Teaching and Learning Mathematical Modelling:

Innovation, Investigation and Applications

Dr Samuel Johnson on lectures and books:

Talking of education, "People have now a-days" (said he) "got a strange opinion that every thing should be taught by lectures. Now, I cannot see that lectures can do so much good as reading the books from which the lectures are taken. I know nothing that can be best taught by lectures, expect where experiments are to be shewn. You may teach chymestry by lectures. — You might teach making of shoes by lectures!"

James Boswell: *Life of Samuel Johnson, 1766*

Teaching and Learning Mathematical Modelling:

Innovation, Investigation and Applications

S.K. Houston
The University of Ulster, Northern Ireland

W. Blum
The University of Kassel, Germany

I.D. Huntley
The University of Bristol, England

N.T. Neill
The University of Ulster, Northern Ireland

Albion Publishing
Chichester

First published in 1997 by
ALBION PUBLISHING LIMITED
International Publishers
Coll House, Westergate, Chichester, West Sussex, PO20 6QL England

QA
401
T435
1997

British Library Cataloguing in Publication Data
A catalogue record of this book is available from the British Library

ISBN 1-898563-29-2

Printed in Great Britain by Hartnolls, Bodmin, Cornwall

Table of Contents

Preface   ix

Section A   **Reflections and Investigations**

1   Curriculum Development and Assessment in Northern Ireland   3
     Catherine Coxhead
     Northern Ireland Council for the Curriculum, Examinations and
     Assessment, Belfast, BT1 3BG, Northern Ireland.

2   Mathematical Modelling and the Way the Mind Works   23
     Susan J Lamon
     Marquette University, Milwaukee, WI 53233-1881, USA.

3   What Makes A Model Effective and Useful (or Not)?   39
     Eva Jablonka
     Technische Universität Berlin, D-10623 Berlin, Germany.

4   A Case Study of Instruction and Assessment in Mathematical   51
     Modelling - 'the delivering problem'
     T Ikeda
     Yokohama National University, Yokohama, 240, Japan.

5   The Quest for Meaning in Students' Mathematical Modelling   63
Activity
João Filipe Matos   *Susana Carreira*
Universidade de Lisboa   *Universidade Nova de Lisboa*
Campo Grande, C1-2° Piso   *Quinta da Torre, 2825 Monte*
1700 Lisboa   *da Caparica*
Portugal   *Portugal*

6   Some Mathematical Characteristics of Students entering Applied   77
Mathematics Degree Courses
P L Galbraith   *C R Haines*
University of Queensland   *City University*
Brisbane, Queensland 4072   *London, EC1V 0HB*
Australia.   *England.*

Section B   **Assessment at Tertiary Level**

7   The Assessment Factor - by Whom, for Whom, When and Why   95
Leone Burton
University of Birmingham, Birmingham, B15 2TT, England.

8   Assessment of Complex Behaviour as Expected in Mathematical   109
Projects and Investigations
John Izard
Australian Council for Educational Research, Hawthorne,
Victoria, Australia.

9   Mathematical Proficiency on Entry to Undergraduate Courses   125
Peter Edwards
Bournemouth University, Bournemouth, BH1 3NA, England.

10   Evaluating Rating Scales for the Assessment of Posters   135
Ken Houston
University of Ulster, Jordanstown, BT37 0QB, Northern Ireland.

11   Deriving Learning Outcomes for Mathematical Modelling Units   149
within an Undergraduate Programme
Andrew Battye and Maggie Challis
Sheffield Hallam University, Sheffield, S1 1WB, England.

Section C

Secondary Courses and Case Studies

12   An Operative Approach to Formal Reasoning   159
      Paola Forcheri and Maria Teresa Molfino
      Istituto per la Matematica Applicata del Consiglio Nazionale
      delle Ricerche, Genova, Italy.

13   The Application of Mathematics - an Essential Component of   171
      New Vocational Qualifications in the UK
      John Gillespie
      Shell Centre for Mathematical Education, University of
      Nottingham, Nottingham, NG7 2RD, England.

14   Mathematics as Orientation in a Complex World   183
      Hans-Wolfgang Henn
      Staatliches Seminar für Schulpädagogik, Karlsruhe, Germany.

15   The 'Mathematical Modelling' Course for Russia's Schools: its   203
      Aims, Methods and Content
      EK Henner and AP Shestakov
      Pedagogical University of Perm, Perm, Russia.

16   On the Use of Open-ended, Real-world Problems   211
      Ted Hodgson
      Montana State University, Bozeman, MT 59717, USA.

17   Slow Learners, Mathematics and Future Profession: the search for   219
      mathematics on the shop floor of the future
      Pieter van der Zwaart,
      SLO, Institute for Curriculum Development, Postbus 2041,
      7500 CA Enschede, The Netherlands.

18   Mathematical Modelling for 16-19 Vocational Courses   229
      Julian Williams and Geoff Wake
      Centre for Mathematics Education, University of Manchester,
      Manchester, M13 9PL, England.

19   Mathematical Modelling and Children's Development of Science   243
      Concepts
      Brian Doig   *Susie Groves*
      The Australian Council for   *Deakin University,*
      Educationa Research,   *221 Burwood Highway,*
      Camberwell 3124,   *Burwood, 3125,*
      Australia.   *Australia.*

                        Julian Williams
                        University of Manchester,
                        Manchester, M13 9PL,
                        England.

20   The Development of a Secondary-School Course in Probability,   255
      Statistics and Modelling that Attracts and Empowers Students
      Thomas L Schroeder and Barry E Shealy
      State University of New York at Buffalo,
      Buffalo, NY. 14260-1000, USA.

Section D   Tertiary Case Studies

21   A Unified Approach to the Mathematical Modelling of   269
      Mechanical Systems
      Krzysztof Arczewski   *Wojciech Blajer*
      Warsaw University of   *Technical University of*
      Technology,   *Radom,*
      Warsaw, Poland.   *Radom, Poland.*

22   Performance Modelling of Parallel Algorithms   281
      D B Clegg
      Liverpool John Moores University, Liverpool, L3 3AF, England.

23   Modelling Cancer Chemotherapy   293
      JR Usher and D Henderson
      The Robert Gordon University, Aberdeen, AB1 1HG, Scotland.

24   Motoring - Modelling in the Fast Lane   309
      M J Herring
      Cheltenham and Gloucester College of Higher Education,
      Cheltenham, GL50 2QF, England.

25   Computer-based Experiments in Mechanics   321
D A Lawson and J H Tabor
BP Mathematics Centre, Coventry University,
Coventry, CV1 5FB, England.

26   Modelling Patient Flow through Hospitals   331
Sally McClean
University of Ulster, Coleraine, BT52 1SA, Northern Ireland.

27   Lagging a Pipe or Bandaging a Limb   343
A J I Riede
Universität Heidelberg, D-69120 Heidelberg, Germany.

Section E   Tertiary Courses

28   The Relevance of Research in the Development of Undergraduate   353
Courses in Mathematical Modelling
Angela Botham and Jean Crowe
Edge Hill College of Higher Education,
Lancaster, L39 4QP, England.

29   An Introductory Course on Mathematical Models and Modelling: 363
A Constructivist Approach for Middle School Teachers
Don Cathcart and Tom Horseman
Salisbury State University, Salisbury, MD 21801, USA.

30   Simulation Modelling for Undergraduate Mathematicians   373
Andrew Fitzharris
University of Hertfordshire, Hatfield, AL10 9AB, England.

31   Experiences With System Modelling in a Social and Business   385
Context
MJ Hamson and MAM Lynch
Glasgow Caledonian University, Glasgow, G4 0BA, Scotland.

32   Using Critical Reviews in Mathematical Modelling Courses   395
Bryan A Orman
University of Southampton, Southampton, SO17 1BJ, England.

Preface

It is now widely accepted, both by practitioners and educationalists, that mathematical modelling is 'the way of life of an applied mathematician'. There is a recognition that undergraduate university students, if they are to earn a living as mathematicians of one sort or another, should have opportunities to learn mathematical modelling during their studies - 'mathematical modelling is a must!' There is a recognition that pupils in schools should learn about applications of mathematics through engagement in modelling, and have their knowledge and skills assessed. There is the belief that much of mathematics is learnt best when it is situated in the context of a pupil's everyday experience, that is, through applications.

Nevertheless, despite this widespread recognition and practice, there are still needs to be met. There is still the need to convert some hearts and minds to the belief that modelling and applications are essential components of mathematics learning. There is the need to devise, implement and evaluate new methods of teaching and assessing. There is the need to conduct research on the teaching and learning of modelling and applications, and there is the need to disseminate examples of good practice and the fruits of research into student learning.

This book attempts to meet these needs, at least to some extent. It contains 32 articles on the teaching, learning and assessment of mathematical modelling and applications which were originally presented as papers at the 7th International Conference on the Teaching of Mathematical Modelling and Applications (ICTMA-7) which was held on the Jordanstown campus of the University of Ulster in Northern Ireland in July 1995. (The ICTMA Conferences are held biennially and the next conference is planned for Brisbane, Australia, in August 1997.) There are contributions by authors from 12 different countries, with 16 of the articles coming from authors in the United Kingdom. This reflects the truly international dimension of the subject matter of this book, and provides valuable insights into practices worldwide.

The book is divided into five sections. Section A contains four articles on Reflections and Investigations and includes an article which gives an account of recent educational developments in Northern Ireland. Assessment of student learning is an essential part of the education system and is currently an active research field. Section B contains five articles which deal with assessment, particularly at undergraduate level. Section C contains seven articles dealing with secondary education and contains both case studies of good practice and examples of courses and how they are taught in a variety of countries including Russia, the Netherlands, the USA and the UK. Section D presents seven case studies suitable for use in university teaching and Section E concludes the book with descriptions of five ideas relating to undergraduate modelling courses.

The book as a whole complements the series of books emanating from the ICTMA conference series (see next page) and should prove to be a useful source of information and ideas for those engaged in the teaching, learning and assessment of mathematical modelling and applications.

This is now the age of electronic communication and publication. Email addresses of most of the authors are included in the chapter headings, while the Preface, Table of Contents and publishing details of the book are available at the internet address
http://www.infj.ulst.ac.uk/~cdmx23/ictma7.html.

The IRA cease-fire of October 1994, which held through 1995, provided a window of opportunity for tourists and conference delegates to come to a relatively trouble free Northern Ireland, and we are grateful to all those who attended the conference, either to present a paper or just to listen. It is rumoured that they had an enjoyable and profitable time! See the review by Duncan *et al.*, Zentralblatt fur Didaktik der Mathematik (ZDM), 28 (2), pp 67 - 69, 1996, or the internet page
http://www.infj.ulst.ac.uk/~cdmx23/report.html.

We are also grateful to the authors of these articles for their excellent contributions and for their care in presenting material in the form we requested.

We thank the other members of the local team - Derek Curran, Jack Cromie, Norman Smith, Keith Simpson (Catering), Fred Ruck (Premises) and Hazel Alderdice (Accommodation) - for organising the conference and making it run smoothly. The technicians and computing officers of the Faculty of Informatics also receive our thanks. Above all we wish to thank the conference secretary, Sharon Kelly, for all her work, before, during and after the conference, and for preparing this manuscript. Finally we acknowledge the encouragement and interest of our publisher, Ellis Horwood of Albion Publishing Ltd.

S K Houston   W Blum   I D Huntley   N T Neill

January 1997

As editor-in-chief, it was my responsibility to ensure that all the i's were dotted and t's crossed, so any residual mistakes are down to me. I express my heartfelt thanks to the work of my fellow editors in producing this book.

S K H.

PREVIOUS BOOKS EMANATING FROM ICTMA CONFERENCES

Berry JS, Burghes DN, Huntley ID, James DJG and Moscardini AO, 1984, **Teaching and Applying Mathematical Modelling**, Ellis Horwood, Chichester.

Berry JS, Burghes DN, Huntley ID, James DJG and Moscardini AO, 1986, **Mathematical Modelling Methodology, Models and Micros**, Ellis Horwood, Chichester.

Berry JS, Burghes DN, Huntley ID, James DJG and Moscardini AO, 1987, **Mathematical Modelling Courses**, Ellis Horwood, Chichester.

Blum W, Berry JS, Biehler R, Huntley ID, Kaiser-Messmer G and Profke L, 1989, **Applications and Modelling in Learning and Teaching Mathematics**, Ellis Horwood, Chichester.

Niss M, Blum W and Huntley ID, 1991, **Teaching of Mathematical Modelling and Applications**, Ellis Horwood, Chichester.

De Lange J, Keitel C, Huntley ID and Niss M, 1993, **Innovation in Maths Education by Modelling and Applications**, Ellis Horwood, Chichester.

Sloyer C, Blum W and Huntley ID, 1995, **Advances and Perspectives in the Teaching of Mathematical Modelling and Applications**, Water Street Mathematics, Yorklyn, Delaware.

Section A

Reflections and Investigations

1

Curriculum Development and Assessment in Northern Ireland

Catherine Coxhead
Northern Ireland Council for the Curriculum, Examinations and Assessment, Clarendon Dock, Belfast, BT1 3BG

ABSTRACT

This paper attempts to provide some insights into Northern Ireland education, with special reference to mathematics. It begins with some background before going on to outline recent developments in respect of curriculum provision and assessment arrangements for mathematics for 4-19 year olds in Northern Ireland. The distinctive features of the Northern Ireland Curriculum as set out in the Education Reform (Northern Ireland) Order 1989 are used to provide the context for comparisons between the Northern Ireland programmes of study and attainment targets for mathematics and those of the National Curriculum in England and Wales, with particular reference to the statutory cross-curricular themes of economic awareness and information technology. The implications of curriculum provision and assessment arrangements for the development of students' skills in mathematical modelling and applications are explored, and the outcomes of a curriculum development project commissioned by the Northern Ireland Council for the Curriculum, Examinations and Assessment (CCEA) are used to illustrate the benefits and difficulties of the recent shift of emphasis in mathematics teaching in the Province. Reference is also made to recent changes in CCEA syllabuses for GCSE and A-level mathematics, and to the development of General National Vocational Qualifications (GNVQs). The paper ends with some reflections on the current debate on the appropriateness of school mathematics as preparation for higher education.

INTRODUCTION

Northern Ireland is an integral part of the United Kingdom and yet, in some ways, it is very different from the rest. Separated from the Great Britain mainland and very much influenced by its own cultures and history, it has its own unique traditions. From its establishment in the 1920s, until the early 1970s, education policy was set by the local devolved parliament in Belfast. Since then, with direct rule from London, the Northern Ireland Office, via its Secretary of State and a number of junior ministers, has held responsibility for areas previously devolved to the Stormont Parliament in Belfast. There is a Department of Education for Northern Ireland (DENI), and the education minister is responsible to the Secretary of State for Northern Ireland (rather than the Secretary of State for Education and Employment). Thus education in Northern Ireland is not bound by Department for Education and Employment decisions taken in England, though it is influenced by them.

Up until the late 1960s, Northern Ireland had its own system of examinations and qualifications. As mobility increased, and more students wanted to go to universities in Great Britain, it was felt that there was a need to align qualifications with those in England or Scotland, and it was finally decided that the English system of General Certificate of Education (GCE) O- and A-levels should be adopted here (Ministry of Education for Northern Ireland, 1969). Many, looking back with hindsight, would now argue that Northern Ireland should have adopted the Scottish system, but it did not. One consequence of this is that, whereas in many ways Northern Ireland is more removed from England than Scotland is, in education it is more closely integrated. While it has an Examinations Board setting General Certificate of Secondary Education (GCSE) and GCE A-level examinations, many schools choose to take these examinations through the English groups such as the Northern Examinations and Assessment Board (NEAB), Cambridge Local, London and so on. On the vocational front, organisations such as the Business and Technology Education Council (BTEC), City and Guilds and the Royal Society of Arts (RSA) operate in Northern Ireland as well as in England and Wales. Thus Northern Ireland is inevitably linked into developments in England which affect 14 to 19 year olds, even if it is able to maintain some distinctions for younger children.

Not all English policies automatically reach Northern Ireland. One very significant factor is that a selective system of secondary education operates in almost all parts of the Province. There are enough grammar school places for about one third of Northern Ireland eleven year olds - those who want to go to a grammar school take a selection test in the last year of primary school and the unsuccessful candidates go to secondary intermediate schools. The term 'secondary school' has thus come to mean non-grammar and those involved in education tend therefore to talk about post-primary education rather than secondary. Although not officially a matter for debate, this selective system causes annoyance and distress to many outside the grammar-school sector. The 11-plus tests undoubtedly exert a massive influence on Northern Ireland primary schools (Sutherland and Gallagher, 1986; Teare and Sutherland, 1988), and while many secondary intermediate schools are very good indeed, there appears to be a considerable cost to the Province as a whole in terms of the negative impact on those who fail to obtain grammar school places (Wilson, 1985).

Another feature of the system which has a significant effect in many ways is the religious divide. Most of the schools in Northern Ireland have either a Roman Catholic or a Protestant background and ethos. The integrated schools movement is a growing sector, but remains quite small. Most children are educated in Catholic schools or Protestant schools. Recent research has shown consistently that a higher proportion of children from Protestant schools get good qualifications. This is particularly true in science and mathematics (Osborne, 1986; Cormack *et al.*, 1992; DENI, 1995).

One final general point worth mentioning concerns the teaching force. For a variety of reasons, including traditions and the state of the economy, teaching is an esteemed profession which can attract reasonably well-qualified applicants. The result is a strong teaching force and, even in science and mathematics, teachers are generally well qualified in their specific subject areas.

THE TEACHING AND LEARNING OF MATHEMATICS IN NORTHERN IRELAND

Northern Ireland has strong traditions and high standards in mathematics. In the late 1970s and early 1980s the Assessment of Performance Unit (NFER, 1984) conducted a range of tests in mathematics throughout England, Wales and Northern Ireland over a number of years. In 1982, at age 11, Northern Ireland children did better than their English and Welsh counterparts in all but two of the thirteen topics tested. Their performance in number skills was particularly outstanding. On some items, Northern Ireland success rates exceeded those in England by 30 percentage points. Differences in performance at age 15 were less obvious, but Northern Ireland scores were highest on ten out of the fifteen topics tested. Details are in Table 1.

An examination of these results suggests that Northern Ireland was strong on the traditional skills of computation and algebraic manipulation, but weaker in some areas such as problem-solving, probability and statistics and number concepts. This would be consistent with the findings of school inspections in the mid 1980s. There was a lot of strong traditional teaching which clearly showed in the level of skills developed. In terms of paragraph 243 of the Cockcroft Report (1982), which was frequently discussed in those days, there was plenty of exposition and consolidation but rather less by way of discussion, practical or investigational work and problem-solving, including the application of mathematics to everyday situations. In those days the Inspectorate was heavily involved in developmental work and in-service training, and much time and effort was invested in disseminating a wider range of teaching styles. This took on an added dimension and greater urgency when GCSE was introduced in 1986, with its commitment to mathematics coursework for all candidates.

At the same time, the Northern Ireland Council for Educational Development, with the support of DENI, began a curriculum development initiative in primary schools, known as the 'primary guidelines'. The objective was to get each primary school to work on developing its own schemes of work in each subject. Each school would choose to work on a particular subject over a year or two in some cases; additional funds were available to support their work. Also available were booklets giving guidance on what needed to

Rank order of countries within each sub-category

At age eleven:

Category	Sub-category	Eng	Wal	NI
Geometry	Shapes, lines, angles	1	3	2
	Symmetry, transformations, co-ordinates	1	3	2
Measures	Money, time, mass, temperature	2	3	1
	Length: area, volume, capacity	2	3	1
Numbers	Concepts: whole numbers	2	3	1
	Concepts: decimals and fractions	3	2	1
	Computation : whole numbers and decimals			
	Computation : fractions	3	2	1
	Applications of numbers	3	2	1
	Rate and ratio	3	2	1
Algebra	Generalised arithmetic	2	3	1
Probability & Statistics	Probability and data representation	2	3	1

At age fifteen:

Category	Sub-category	Eng	Wal	NI
Numbers	Concepts	2	3	1
	Skills	2	3	1
	Applications	2	3	1
Measures	Unit	2	3	1
	Rate and ratio	2	3	1
	Mensuration	2	3	1
Algebra	General algebra	2	3	1
	Traditional algebra	2	3	1
	Modern algebra	2	3	1
	Graphical algebra	2	3	1
Geometry	Descriptive geometry	1	3	2
	Modern geometry	1	3	2
	Trigonometry	1.5	3	1.5
Probability & Statistics	Probability	1	3	2
	Statistics	1	3	2

Table 1. Pattern of performance in England, Wales and Northern Ireland, 1982 - Mathematical Concepts and Skills Written Tests

be considered: aims, objectives, range of content, teaching styles, and so on. One of the key points was that nothing was imposed - schools did their own development work. In mathematics, they were very much encouraged to think about teaching styles and practical work.

Perhaps the most significant mathematical development in the late 1980s, prior to the National Curriculum, was GCSE, and in particular the coursework element. GCSE was examined for the first time in 1988, as in England and Wales, although coursework in mathematics did not become a compulsory element of the examination until 1991. There have been, and remain, criticisms of the GCSE, and these will be referred to later. However there was a determined effort to introduce opportunities for pupils to engage in more extended pieces of work. The Northern Ireland Examination Council's syllabus required candidates to submit work of two kinds: a 'pure' open-ended investigation, and a more 'applied' project in which mathematics was used in a practical situation. The latter category included applications of number in, for example, finance in Northern Ireland, geometry in designing room layouts or car-parks, and statistics.

EDUCATION REFORM

Curriculum

Even as the first cohort of GCSE candidates was revising for its examinations, the government announced its intentions for a ten-subject National Curriculum, with associated assessment at ages 7, 11, 14 and 16. Shortly afterwards the Department in Northern Ireland began consulting about its own similar plans. Legislation, the Education Reform (Northern Ireland) Order (1989), was passed and the process of implementation began about a year behind England and Wales. This lagging behind did have some considerable advantages - mostly the opportunity to take account of the debate and immediate criticisms of the developments in England. In terms of general structure, there was a concern to have a clearer rationale for the curriculum package, and the Northern Ireland version has some distinctive features.

In the first place, the Northern Ireland Curriculum is prescribed in terms of six Areas of Study which group together cognate subjects. The Areas of Study of English, mathematics, science and technology, the environment and society, creative and expressive studies and language studies provide a framework for ensuring a broad and balanced curriculum for all pupils. Pupils aged 11 to 16 are required to study subjects from each of the areas of study.

Then the 1989 Order also identified six educational, cross-curricular themes which were to be delivered mainly through the compulsory subjects of the curriculum. These themes are statutory requirements. They differ significantly from their non-statutory counterparts in England and Wales in terms of titles and descriptions, reflecting differing curriculum priorities and emphases.

Two of these cross-curriculum themes, Education for Mutual Understanding and Cultural Heritage, are unique to Northern Ireland and were intended to play a part in addressing the peculiar problem of the Province, namely the problem of a community divided by religion and by national aspiration. Recognising that segregation in education could lead to a lack of knowledge and understanding of other traditions within the community, the Government was anxious to ensure that some effort was made to counteract ignorance and prejudice. The other four themes: Health Education, Economic Awareness, Careers Education and Information Technology are likely to be more familiar. In order to ensure that these themes were adequately addressed, requirements were drawn up before the details for each subject were finalised.

As far as mathematics is concerned, the requirements were very similar to those in England and Wales. The broad structure was the same, with attainment targets and statements of attainment set out at 10 levels. The main difference was that in Northern Ireland detailed programmes of studies were provided, rather than simple re-formulations of the statements of attainment. This gave clear guidance to teachers about general teaching approaches and the kinds of opportunities that should be provided. For example, the programme of study for 11 to 14 year olds begins with the statements in Table 2:

Pupils should regularly experience success in mathematics to help them to develop confidence and a positive attitude towards mathematics.

Throughout the mathematics programme pupils should be engaged in a wide range of purposeful activities. These activities should:

- involve pupils in different modes of learning: doing; observing; talking and listening; discussing with other pupils and the teacher; reflecting; drafting; reading and recording,
- develop pupils' personal qualities,
- include both independent and co-operative work,
- develop mental skills,
- be balanced between those which are short in duration and those which have scope for development over an extended period,
- be both of the kind which have an exact answer or result and those which have many possible outcomes.

Table 2. Extract from the Curriculum (Programmes of Study and Attainment Targets in Mathematics) Order (Northern Ireland), 1990

The National Curriculum in England and Wales and the Northern Ireland Curriculum have both recently been the subject of detailed reviews. The revisions agreed in England and Wales will apply from August 1995 for 5-14 year olds and from August 1996 for 14-16 year olds. The proposals for revisions to the Northern Ireland Curriculum are currently the subject of consultation: the revisions, when finalised, will be introduced for the start of the 1996/97 school year when statutory key stage assessment will begin in Northern Ireland for the first time. The proposed revisions to the Northern Ireland Curriculum involve a restructuring of the existing orders and, in some subjects, a substantial slimming down of content to make the curriculum more manageable. As in England and Wales, the revised subject proposals consist of programmes of study and level descriptions which are general descriptions of pupils' attainment in terms of the key elements of the subjects. Their purpose is to assist teachers in making holistic summative judgements about the level at which a pupil is working. Level descriptions are intended for assessment purposes and replace detailed statements of attainment which had tended to lead to fragmentation and atomisation of teaching and assessment.

Within both the current and the proposed programmes of study for mathematics in the Northern Ireland Curriculum (CCEA, 1995a) opportunities are provided for the development of pupils' skills in mathematical modelling and applications, starting in the early years of schooling. There is an attainment target devoted to 'Processes in Mathematics' and relevant sections of the programmes of study contain the following requirements:

Pupils should have opportunities to:

(in the early years)
- develop different approaches to solving problems and look for ways to overcome difficulties;
- recognise simple patterns and relationships and make predictions about them based on experience;

(in the upper primary school, in addition,)
- develop their own mathematical strategies for solving problems.

(in post-primary schools)
- select, trial and evaluate a variety of possible approaches; identify what further information may be required in order to pursue a line of enquiry;
- develop a range of mathematical strategies for solving problems; review, evaluate and refine the effectiveness of these strategies;
- use patterns and relationships produced from a task to make and test predictions; make and test generalisations, initially in words later translating these into symbolic form.

Table 3. Extracts from Recommendations for Revised Subject Requirements for Mathematics, (CCEA, 1995a)

In the mathematics programmes of study in England and Wales (DFE, 1995), some similar opportunities for problem solving are presented within the sections relating to the attainment target for Using and Applying Mathematics. However less emphasis appears to be given to the recognition of patterns and relationships, and the making and testing of predictions.

The proposed Northern Ireland programmes of study also give greater emphasis to the contextualising of pupils' mathematical activities in real-life situations, and to the use of information technology. For example, in the programme of study for key stages 3 and 4, pupils are required to:

> work with money in a variety of real life situations, for example, shopping, banking and travel

and

> to collect represent and interpret real data from a range of sources, for example, road traffic statistics, health and education, population census and environmental statistics.

These and other similar contexts provide ideal opportunities for mathematics to contribute to the statutory cross-curricular themes, such as economic awareness, in which pupils are required to use appropriate knowledge, understanding and skills:

> to evaluate economic information and ideas including their presentation in words, images graphs and statistics

and

> to make balanced and informed judgements and, where necessary, know the appropriate action to be taken on issues, problems and events where there is an economic dimension.

Similarly, whereas the National Curriculum programme of study simply contains the very general requirement that

> Pupils should be given opportunities where appropriate, to develop and apply their information technology (IT) capability in their study of mathematics,

the Northern Ireland Curriculum proposals for 11 to 16 year olds (CCEA, 1995a) require that pupils should

> explore, manipulate, represent and communicate mathematics through a variety of appropriate computer applications/packages. They should have opportunities to prepare data structures including spreadsheets and databases, entering information into them, investigating patterns and relationships, and framing and testing questions. They should have opportunities to create and use procedures which include variables, using a programmable language for example, LOGO.

Of course the setting down of opportunities to develop pupils' skills in mathematical modelling and applications in curriculum specifications does not, in itself, mean that the development of these skills is always realised in practice. CCEA's monitoring of the implementation of the Northern Ireland Curriculum programme of study for mathematics suggested that many teachers were finding it difficult to devise appropriate learning activities to develop pupils' processes in mathematics and provide opportunities for assessment. Evidence indicated that many teachers were concentrating their attention on the content-based attainment targets of number, measure, shape and space, and were assuming that pupils' competence in mathematical processes would be at the same level as their attainment in these other areas.

Concern about teachers' difficulties in teaching and assessing mathematical processes prompted CCEA, in conjunction with the Western Education and Library Board, to initiate a curriculum development project in process-based mathematics (CCEA, to be published). The project aimed to develop teaching and learning materials which would encourage and guide teachers in integrating the promotion of pupils' process skills into the broad content of the mathematics curriculum and support teacher assessment. The project developed some 20 tasks, suitable for pupils of a wide range of ages and abilities. As the tasks were devised they were trialled in schools and evaluated by the project officer, with the assistance of teachers and pupils; the tasks were then refined in the light of the evaluation. Associated with the tasks are exemplars of pupils' work annotated to support teacher assessment. The outcomes of the project are presently in draft form awaiting publication once the final shape of the revised mathematics curriculum requirements is known. However, brief descriptions of two of the tasks will serve to give a flavour of the materials.

In one task, pupils are required to help a shop assistant, who provides a gift-wrapping service for customers, to work out quickly how much ribbon is required to wrap parcels of various sizes. In another, pupils are asked to help a child design the largest possible enclosure for a pet rabbit from a given amount of wire netting. They are then required to investigate the problem further, using other lengths of netting, to suggest a quick way of working out the maximum area for any length of fencing, and to put this method into the form of a simple equation.

Assessment

An important feature of the Government's education policy is the assessment of pupils at regular intervals. Having listened to the debate in England about testing at 7 years of age, DENI decided to legislate for assessment at 8, 11, 14 and 16. As yet, assessment has not been made statutory, and there has recently been a wide-ranging consultation about the form that assessment should take when it becomes statutory in the 1996/7 school year. There is a strongly held view in Northern Ireland that statutory assessment should support and enhance teachers' normal assessment of pupils rather than either conflicting with it or appearing completely separate from it. For the past three years CCEA has been piloting moderated teacher assessment, trying to find a process which is both consistent and reliable across schools, and manageable by teachers. More than half of primary and a quarter of post-primary schools in the Province were involved in the 1994/95 pilot. As part of this pilot, CCEA (1994) has developed assessment units to help teachers confirm their assessment of pupils' progress in an attainment target. Several of the units developed so far address processes in mathematics.

In one unit, information is provided on the time taken to get to school, by a class of pupils, using various methods of travel. From this information pupils are required to make a number of calculations including total and average travel times, to present information in the form of a bar chart, to use this information to decide on the most popular method of transport, and to make a prediction about the travel arrangements of the whole year group based on information known about one class.

In another unit, designed to assess process skills at a higher level, pupils are provided with information on a spreadsheet showing the prices of various toys in 1992, and the percentage increase in two subsequent years. They are asked to use the spreadsheet to find out information about prices and price changes, to respond to various 'What would happen if?' questions, to look for patterns in their results, to make and check predictions, and to make generalisations about relationships between variables.

IMPLICATIONS FOR TEACHER TRAINING

Providing practical classroom materials which teachers can use with their pupils, and adapt to their own needs, is one contribution that CCEA is making to developing teachers' and pupils' confidence in handling mathematical processes. Of course, this on its own will not be enough and in-service training is essential if the long-term attitudinal changes required to introduce new teaching methods are to come about. This was recognised by DENI, and the Education Reform Order (1989) placed a duty on the five Education and Library Boards across Northern Ireland to provide support for schools. Since 1990, extensive advisory and support services have been developed. This is in direct contrast to the position in England, where local education authority support services have, on the whole, been weakened.

GCSE AND GCE COURSES

The impact of GCSE and GCE syllabuses and examinations on teaching methods is always significant. The demands of society, and of the Government and parents in particular, to see students achieving better qualifications puts teachers under great pressure, especially when seen in the context of open enrolment and the publication of results. Thus the need to cover the syllabus is often considered paramount, and the use of more enlightened teaching methods to increase pupils' facility in mathematical processes may often be strangled by the need to 'get through the content'.

The role of syllabus development and examination question setting is therefore central to the development of students' modelling and application skills. Even if the Northern Ireland Curriculum programmes of study eventually have their desired impact on mathematics teaching from 4-14, this may be dissipated from 14-19 if syllabuses and examinations do not encourage and assess students' process skills. Experience suggests that it is not sufficient for examination boards such as CCEA merely to offer alternative courses which put a greater emphasis on such skills whilst still offering 'traditional' mainstream courses. When the Northern Ireland Schools Examinations and Assessment Council (NISEAC), one of the predecessor organisations from which CCEA was formed, introduced an innovative Further Mathematics A-level syllabus in 1988, with an emphasis on mathematical modelling, coursework and pre-seen examination material, it was much praised in academic circles throughout the UK (Houston, 1989). However, the syllabus was short-lived and was withdrawn in 1992 due to a poor take-up in Northern Ireland schools (Houston, 1992). Despite a very good general level of performance being achieved by those candidates who did enter, the examination proved unpopular with Northern Ireland teachers, either owing to their reservations about taking on anything new or as a result of their practical concerns about the heavy workload assumed to be associated with coursework and a perceived lack of suitable published teaching resources.

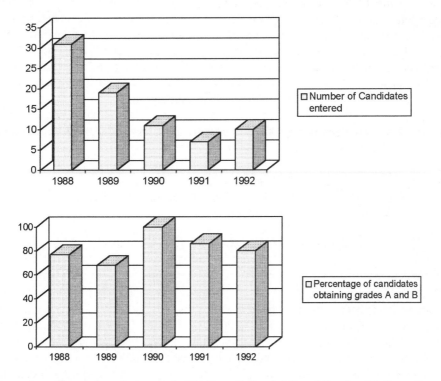

Figure 1: entries and performance in Northern Ireland A Level Further Mathematics Mode 2, 1988-92

Thus the evidence appears to be clear - if the development of students' mathematical modelling and application skills is to be taken seriously by teachers, these skills need to be an integral part of the requirements of mainstream courses leading to recognised qualifications. As will be seen, this lesson appears to have been understood and steps are now being taken to ensure mathematical applications and modelling are included in all GCE A- and AS-level syllabuses.

Currently CCEA syllabuses are being revised to take account of revised regulations (SCAA, 1995a) and subject criteria at GCSE (SCAA, 1995b) and the new agreed core for A- and AS-levels (SCAA, 1993).

The GCSE criteria for mathematics require that all syllabuses include provision for candidates to demonstrate their ability to use and apply mathematics. This encompasses:

- making and monitoring decisions to solve problems,
- communicating mathematically,
- developing skills of mathematical reasoning.

The existing CCEA GCSE syllabuses (CCEA, 1995b) reflect the Northern Ireland Curriculum programme of study for Key Stage 4, and the revised syllabuses will also take account of the revisions made to the programme of study as a result of the curriculum review. The draft syllabuses for 1998 (CCEA, 1995c) contain a scheme of assessment which includes the following among its objectives (see Table 4).

The scheme of assessment will test the ability of students to:

- analyse a problem, select a suitable strategy and apply an appropriate technique to obtain its solution,
- apply combinations of mathematical skills to techniques in problem solving,
- make logical deductions from given mathematical data,
- respond to a problem relating to a relatively unstructured situation by translating it into an appropriately structured form.

Table 4. Extracts from the GCSE Mathematics 1998 Draft Syllabuses A and B, (CCEA, 1995)

Candidates' competence in mathematical processes is assessed in the main through the coursework element of the examination. In this respect candidates are required to submit two tasks, one of a practical nature and one of an investigative nature. Practical tasks are intended to provide candidates with opportunities to put mathematics to use in the real world. Investigative tasks are intended to provide candidates with opportunities to engage in investigations which are 'pure', in the sense that their interest and importance do not depend on possible applications. CCEA provides examples of starting points for structured and open-ended practical and investigative tasks to assist teachers in devising their own coursework activities for candidates.

One suggestion for a practical task involves the organisation of a sports tournament. Students are provided with information about the number of teams and the length of matches and breaks, and are required to work out the number of matches to be played and the length of the tournament. They are then asked to examine the effect of changes in the number of pitches and teams, to look for patterns in their findings, to describe a method for calculating the number of matches needed for a given number of teams, and to present their conclusions.

CCEA also provides an Additional Mathematics GCSE syllabus which is designed to broaden the mathematical experience of high attaining students. This remains relatively popular, with an entry of some 3600 in 1995 compared with 18700 in Mathematics.

At GCE A- and AS-level, CCEA works with the other GCE boards and SCAA in the development of agreed subject cores and codes of practice. The new subject core (SCAA, 1993), which must be included in all A- and AS-syllabuses with the title of Mathematics, contains important innovations, The most significant of these is the requirement to study at least one area of the application of mathematics. No particular area is specified in terms of content, but the core contains skills in modelling and application which must be developed in all syllabuses. The proportion of any syllabus devoted to application depends in part on how much of the non-core syllabus is devoted to it, but it is stipulated that around 20% of the core content for an A-level syllabus and about 30% of the core content of an AS-level syllabus should be related to this aspect of mathematics.

Within the knowledge and understanding that must be included in any GCE syllabus, the core section on the application of mathematics has the following requirements (see Table 5).

Syllabuses must develop

> an understanding of the process of mathematical modelling with reference to one or more application areas,

which includes

- abstraction from a real world situation to a mathematical description,
- the selection and use of a simple mathematical model to describe a real world situation,
- approximation, simplification and solution,
- interpretation and communication of mathematical results and their implications in real-world terms,
- progressive refinement of mathematical models.

Table 5. Extract from the GCE A/AS Subject Core for Mathematics, (SCAA, 1993).

Within the skills and assessment objectives that must be included, the core states that students must be able to demonstrate their ability to evaluate mathematical models, including an appreciation of the assumptions made, and interpret, justify and present the results from a mathematical analysis in a form relevant to the original problem.

GNVQs AND CORE SKILLS

Recent years have seen the development of alternatives to the traditional academic qualifications of GCSE and GCE. In particular the development of General National Vocational Qualifications (GNVQs) has provided an opportunity for students in schools and colleges to take vocationally recognised courses - in areas such as art and design, business, health and social care, leisure and tourism, and manufacturing - which are designed to equip them with a wide range of working skills. In a large measure, this movement towards a vocational dimension in 16+ qualifications has been the result of political pressure aimed at correcting a perceived gulf in educational and training standards in the UK compared with those of major competitor countries in Europe and elsewhere. In Northern Ireland many schools and colleges have introduced GNVQs post-16 and, from September 1995, a small number of schools will be involved in piloting a new Part One GNVQ qualification for 16 year olds. The attraction of GNVQs lies in their emphasis on assignment-based work and continuous assessment, as opposed to traditional terminal examinations to which many students find themselves unsuited.

A key element of GNVQs is the inclusion of core skills units in application of number, communication, and information technology. It is intended that these core skills should be transferable to a variety of settings that students may encounter within and beyond their learning programmes. Attainment in all three core skills is required at level 1 in Foundation GNVQs, at level 2 in Intermediate GNVQs, and at level 3 in Advanced GNVQs. It is the Advanced GNVQ, the 'vocational A-level', that has seen the greatest up-take in Northern Ireland schools and colleges. To gain this qualification students have to demonstrate their attainment of level 3 in application of number in three elements - the collecting and recording of data, the tackling of problems and the interpretation of data. In the second of these elements, students have to demonstrate their ability to choose and use appropriate techniques to solve a range of problems, including those in three dimensions, to use mathematical terms and carry out calculations correctly, and to check that results make sense in respect of the problem being tackled.

It is too early to come to a definitive judgement about the effectiveness of GNVQs, and the development of core skills in particular, and it remains to be seen whether higher education institutions and employers will grant equal recognition to students seeking access to courses and careers from the vocational education pathway. Recent statistics from BTEC indicate that seven out of ten GNVQ students went on to do a degree course or higher diploma course in 1994 (BTEC, 1995). However, despite some concerns about the quality of some courses and the validity of assessment, it appears that the potential exists to make a real contribution to the development of mathematical modelling and application skills in vocational contexts for a much wider range of students than had hitherto been possible. In an introduction to core skills units by the National Council for Vocational Qualifications (NCVQ, 1995), reference is made to evaluations of core skills in GNVQs by NCVQ, the Further Education Funding Council (FEFC) and the Office for Standards in Education (OFSTED). NCVQ draws attention to the findings of these evaluations, which emphasise the importance of contextualising core skills so that students are encouraged to see how they can be used to tackle real problems and issues in vocational settings. This point is illustrated by reference to examples from Health and Social Care GNVQs in which students

> initially intimidated by what they see as 'maths'…have since become highly motivated to learn the skills required in the core skills units on being shown how collecting data, analysing it and presenting it, can help them understand the profile of health care in a locality, or determine whether a particular disease affects one group in a locality more than another.

Such examples point to the wider value of core skills in application of number. As Houston (1993) has suggested, "…to be fully informed and functioning citizens, students of all disciplines need to be aware, at least to some extent, of the mathematical modelling process…". Core skills units in mathematics are beginning to feature in a number of other settings including, for example:

- as additional units in competence-based, workplace - assessed National Vocational Qualification (NVQs),
- as curriculum enrichment within GCSE and GCE A- and AS-level courses,
- in university teaching programmes.

Thus, though the provision of core skills mathematics education post-16 may be rather uneven and is certainly far from universal, the first steps have been taken. The best way of promoting further developments in this area might be one issue which Sir Ron Dearing may wish to address in his current review of post-16 education and qualifications in England, Wales and Northern Ireland.

SCHOOL MATHEMATICS AS PREPARATION FOR HIGHER EDUCATION

It should now be clear that much of what is happening in Northern Ireland in mathematics mirrors what is going on in England and Wales. The preceding analysis has sought to demonstrate that, since the Cockcroft Report of 1982, in Northern Ireland as elsewhere in the UK there has been an attempt to broaden teaching styles and offer wider opportunities for using and applying mathematics in realistic contexts and in non-standard ways.

Inevitably, at a time when there are increasing demands on the whole school curriculum, this broader range of activities could only be provided at the expense of something else in mathematics, and there was a deliberate step taken to reduce the content required for GCSE; there has since been some reduction at GCE A-level also. This has led, quite recently, to an upsurge of concern about the mathematical standards of students entering higher education, not only in mathematics but also in science and engineering. There have been letters to the press, and working groups have been established by bodies such as the London Mathematical Society.

Concerns expressed recently have tended to suggest that many students leaving school to study mathematics at a higher level are:

- weak in skills and techniques such as algebraic manipulation, integration, and suchlike,
- have no concept of proof in mathematics,
- are generally lacking in understanding of mathematical concepts.

The analysis that some present of the causes of these problems is alarming (London Mathematical Society, 1995). It is argued that the fault lies in recent developments: the attempts to broaden the mathematical curriculum, to teach process skills as well as content, to set mathematical ideas in real life contexts and to demonstrate how they can be applied. It has even been suggested that it is absurd to introduce problem-solving to young children, when they have few techniques to draw upon.

Of course, there has always been a debate among teachers as to the relative merits of teaching skills and techniques first, out of context, and then showing how they can be applied, and motivating the initial learning and practice by placing it in context. In mathematics, there are certainly some pupils who enjoy learning not only the abstract ideas, but also skills and techniques for their own sake. However experience would suggest that they are in a minority, and that mathematics teachers frequently get asked "What are we learning this for? What's the point?" The usual answer is that "it is on the syllabus".

Furthermore, there would appear to be a particular difficulty in mathematics. If a football coach spends an hour or two on basic skills, those involved can see the point because they know what the game is all about and they can see what they are aiming for. However, children learning mathematics have no way of knowing what the end product can be - it can really only be appreciated looking backwards. Surely then, it is right to try from the start to demonstrate the applicability of mathematics, and approaches to mathematical exploration and problem-solving, however simple.

However this should not be taken to imply that the concerns can be ignored. Clearly the evidence needs to be examined, and there should be a wide-ranging debate about the requirements of higher education, while recognising that other requirements also need to be addressed. It is to be hoped that the current work of the London Mathematical Society will facilitate a healthy debate in which everyone listens carefully to what others are saying, not one in which battle lines are drawn and the two sides snipe at one another.
In this context there are three points which should be made. First, if there is a genuine decline in standards - and not all the evidence points that way - the blame cannot yet be

attributed to the National Curriculum. The first pupils whose GCSE courses were based on National Curriculum requirements are just completing their first year in the sixth form in the summer of 1995. Here in Northern Ireland, which is a year behind, the first students have taken GCSE this summer, 1995.

Second, if a general lowering of standards is demonstrable, it needs to be considered in a broad context, not just in terms of mathematics syllabuses. There is - allegedly - a similar decline in standards of written English. Maybe the decline has more to do with general cultural changes, and the attitudes and aspirations of today's young people, than it has to do with syllabuses. Good technique requires hard graft, and the results depend in large measure upon the efforts of the pupil rather than those of the teacher.

The third point is that the debate about standards should not be approached with a view to turning the clock backwards, for two reasons. The first reason is that not all was well in the past. Sir Wilfred Cockcroft conducted his enquiry in the early 1980s because of widespread concern about mathematics standards, so there must have been problems then. The second reason is that the world is a different place. There are now hand-held computers with built-in software that would enable students to complete a 3 hour A-level mathematics paper in 10 minutes. Very careful thought should be given to what should be taught in these wholly new circumstances. There has, of course, been a similar debate about arithmetic in the context of simple calculators for some time now. It is often argued that skills need to be practised in order for understanding to develop. This can be the case but it depends how transparent the algorithms are. However, it is by no means apparent that many children's understanding of number has been enhanced by extended practice in long division. Calculators make no difference to the need for mental skills, but there is a need to reconsider pencil and paper algorithms. There needs to be some rigorous thinking about the A-level curriculum, and indeed university curricula, trying to disentangle those elements of technique which enhance understanding and those that merely get in the way and could now be facilitated through technology. This thinking should take the debate forward, not backwards.

CONCLUSIONS

Mathematics education in Northern Ireland is closely linked to that of England and Wales. The statutory curriculum and assessment framework in Northern Ireland provides a number of opportunities for teachers to begin to develop their pupils' skills in mathematical modelling and applications. The major constraint on the achievement of this goal appears to be a lack of teacher familiarity and confidence with new ways of teaching and assessing process skills. CCEA is working, in partnership with the advisory and support service of the Education and Library Boards, to help overcome this through the provision of appropriate support materials. Nevertheless there is still a great need for pre-service and in-service training for teachers in this area.

From 14-19, the framework of courses, examinations and qualifications has an important influence on classroom practice. This is magnified by government and parental pressure to raise the standard of educational achievement. If the development of students' modelling and application skills is to be fully realised, these skills need to be embedded in the requirements of mainstream qualifications. Developments at GCSE and particularly at GCE level are addressing this issue, and providing greater opportunities for those following mathematics courses. The incorporation of core skills units in application of number in GNVQ and other courses is also helping to provide a basis for the development of mathematical process skills in real world contexts for those not studying specialist mathematics courses.

The current debate about the appropriateness of school mathematics as preparation for higher education should be based on a full examination of the evidence, and should contribute to a clear vision of the curricular needs of young people in a period of rapid social, economic and technological change. In Northern Ireland, standards remain high and the potential exists for the Province to make a strong contribution to future developments.

REFERENCES

Business and Technology Education Council (BTEC), 1995, **Shaping the Future : The BTEC Report, 1994**, BTEC, London.

Cockcroft, WH, 1982, **Mathematics counts : Report of the Committee of Enquiry into the Teaching of Mathematics in Schools**, HMSO, London.

Cormack RJ, Gallagher AM, Osborne RD and Fisher NA, 1992, *Secondary Analysis of the School Leavers' Survey (1989)*, Annex D, **Seventeenth Report of the Standing Advisory Commission on Human Rights**, House of Commons Paper 54, HMSO, London.

Department for Education (DFE), 1995, **The National Curriculum Order for Mathematics**, HMSO, London.

Department of Education for Northern Ireland (DENI), 1989, **The Education Reform (Northern Ireland) Order**, HMSO, Belfast.

Department of Education for Northern Ireland (DENI), 1990, **Curriculum (Programmes of Study and Attainment Targets in Mathematics) Order (Northern Ireland)**, HMSO, Belfast.

Department of Education for Northern Ireland (DENI), 1995, *School Leavers 1992/93*, **Statistical Bulletin SB1/1995**, HMSO, Belfast.

Houston SK, 1989, *The Northern Ireland Further Mathematics Project*, **Teaching Mathematics and its Applications, 8**, pp 115 - 121.

Houston SK, 1992, *The Northern Ireland Further Mathematics Project : The End of the Experiment*, **Teaching Mathematics and its Applications, 11**, pp 155 - 158.

Houston SK, 1993, *Comprehension Tests in Mathematics*, **Teaching Mathematics and its Applications, 12**, pp 60 - 73.

London Mathematical Society, 1995, **Tackling the Mathematics Problem**, London Mathematical Society, London.

Ministry of Education for Northern Ireland, 1969, **Education in Northern Ireland in 1968 : Report of the Ministry of Education for Northern Ireland for the year ended 31 December 1968**, HMSO, Belfast.

National Council for Vocational Qualifications (NCVQ), 1995, **Core Skills Units, Application of Number, Levels 1-3**, NCVQ, London.

National Foundation for Educational Research (NFER) 1984, Assessment of Performance Unit (APU), **A Review of Monitoring in Mathematics, 1978 to 1982**, NFER, London.

Northern Ireland Council for the Curriculum, Examinations and Assessment (CCEA), 1994, **Assessment Units in Mathematics**, CCEA, Belfast.

Northern Ireland Council for the Curriculum, Examinations and Assessment (CCEA), 1995a, **Proposals for Revised Subject Requirements : Mathematics**, CCEA, Belfast.

Northern Ireland Council for the Curriculum, Examinations and Assessment (CCEA), 1995b, **GCSE Mathematics 1997, Syllabus A and B**, CCEA, Belfast.

Northern Ireland Council for Curriculum, Examinations and Assessment (CCEA), 1995c, **GCSE Mathematics 1998, Draft Syllabus A and B**, CCEA, Belfast.

Northern Ireland Council for the Curriculum, Examinations and Assessment (CCEA), to be published, **Processes in Mathematics**, CCEA, Belfast.

Osborne RD, 1986, *Segregated Schools and Examination Results in Northern Ireland : Some preliminary Research*, **Educational Research, 28**.

School Curriculum and Assessment Authority, (SCAA), 1993, **GCE A/AS Subject Core for Mathematics**, SCAA, London.

School Curriculum and Assessment Authority (SCAA), 1995a, **Regulations for the General Certificate of Secondary Education (GCSE***)*, SCAA, London.

School Curriculum and Assessment Authority (SCAA), 1995a, **Criteria for GCSE /
Key Stage 4 Mathematics**, SCAA, London.

Sutherland AE and Gallagher AM, 1986, **Transfer and the Upper Primary School**,
Northern Ireland Council for Educational Research (NICER), Belfast.

Teare SM and Sutherland AE, 1988, **At sixes and Sevens ; a Study of the Curriculum
in the Upper Primary School**, Northern Ireland Council for Educational Research
(NICER), Belfast.

Wilson JA, 1985, **Secondary school organisation and pupil progress**, Northern Ireland
Council for Educational Research (NCER), Belfast.

2

Mathematical Modelling and the Way the Mind Works

Susan J Lamon
Marquette University, Milwaukee, Wisconsin, USA 53233-1881
e-mail: sue@mscs.mu.edu

ABSTRACT

There is a confluence of ideas from medical research, cognitive science and educational mathematics indicating that the process of mathematical modelling facilitates the construction of conceptual models which, in turn, develop the mind's power to structure experiences. This paper uses interviews with elementary, middle school and adult students to illustrate the characteristics of conceptual models and the nature of their evolution into more stable systems during the mathematical modelling process. It concludes that mathematical modelling should be a part of the school mathematics curriculum beginning in the elementary years.

INTRODUCTION

Educational psychologists, teachers and mathematics educators argue that preparing the next generation to meet the challenges of our technological environment is not as simple as teaching more people to do mathematics. We do not teach mathematics - we teach children mathematics. Without some regard for the way in which children are thinking about, integrating and using mathematical ideas, teaching can be a fruitless endeavour. Furthermore, we do not teach children merely to do mathematics - we teach them about mathematics, about values and about scientific habits of mind. We initiate them into a mathematical culture (Bishop, 1988). For these reasons modelling may be viewed as an interface for mathematics content, realistic context, mathematical culture and human cognition.

New research methodologies in modern cognitive science open windows to the workings of the human mind and enable mathematics education researchers to do conceptual analyses of children's and adults' mathematical activity rather than purely procedural analyses of problem solving. We are studying in greater depth than ever before the development of conceptual models. As children and adults work in real situations, they make sense of them by imposing some organisation or structure that takes the form of language, pictures, written symbols or real objects. These representations give witness to the person's underlying cognitive model. The process of modelling sustains a dialectic between cognitive models and mathematical models. The more co-ordinated and internally consistent a person's conceptual model is, the better he or she can solve real problems; at the same time, conceptual models often evolve to greater levels of stability and sophistication during the process of solving problems by mathematical modelling.

Our understanding of the connections between mathematics, modelling and the human mind is further enhanced by technological advances. One of the most amazing accomplishments of the last decade has been a technology called positron-emission tomography (PET) scanning that allows researchers to see inside living brains and to examine brain functions at their subcellular level. As the human brain works on various tasks, increased metabolism in the active cells produces a spectrum of colours to mark the areas of the brain involved in the activity and the intensity of their involvement (Raichle, 1995). What scientists are discovering about the brain and its neurological communication system complements and supports the theories and results of cognitive research concerning thinking, learning and problem solving.

I will begin by examining some of the concordant results of cognitive science and medical research about the human brain and the way it functions. Later I will offer some compelling protocols of children's and young adults' modelling activity to illustrate some of the properties of conceptual models currently under investigation in mathematics education research.

COGNITIVE SCIENCE AND MEDICAL RESEARCH

Beliefs that only observable stimuli and responses were legitimate objects of study stifled research on thinking until the beginning of the 20th century (Mayer, 1983). In a relatively short period of time, cognitive science research - both from cognitive psychology and artificial intelligence perspectives - together with the results of PET scan brain research, have considerably enhanced our understanding of human cognitive functioning. The following examples consider a few of the ways in which substantially different types of research, each using different methodologies, arrive at similar conclusions.

- Cognitive psychology holds the position "that cognition is a process, whereas the psychometric view makes it a collection of abilities" (Hunt, 1995: page 359). The neuroscientist would contend that both perspectives are useful. Cognition is an electrochemical communication process among neurons in the brain, but people also have innate abilities. For example, the planum temporale in the left hemisphere of the brain is associated with auditory processing. It is larger in musicians than in nonmusicians and is larger still in musicians with perfect pitch (Swerdlow and McNally, 1995).

- One of the major contributions of the information-processing models of human problem solving was in requiring the specification of the initial state of the problem solver. Today, cognitive psychology has abandoned the search for learning models which assume a universal simplicity in the mental processes and capacities used to assimilate knowledge, in favour of emphasizing the richness and complexity of the experience students bring into the classroom. Brain researchers have found that richness of experience results in more connections between the hundred billion neurons in the brain which, in turn, means a better functioning brain. For example, rats raised in cages full of toys have a greater brain mass than those raised in empty cages, probably due to the growth of dendrites or threadlike extensions that grow out of the neurons to enhance neural communication (Swerdlow and NcNally, 1995).

- One of the earliest constructs from cognitive psychology was the 'schema' (Bartlett, 1932). A schema refers to a cognitive structure or configuration of existing pieces of knowledge associated with our understanding of a particular situation. Meaningful learning is attaching new information to existing cognitive configurations (Skemp, 1987). Likewise, in information processing, the ability to 'chunk' pieces of data rather than remember them individually extends the capacity and improves the performance of the short term memory and is thought to increase with age, presumably because it depends on a person's previous experience (Case, 1978). Neuroscience confirms the validity of connectivity models when it defines learning as "forming new connections between neurons". Memory, rather than residing in a particular location, consists in "changed connections between different sets of neural networks" (Swerdlow and NcNally, 1995: pages 7, 35) or bundles of associated signals.

- Psychometricians have noted that creative problem solving decreases from early adulthood onward; young people are much better at dealing with unfamiliar tasks (Hunt, 1995). Cognitive psychologists would say that this is because novices have to develop new problem representations, whereas with age and experience problem representation switches more to pattern recognition or the activation of already established connections. Correspondingly, neuroscientists have determined that the process of producing new connections between neurons (learning) is most intense in childhood and that dendrite production rises rapidly after birth and remains at a peak from about age four to age ten.

CONCEPTUAL MODELS

Researchers now examine student thought processes in fine detail using qualitative methods such as clinical interview techniques, introspection and protocol analysis. Going beyond mere descriptions of what children know or what their deficiencies are, beyond litanies of their misconceptions and bugs and beyond procedural analyses of problem-solving episodes, some current research efforts are now aimed at determining the characteristics of cognitive models. A **conceptual model** or C-model is a cognitive system of mathematical ideas and their associated processes that a student brings to bear on a phenomenon of interest, as inferred from the student's representations in words, pictures, diagrams, symbols or other external media (models of models).

'Model-eliciting' problems (Lesh and Lamon, 1992) are used to explore the characteristics and organisation of a student's thinking in mathematically rich contexts. While interacting with a realistic situation that may be organised using a variety of mathematical ideas, student thinking is analysed to determine what mathematical structures the student imposes on the situation in an effort to make sense of it. C-model analysis is used not only to determine what it means to understand a mathematical idea, but also to determine which viable models constitute more sophisticated understandings. Which models are 'unstable' (Lesh, 1885), *ie* poorly co-ordinated; which are unsuited to the task or are immature? Which models are most resistant to change and in which direction are they changing? What are some of the intermediate stages in the development of a mathematical idea and how can the teacher recognise them in a student's representations?

CHARACTERISTICS AND DEVELOPMENT OF C-MODELS

The following examples illustrate some of the most salient characteristics of unstable but evolving conceptual models as well as some of the mechanisms by which they develop.

- Models develop over time. They are primitive at first - sometimes not even mathematical - and they gradually increase in sophistication.

The activity, shown in Fig. 1, is an adaptation of a problem first proposed by Hans Freudenthal (1983). It is particularly good in eliciting a wide range of solutions, not all of them mathematical. Activities with multiple levels of responses provide good assessments of a student's primitive ideas whilst allowing room during the investigation for that student to adopt more sophisticated perspectives, in essence making learning and assessment a seamless process. A range of responses elicited from many people often suggests directions in which personal models will change over time to reach new levels of sophistication (adequacy or usefulness).

As shown in Fig. 2, Student A employs the most primitive ideas, using a model that makes true/false, right/wrong decisions based purely on perceptual cues and employing no mathematical ideas at all. Frequently a logical model is used to organise the situation. Student B's logical model applies some faulty reasoning. Next, students move to a measurement model. The measurement model used by Student C showed some problems in choosing appropriate units of measure, while Student D used a

Figure 1. The tower problem.

multiplication-as-repeated-addition model with faulty recall of multiplication facts. Students E and F appear to use some elements of a proportionality model.

- C-models are complicated and multi-phased.

- Primitive models often contain misconceptions that need to be challenged.

- In the development of a model, obvious, surface-level features of reality are usually noticed first. Several rounds of re-examination of the given situation are needed to extract less obvious relationships and details.

- Models sometimes go through several rounds of revision during a student's engagement with an activity so that we are able to see progression toward more sophisticated ways of thinking within a single problem-solving episode.

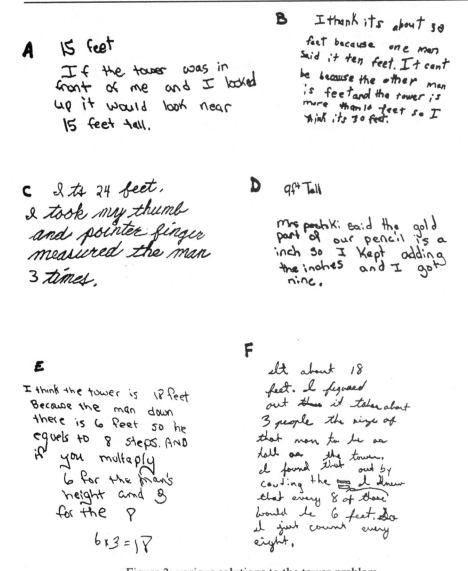

A 15 feet

If the tower was in front of me and I looked up it would look near 15 feet tall.

B I thank its about 30 feet because one man said it ten feet. It cant be because the other man is feet and the tower is more than 10 feet so I think its 30 feet.

C It ts 24 feet.
I took my thumb and pointer finger measured the man 3 times.

D 9ft Tall

mrs pashki said the gold part of our pencil is a inch so I kept adding the inches and I got nine.

E

I think the tower is 18 feet Because the man down there is 6 feet so he equels to 8 steps. AND if you multaply 6 for the man's height and 3 for the P

$6 \times 3 = 18$

F

It about 18 feet. I figuaed out that it takes about 3 people the size of that man to be as tall as the tower. I found that out by couting the I knew that every 8 of those would be 6 feet. So I just count every eight.

Figure 2: various solutions to the tower problem

The following interview was part of a teaching experiment with middle school children, designed to help them use their informal knowledge about the human body to develop a sense of proportion. Michael, a seventh grader, revised his model several times in a short period of time.

I: If your arms were six feet long, how tall would you be?

M: *Long wait.* Six feet one.

I: Why do you say that?

M.   'Cause you don't want your fingers to drag on the ground.

I: Show me with your thumb and your pointer finger how long an inch is.

M: *Shows a length of approximately two inches.*

I: Seems to me, if your fingers were only an inch off the ground, you would bump them on every stone you passed.

M: Yeah, I guess you would.

I: Well what do you think about that?

M: I guess you better have them up a few more inches. Probably five or six.

I: Draw me a picture. What would you look like then?

M: *Long pause.* Nah. I can't do that. My arms would be coming out my head. Now wait a minute. My arms are as long as my head plus the rest of me down to my stomach. My waist is half of me. My waist is about the middle of my body and my arm is about half my body. So I would say that if I had arms six feet long, I would have to be about 12 feet tall.

    (The next day, Michael reported that he had had a doctor's appointment for a physical and this caused him to return to the question we had been discussing.)

M: I just had a physical after school yesterday, so I found out I am exactly five feet tall. *M extended his arm in front of him.* My arm is about two feet long I would say. So if I grew twice as tall, my arms would be twice as long. That would be four feet. Somebody with arms six feet long would be about two feet taller than that, or 12 feet tall. So the answer I got yesterday must be about right.

In the end Michael showed a fairly sophisticated strategy, a combination of building up and doubling. Unfortunately he resorted to additive thinking when it was impossible to double again, revealing that he is not yet a multiplicative thinker. Even worse, the fact that he arrived at the same answer twice convinced him of the goodness of his additive model. Nevertheless, within a single episode, we can see model development from nonsensical to mathematical thinking.

- The stability (maturity, internal consistency) of a C-model is demonstrated by translations within and between representations.
- When a C-model is stable, students do not apply procedures as rote.
- When the mental model is stable, successive iterations of the modelling cycle are able to incorporate more of the relevant information .
- In successive iterations of the modelling cycle, conceptual models move to higher levels of abstraction.

University Algebra students were given a traffic-flow problem (de Lange, 1992) to investigate in a two-stage examination. They were given one week to investigate fully the information given in Fig. 3 before they were asked the following questions.

(1)   How much time is needed to have all the lights turn green exactly once?

(2)   Determine $G = G_1 + G_2 + G_3 + G_4$ then $T = 30\ G$. What do the elements of T signify?

(3)   Ten cars per minute can pass through the green light. Show, in a matrix, the number of cars that can pass in each direction in one hour.

(4)   Compare this matrix to matrix A. Are the traffic lights regulated accurately? If not, can you make another matrix G in which traffic can pass more smoothly?

A young woman entered the final examination with the main question already answered. 'In my mind, this is a maximisation problem - actually, a maximisation and a minimisation problem', she explained. 'You have to maximise the safety of all the drivers, and minimise the amount of time they sit at the red lights so that they don't lose their tempers and jeopardise their safety.'

Here you see a cross-roads in Geldrop, the Netherlands, near the Great Church.

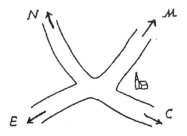

In order to let the traffic flow as smoothly as possible, the traffic lights have been regulated to avoid rush hour traffic jams.

A count showed that the following number of vehicles wished to pass the cross-roads during rush hour (per hour):

$$
A:\ \text{from}\quad
\begin{array}{c}
M \\ N \\ E \\ C
\end{array}
\begin{bmatrix}
0 & 40 & 200 & 30 \\
30 & 0 & 80 & 50 \\
210 & 60 & 0 & 60 \\
30 & 40 & 80 & 0
\end{bmatrix}
$$
$$\quad\quad\quad M \quad\ N \quad\ E \quad\ C$$

The matrices G_1, G_2, G_3 and G_4 show which directions had a green light and for how long. In G_1, for instance, you can see that the directions M $\to$ E, E $\to$ M, M $\to$ N and E $\to$ C can flow simultaneously for a period of $\frac{2}{3}$ minute. The other directions are then blocked by a red light.

$$G_1 : \quad \begin{array}{c} \\ M \\ N \\ E \\ C \end{array} \begin{array}{cccc} M & N & E & C \\ \left(\begin{array}{cccc} 0 & \dfrac{2}{3} & \dfrac{2}{3} & 0 \\ 0 & 0 & 0 & 0 \\ \dfrac{2}{3} & 0 & 0 & \dfrac{2}{3} \\ 0 & 0 & 0 & 0 \end{array}\right) \end{array} \qquad G_3 : \quad \begin{array}{c} \\ M \\ N \\ E \\ C \end{array} \begin{array}{cccc} M & N & E & C \\ \left(\begin{array}{cccc} 0 & 0 & 0 & 0 \\ 0 & 0 & \dfrac{1}{2} & \dfrac{1}{2} \\ 0 & 0 & 0 & 0 \\ \dfrac{1}{2} & \dfrac{1}{2} & 0 & 0 \end{array}\right) \end{array}$$

$$G_2 : \quad \begin{array}{c} \\ M \\ N \\ E \\ C \end{array} \begin{array}{cccc} M & N & E & C \\ \left(\begin{array}{cccc} 0 & 0 & 0 & \dfrac{1}{3} \\ 0 & 0 & 0 & 0 \\ 0 & \dfrac{1}{3} & 0 & 0 \\ 0 & 0 & 0 & 0 \end{array}\right) \end{array} \qquad G_4 : \quad \begin{array}{c} \\ M \\ N \\ E \\ C \end{array} \begin{array}{cccc} M & N & E & C \\ \left(\begin{array}{cccc} 0 & 0 & 0 & 0 \\ \dfrac{1}{2} & 0 & 0 & 0 \\ 0 & 0 & 0 & 0 \\ 0 & 0 & \dfrac{1}{2} & 0 \end{array}\right) \end{array}$$

Figure 3: the stoplight problem

She had constructed an algorithm for progressively maximising safety and minimising wait time for all the drivers. Her matrices, shown in Fig. 4, were accompanied by the following explanation.

'If I arrange the times as in the New G_1 matrix, then it leads to the New C_1 matrix for the number of cars that may pass through the intersections. This arrangement just barely handles the traffic jam that would occur in travelling from E to M, so I experimented to see what other times would give. But, as you can see, all I need to do is adjust the fractions to get different results. If I add a little time to increase the margin of safety at the trouble spot, I get the New G_2 matrix and the New C_2 matrix. Everyone can then pass through the intersections safely, but there is too much waiting time in the directions where not as many cars need to pass. I could keep this up, but I would really prefer to make one other change. If I could change the original condition that the traffic lights go through a full cycle every two minutes, then I would suggest reducing the cycle to 1.5 minutes. I would start assigning the times as follows:

$$G_1 : \frac{14}{20} \qquad G_2 : \frac{4}{20} \qquad G_3 : \frac{6}{20} \qquad G_4 : \frac{6}{20}$$

Then I would proceed to adjust the fractions as I did before until I could get the best results'.

This student mathematised the given situation by first organising and structuring the critical requirements and constraints into her fraction model. In her initial, intuitive investigations, she manipulated the fraction model to see how it affected the matrix A. Finally she formalised and abstracted the process of adjusting the lights with an algorithm. Having simplified the adjustment process, she then saw a way to incorporate another variable: the length of the cycle. In her next iteration of the modelling cycle, her C-model assumed a higher level of abstraction when she operated on the algorithm she had constructed in previous stages.

- To help accommodate new or slightly different circumstances, cognitive processes appear to include analogical sensors.
- C-models are not really discrete entities. They often incorporate other models that are used in series and in parallel.
- C-models often find expression in a variety of interacting representations.
- Sometimes the C-model is highly personal; it might not be a model for someone else.

The following questions about ratios were devised to help children focus on the size of a ratio number, the relationship between a and b, rather than on its constituent parts. The questions required qualitative reasoning. 'Today I have fewer friends but the same number of candy bars as I had yesterday. Will each of us get more today, fewer, the same number, or can't you tell?'

A week after discussing the candy bar problems, Derek was asked structurally similar questions about pitchers of lemonade and glasses of different sizes.

I: Suppose you have a larger pitcher of lemonade and smaller glasses than you had before. Can you pour more glassfuls, fewer, the same number as yesterday, or can't you tell?

D: Well, the water got more and the glass got smaller. *Thinks for a long time.* You would get more glasses.

I: How did you figure that out?

D: Well, it's like this. The uh, well, it's like the kids and the cookies. You got more cookies and less friends, so you get more for everybody.

I: Could you explain a little more?

$$G = G_1 + G_2 + G_3 + G_4 \qquad\qquad T = 30G$$

$$G = \begin{bmatrix} 0 & \frac{2}{3} & \frac{2}{3} & \frac{1}{3} \\ \frac{1}{2} & 0 & \frac{1}{2} & \frac{1}{2} \\ \frac{2}{3} & \frac{1}{3} & 0 & \frac{2}{3} \\ \frac{1}{2} & \frac{1}{2} & \frac{1}{2} & 0 \end{bmatrix} \qquad T = \begin{bmatrix} 0 & 20 & 20 & 10 \\ 15 & 0 & 15 & 15 \\ 20 & 10 & 1 & 20 \\ 15 & 15 & 15 & 0 \end{bmatrix} \quad C = 10T = \begin{bmatrix} 0 & 200 & 200 & 100 \\ 150 & 0 & 150 & 150 \\ 200 & 100 & 0 & 200 \\ 150 & 150 & 150 & 0 \end{bmatrix}$$

$$\frac{2}{3} \text{ minutes} \qquad\qquad 200 \text{ cars per hour}$$

$$? \qquad\qquad 210 \text{ cars per hour}$$

$$\frac{\frac{2}{3}}{200} = \frac{x}{210}$$

$$x = \frac{7}{10} \text{ minutes}$$

$$\frac{7}{10} + \frac{3}{10} + \frac{5}{10} + \frac{5}{10} = 2 \text{ minutes}$$

$$\frac{7}{10} + \frac{3}{10} + \frac{5}{10} + \frac{5}{10} \qquad\qquad\qquad \frac{9}{10} + \frac{3}{10} + \frac{4}{10} + \frac{4}{10}$$

$$NG_1 = \begin{bmatrix} 0 & .7 & .7 & .3 \\ .5 & 0 & .5 & .5 \\ .7 & .3 & 0 & .7 \\ .5 & .5 & .5 & 0 \end{bmatrix} \qquad\qquad NG_2 = \begin{bmatrix} 0 & .9 & .9 & .3 \\ .4 & 0 & .4 & .4 \\ .9 & .3 & 0 & .9 \\ .4 & .4 & .4 & 0 \end{bmatrix}$$

$$NC_1 = \begin{bmatrix} 0 & 210 & 210 & 90 \\ 150 & 0 & 150 & 150 \\ 210 & 90 & 0 & 210 \\ 150 & 150 & 150 & 0 \end{bmatrix} \qquad\qquad NC_2 = \begin{bmatrix} 0 & 270 & 270 & 90 \\ 120 & 0 & 120 & 120 \\ 270 & 90 & 0 & 270 \\ 120 & 120 & 120 & 0 \end{bmatrix}$$

Figure 4: student matrices for the traffic flow problem

D: *Showing with his hands.* The cookies go up and the kids go down.

I: Tell me which is which. Are the glasses like the kids or like the cookies?

D: Well, you give out water to glasses. So it's like givin' cookies to kids.

I: Suppose you have a smaller pitcher of lemonade and you are using smaller glasses.

D: *Thinks for a long time.* Well it depends how many cookies I have.

I: What do you mean?

D: I don't know. This one tough 'cause you can't tell what it doin'. *Long pause.* If you got a whole lot of cookies like a whole package, you could just take one away an' me an' my two friends would still get a lot. But you only got a couple, and you take one away, we might have to split. So you don't know if you got enough water for them glasses or not.

Derek's C-model was complex. Using several analogies in tandem, he moved from a bottles-and-glasses schema to a candy-bars-and-children schema to a cookies-and-children schema, effectively using several models simultaneously to make sense of the situation. The cookies-and-children model was so much like the candy-bars-and-children model that he spoke of it as if it were identical. It appears that some personal preference caused this translation and thus the model and the reasoning it entailed were evidence of a unique, personal construction. Derek's ability to move between models and to identify an ambiguous situation showed that he had a strong conceptual model, well adapted to the situation.

My work with children and adults as their conceptual models evolve during sense-making activities confirms perspectives that have come from artificial intelligence research: "People are more fundamentally model builders than they are formal system builders. They reason by analogy. They induce" (diSessa, 1979: page 251).

GETTING BACK TO BASICS

Understanding the real problem, the first step in the modelling process, entails differentiating the problem from the question. In part this is what is meant by identifying the mathematical structure within the context (Freudenthal, 1991). Most often the question being asked does not immediately highlight the real issue that needs to be resolved. For example, in the traffic flow problem, while the question asked students to consider the timing of the lights, it took some 'mucking around' in the situation to discover that the real problem was in maximising safety while minimising wait-time. This suggests that quick-answer questions in the mathematics classroom deny students access to the critical non-answer-giving stages of problem solving, where the real business of model building and building understanding takes place.

As many notable modellers have advised (Berry and Houston, 1995; Edwards and Hamson, 1989; Cross and Moscardini, 1985), modelling is a purposeful process, not an exercise in which mathematics is an end in itself. We construct mathematical models in order to solve problems. Similarly the chemical processes in the brain that build dendrites are inactivated unless the disequilibrium caused by a problem is sensed by the individual and he or she is motivated to consider alternative perspectives or resolve anomalies. This suggests that the drill and practice questions currently asked of students in the classroom, and not necessarily perceived as being problematic, may not engage the brain in knowledge-building activity.

Modifications may be needed in instructional sequencing that starts with simple situations and spares children the complexity of co-ordinating several variables and relationships. Brain architecture suggests, as does psychometric research on fluid (creative) intelligence and crystallised (previously acquired) intelligence, that "learning to do an intellectual task will generally be harder than doing it" (Hunt, 1995: page 360) or, in other terms, electrochemical communication is easier than growing new dendrites. These results suggest that up-front, in-depth struggles will make later work easier. The sterility of school-type problems denies students adequate access to the challenge and complexity needed to build cognitive structures that might facilitate later mathematical thinking.

Students' first pass at a problem is often superficial. Off-the-cuff responses, like those students are often expected to give in their mathematics classes, are different from ones that result from careful thought. Since modelling focuses on the purpose of the mathematical activity and considers all of the constraints involved, sensible judgements can be made about such issues as how accurate a solution needs to be, what are the consequences of making an error and how quickly a response needs to be generated. Textbook problems fail to provide enough context to encourage student sense making, planning, and judgement.

CONCLUSION

The biggest problem in mathematics education is how to package mathematics in an instructional setting in a way that best preserves the integrity of the powerful, logical structure of the subject and conveys our beliefs about what it means to do mathematics, whilst taking into account the cognitive reality of human beings. For many reasons mathematical modelling offers a viable methodological alternative to direct instruction in the mathematics classroom. From what we know about the human brain and its functioning, it appears that modelling engages human cognition in a process to which it is best suited. In many ways, current classroom practices have robbed students of what should be a natural process. Mathematical modelling should not be merely 'incorporated' into the mathematics curriculum as an occasional activity. For our students, it should become synonymous with doing mathematics.

Neurological evidence delivers the powerful message that young children's brains are more powerful than they will be later in life. This fact should create an urgency about harnessing that power and initiating young children into the mathematical modelling process. Beginning in the early elementary years, children should learn to handle the complexity of real problems, to make explicit assumptions, to mathematise the situation at hand, to draw conclusions, and to relate their conclusions to the initial conditions and assumptions.

REFERENCES

Bartlett FC, 1932, **Remembering**, Cambridge University Press, London.

Berry J and Houston K, 1995, **Mathematical Modelling**, Edward Arnold, London.

Bishop AJ, 1988, **Mathematical Enculturation: A Cultural Perspective on Mathematics Education**, Kluwer, Dordrecht.

Case R, 1978, *Intellectual Development from Birth to Adulthood: A Neo-Piagetian Approach*, in Siegler RS, (ed.), **Children's Thinking: What Develops?**, Erlbaum, Hillsdale, NJ.

Cross M and Moscardini AO, 1985, **Learning the Art of Mathematical Modelling,** Ellis Horwood, Chichester.

de Lange J, 1992, *Assessing Mathematical Skills, Understanding, and Thinking*, in Lesh R and Lamon SJ, (eds.), **Assessment of Authentic Performance in School Mathematics**, Association for the Advancement of Science, Washington, DC, pp 195-214.

di Sessa AA, 1979, *On 'Learnable' Representations of Knowledge: A Meaning for the Computational Metaphor*, in Lochhead J and Clement J, (eds.), **Cognitive Process Instruction**, Franklin Institute Press, Philadelphia, PA, pp 239-266.

Edwards D and Hamson M, 1989, **Guide to Mathematical Modelling**, Macmillan, London.

Freudenthal H, 1983, **Didactical Phenomenology of Mathematical Structures,** D Reidel, Dordrecht.

Freudenthal H, 1991, **Revisiting Mathematics Education: China Lectures**, Kluwer, Dordrecht.

Hunt E, 1995, *The Role of Intelligence in Modern Society*, **American Scientist**, July-August, pp 356-368.

Lesh R, 1985, *Conceptual Analyses of Problem-Solving Performance,* in Silver EA (ed.), **Teaching and Learning Mathematical Problem Solving**, Erlbaum, Hillsdale, NJ pp 309-329.

Lesh R and Lamon, SJ, 1992, (eds.), **Assessment of Authentic Performance in School Mathematics**, Association for the Advancement of Science, Washington, DC.

Mayer RE, 1983, **Thinking, Problem Solving, Cognition**, Freeman, New York.

Raichle ME, 1995, quoted in Swerdlow JL and McNally J, *Quiet Miracles of the Brain*, **National Geographic Magazine**, 187, 6, pp 2-41.

Skemp R, 1987, **The Psychology of Learning Mathematics**, Erlbaum, Hillsdale, NJ.

Swerdlow JL and McNally J, 1995, *Quiet Miracles of the Brain*, **National Geographic Magazine**, 187, 6, pp 2-41.

3

What Makes A Model Effective and Useful (or Not)?

Eva Jablonka
Technische Universität Berlin, Strasse des 17. Juni 136, D-10623 Berlin, Germany
e-mail: madid@math.tu-berlin.de

ABSTRACT

Within general education, mathematical modelling can do much more than merely provide experience of solving specific problems, especially if these problems will only be problems for a few of the students in their future lives. Mathematical modelling can contribute to developing a comprehensive picture of mathematics. Genuine examples of mathematical models can provide a source for reflecting and assessing the use of mathematics within diverse areas of application, and can draw attention to the importance of evaluating proposed and underlying models. Whereas the evaluation of the effectiveness of a mathematical model is based on methodological considerations, critical appraisal of its usefulness includes a moral or political evaluation of the purpose, of the criteria for an effective solution, and of the actual and possible consequences of the implementation or use.

INTRODUCTION

Mathematical Models and Modelling

Solving problems by means of mathematical modelling, as is well-known (see, for example, Niss, 1989), is a purposeful activity and is not merely an end in itself. The aim may be to answer a certain question (for description, prediction, explanation or justification), or within a larger context, it may be part of a technological process in which the model then is used to aid decision making, planning and such like. Thus a model can serve as representation or presentation. Developing a model can be seen as a search for a description or even a 'cause' for an observed fact, as constructing a design for something which does not already exist, or for developing a plan for an action. The object of modelling need not be a piece of physical reality, whether natural or artificial. It may also be a social system, a phenomenon of psychology, a procedure of distributing power or money, a calculation of earning or costs, or some other regularity found in social actions.

Modelling can hardly be described as a straightforward activity because problem specification, observation, classifying, identifying relevant factors, hypothesising, choosing assumptions, interpreting and other sub-processes are mutually related and driven by the aims and underlying interests of the modeller. Evaluation takes place throughout the whole process. The mathematical concepts and methods at hand, the technical means (measuring devices, procedures for collecting data, calculators, computers) including rules for their use, domain-specific conceptual frames or (scientific) theories, all influence the form of the model in addition to the constraints of the situation. The mutual relationship between problem definition, the available toolkit of mathematical techniques and the possible form of the solution or model is also emphasised in the descriptions of the modelling process given by Finlay and King (1987) or by Moscardini (1989).

One of the fundamental shortcomings of a simplistic description of the modelling process – underpinned by an objectivistic philosophy of science – is suggesting that the evaluation of the model is unproblematic and can be based on objective methods, simply by confronting it with observed facts (the 'reality'), leading either to a solution of the original problem or to a refinement of the model which is then a more accurate description. However, depending on the aims, there may result qualitatively different descriptions of the same 'reality', none being just a refinement or generalisation of the other. Since the problems are set from outside and since the criteria used in selecting them and in defining what is considered a solution are external to mathematical and methodological considerations, the evaluation cannot be based on these considerations alone. The evolution of the problem has to be studied together with the evolution of its solution. A detailed description of this view of the modelling process is given by Jablonka (1996).

Views and Misconceptions

Influencing the assessment of the results and the interpretation of mathematical models, the following different types of simplistic views and misconceptions may occur (albeit as extreme cases):
- a transmission of the prestige of certainty, exactitude, irrefutability and uncontroversial nature which pure mathematics enjoys (whether justifiably or not) to applications,
- a biased image of usefulness and direct applicability of mathematics in the form of unjustified evaluation or exaggerated optimism,
- mistaking mathematical sophistication as an indicator of the quality of a model,
- a belief in the reliability of mathematical models in general as if they all were constructed following an ideal process of scientific inquiry,

- ignoring the conditions and constraints under which a mathematical model is valid,
- overlooking the different methodological and epistemological status of models with the same mathematical structure,
- a lack of knowledge about refined or qualitative different models of the same 'reality',
- identifying the structure of the model with that of reality, especially with respect to very elementary or to standardised models ('naive realism').

Competence for evaluation

The need for broader reflection on the subject matter (applications of mathematics) and thus for higher levels of knowledge including evaluative competence, follows from the pedagogical argument regarding the mission of mathematics education as something that prepares students for full citizenship, and is underpinned by the fact that mathematics is used within all areas of science, industry, commerce, and administration. Thus some of the 'cultural aims' (Blum, 1993) are:

- to gain some knowledge of which kind of mathematics are successfully used in which areas,
- to educate awareness of applications that affect society's aiming at a citizenry contributing to our mathematised world with intelligence (Davis, 1989),
- to gain individual autonomy – which tends to be suppressed by specialisation and segmentation of knowledge – to assess mathematical models developed by experts,
- to develop consciousness of the limits of reliability, especially of non-transparent and uncontrollable IT-based mathematical models (Booß-Bavnbek and Pate, 1989),
- to be prepared to interpret information presented in a more or less scientific way (de Lange, 1993),
- to contribute to the formation of democratic competence by providing a cluster of exemplary mathematical, technological and reflective knowledge and a democratic attitude (Keitel *et al.*, 1993).

Lack of reflection can even undermine the proper use of skills and techniques. Consequently higher-order thinking or evaluation is not just an additional activity or by-product but the core of expertise which includes reformulating problems, comparing different solutions, justifying and communicating and should be performed with responsibility and commitment (Lewis and Gagel, 1992).

A key question, with respect to evaluative processes, is to what extent lack of specialist knowledge can be compensated by general considerations. According to the analytical distinction discussed below, the issues of usefulness can be addressed without relying on technical knowledge because it is possible to think of any model as if its effectiveness were proved in some way. That does not mean that judgement and evaluation of usefulness is not an even more demanding task. If, on the other hand, the potential usefulness is shown and the underlying modelling process seems reasonable, then specialist domain-specific and mathematical knowledge will be needed to evaluate details.

The following catalogue of elementary questions is meant to guide the analysis of mathematical models from the perspective of a user or 'consumer' of a model, for instance as a basis of a statement within a political discussion, as a rule of financial mathematics, in the form of a computer application, as a piece of scientific literature or as an example in a mathematics textbook. General answers to these questions can be derived partly from criteria for good science – which may be informed by different paradigms – and partly from criteria for reasonable technological solutions, or be based on a theory of rational communication, depending on the context within which the model is used. Similar questions, probably with the exception of those related to criteria for solution and likely consequences, are in a much more specific form thus guiding the process of developing a mathematical model in practice (Blechmann Myškis and Panovko, 1984, or Gross and Knauer, 1989; see also Feichtinger, Jablonka and Winter, 1989 for the results of an investigation into the need and availability of mathematical awareness in Austria).

EVALUATION

Effectiveness and usefulness

When evaluating a mathematical model there are two mutually related aspects: effectiveness and usefulness. Evaluation of effectiveness in the following discussion refers to a context-bound means-end analysis with respect to the specified problem: is the model likely to fulfil the special purpose for which it was constructed? Mathematical knowledge can only help in studying the behaviour and the mathematical implications of the model or solution methods. Besides potentially sophisticated mathematical and methodological considerations concerning empirical tests and other possibilities of validation, the evaluation of effectiveness is based on knowledge about the process of collection and recording of data, as well as knowledge from theories about the domain of interest. The evaluation of usefulness includes a critical appraisal and evaluation of the purpose, of the criteria for an effective and efficient solution, and of the intended, actual and possible consequences of the implementation or usage.

An effective model can be useful as basis for an action: for instance a reliable description of the stopping distance for a car can be used to determine a minimum safety distance. The effectiveness of models which are conventional stipulations prescribing an action (rules for organising social systems, so called normative models) and thus not aiming at describing a piece of experimental or observational reality, can only be judged by analysing and – to a certain degree – formalising the aims and comparing these with the mathematical properties of the model. There are also many models providing only a crude caricature of experimental or observational reality which, for this reason, cannot be interpreted as descriptions. So there might be no distinct interpretation of their effectiveness because of the lack of a clear statement about the problem to be solved, yet they might be useful within the process of scientific inquiry. Sometimes scientists claim a wide range of applicability for a mathematical model leading to exaggerated expectations which later turn out to be unfulfilled. On the other hand, an effective model is not automatically of use to somebody or something because the mode of use might not yet have been invented. The critical appraisal of the usefulness of a mathematical model is only possible with respect to the factual context of usage. For example, if a pacifist uses a model of two warring armies, aiming to demonstrate that fighting is not necessary (as in the case of Richardson), this intention may be achieved, though the model is not effective. Thus, evaluation is not possible without addressing the interests, aims and values associated with the context of usage.

EVALUATING EFFECTIVENESS

Behaviour of model and solution method

Exploration of mathematical properties

Studying the mathematics of a model may have many implications for the assessment of the effectiveness of the particular model, as well as for the theoretical considerations which lead to the construction of the model. Deriving consequences from a given model may reveal the reasonableness or absurdity of some of the assumptions on which it is based, may contribute to clarifying the domain of validity, may help in controlling the consistency of theoretical considerations, can show under which conditions a certain behaviour occurs, may lead to categories for describing a system, may aim at comparison of different models, and so on. If a model is a formalisation of rules, investigating the mathematical properties is the only way of checking its effectiveness with respect to some basic statements about its requirements, or for reconstructing these from the symbolic representation of the model.

Relations between input and output

If a model is relying on data, there are a lot of possibly demanding questions which could be asked concerning the estimation of consequences of different kinds of errors in order to determine the range of values of the measured quantities to guarantee a given accuracy of the results. This may cause unsolvable problems, especially if a lot of components or a lot of data are involved, or if software is used which is always non-transparent and not open to scrutiny by the practitioner.

There are many examples to be found in newspapers which provide an opportunity to address the question of suggestive accuracy of statements derived from mathematical models or simple calculations.

Data

Quality and origin

The considerations above are useless without some knowledge of the data. The theoretical and factual process of quantification is crucial to the effectiveness of a model. In many examples found in textbooks any information on the source of given observational or experimental data is missing. Where do the data (the actual numerical values) come from and how were they measured? Are there other sources available? What factors could have influenced the phenomenon being measured, the measuring instrument or its use? For which scale are they defined. Which theory or which procedure was used for indirect estimates or for condensing complex data? The task of assessment of the quality of data may be extremely complex, varying from case to case - there are big differences between estimating a physical quantity, a rate of unemployment, a sex distribution, a death rate from vital registration data, an index of social cohesion. In surveys people possibly lie, are influenced by the way the question is posed or answer what seems desirable, and so on, so that validity is not easily guaranteed, though the measurement may be reliable. For assessment knowledge from statistics, from hypotheses testing, scaling, empirical social research - as well as knowledge about the process of collection and recording of the data- may be needed, but it is worth making a serious attempt to recognise some of these problems.

Definition of quantities and measuring procedure

Does the definition of the quantities involved state anything about their factual measurability? If some parameters are not accessible to any kind of measurement, the model allows qualitative interpretations only. Sometimes unobservable variables are replaced by numbers so that confusion may arise. Another problem may arise from an inadequate definition of the concept intended to be measured, for example psychological properties or sociological categories, or by choosing arbitrary indicators. Even if standardised, there remains the question whether there is more meaning attached to it than the description of the operations used to measure it.

Checking the results and assumptions

Empirical validation

Any model which is (metaphorically) interpreted as a representation of an observational or experimental reality should be exposed to some attempt at supporting this interpretation. The modes and feasibility of empirical validation vary greatly with respect to different contexts. There is a complicated interplay of data collection, hypotheses in the form of the model, construction of tests, experiments and purpose.

Given a mathematical model aiming at a description - without reference to any factual or possible mode of validation - should lead to an extremely cautious interpretation - the fact that a model fits to some data need not say much. This can be a result of *ad-hoc* manipulations, or of the construction of the test. Checking should comprise a comparison with other mathematical models and other descriptions, or theoretical results if available.

Relation to theory

In terms of the relationship to some theory, one can distinguish *ad-hoc*-models (modelling the performance by arbitrary fitting, not based on a description in terms of a theory) and models based on a theory (those which generate a class of systems allowing the construction of models effective for prediction and control). Booß-Bavnbek, Bohle-Carbonell and Pate (1988) give some alarming examples of the risk of 'feasibility over control'. Construction and implementation of a theory-based model may be too expensive and time consuming, the model can turn out to be too complex, or a theory does not exist. Thus, an *ad-hoc* construction may be more efficient or indeed the only possibility.

A source of misinterpretation of mathematical models is the confusion of their different statuses. Extrapolations of trends are often seen as previsions of future events and not as mere illustrations of future scenarios for instance, world population projections are usually *not* interpreted as scenarios according to different hypothesised future levels of fertility and mortality.   Widely accepted models of economics are interpreted as descriptions based on a theory (in the above sense), even if they lack any reference to observational or experimental data, and even if there is no reason to think of an economic system as not local and time-bound.

A lot of mathematical models cannot be tested because they involve too many restricting conditions so that their domain of validity does not exist or cannot even be constructed. Whether these models are useful can be assessed only with respect to their fruitfulness within a process of inquiry and this is a controversial issue.

EVALUATING USEFULNESS

Contribution to solution

Fruitfulness within scientific inquiry

Assessing the usefulness of a mathematical model which is not effective - for describing a certain process or event, for processing data, for representation of complex information, or for some other clear defined problem within a scientific inquiry - leads to questioning the heuristic value of such a model. It might be a tool for clarifying or reorganising concepts, or be useful for identifying key points, or it may lead to other more suitable models. It can also be a formulation of something already known, a rhetorical artefact, or the consequence of a certain style or fashion. This perception is always controversial, depending on the general role of mathematics which one is willing to ascribe to the field of research.

Contribution to action

If a mathematical model is used within a technological process (see Bunge, 1985: 229-230 for a suitably comprehensive definition) it may serve only as a means of documentation, presentation of data, or for communicating a description of the problem. If it is more than this, it may provide a basis for rational action, for example, if it gives an estimation of chances or risks, or an extrapolation of trends it may suggest beneficial changes than can be made. On the other hand, it may prescribe the course of action to be carried out or a systematic method of achieving a particular end. Examples may include giving concrete strategies for optimising time, room, or profit, or for saving money, being part of a construction design, or prescribing rules for taxation, paying interest rates, organising elections, choosing the winner of an Oscar and so on.

Criteria for solution

Who sets the goal?

Who derives an advantage from the results of the model? Who defines the criteria for the solution? Are there other conceivable criteria for an acceptable solution? Are there conflicts of perception of the situation? In case of applications related to military technology, the evaluation of the goal is crucial - in the course of a seminar on modelling at the University of Oldenburg, the group of students refused to work on a problem because of the potential usage of the outcomes for military purposes (Gross and Knauer, 1989).

Conflicts of perception of a problem may arise in almost any situation. Modelling traffic flow or parking space design will not reduce the amount of cars, maximising advertising exposure will probably not improve the quality of the radio or TV programme, minimizing costs of diet or room in a hospital will probably increase the unease of the patients while minimising production costs need not minimise energy consumption.

Framework

What are the assumptions of the theory and the constraints of practice within which the problem is formulated and the model is constructed? The constraints and assumptions imposed by the framework influence the options for a solution on a lower level. For example, if the problem of a bank employee advising a client is the comparison of financing offers for a loan, for the manager of the bank this is a problem of profitability, and for the customer it is one of planning, embedded in a certain economic system.

Is there a need for a mathematical solution?

Is seeking a mathematical solution a desirable way to find a solution? With everyday practical problems people often feel that there is no need for a mathematical solution. This may be because of the feasibility and efficiency of non mathematical solutions, or lack of transfer and appropriate concepts, or perceived incompatibility of social aspects of the situation with mathematics. To what extent should political, philosophical, social and legal arguments and decisions be substituted by or based on mathematical arguments? The question may arise, for instance, in calculations of social costs, environmental accounting based on monetary assessment and other market-type solutions of social choice problems, or discussions about missiles in terms of reliability, kill factors and so on. If the framework predetermines the use of mathematics, then this may only be questioned by questioning the framework itself.

Consequences

Risks

What are the possible ways of using the model? What are the likely consequences of use or of abstention? It seems reasonable that a model should not be used as a basis for action in a specific context if the consequences of an ineffective model are not acceptable. There are examples of transfer of context-bound reliability of mathematical models to new situations, ignoring the specific constraints and conditions. There are also examples of the application of models whose effectiveness is not known at all (the examples given by Booß-Bavnbek, Bohle-Carbonell and Pate, 1988 and Booß-Bavnbek, 1990). The danger of expanding the domain of application of *ad-hoc* constructions (much like the case of opaque software) lies in the fact that differences perceived as minor aspects may cause totally invalid results. However, in some cases not using unreliable models may be less desirable than their use, for instance when simulations showing an unnatural climatic change are used as arguments calling for desirable actions.

Transformation of meaning

A mass of more or less simple quantifications is institutionalised by 'social contract' (Davis, 1989) or from force of habit. As a consequence, these 'models' emanate a meaning without a conscious act of interpretation and tend to rule out alternative conceptions. On a higher level, the use of a mathematical model as part of a technological solution restricts the space of problem identification, the structure of argumentation, the basis for critique and the scope of possible actions (Skovsmose, 1994: pages 111-114).

CONCLUSIONS

Giving opportunities for discussion of different genuine examples of mathematical models, customs, and practices with respect to their reliability, degree of success and credibility, methodological and epistemological status and aims and consequences, could form an important part of activities aiming at critical competence which means to develop the power of discernment (Herget, 1984; Medley, 1984; Henn or Schroeder and Shealy in this volume). Concerning the use and misuse of statistics, there are a lot of well known books available as well as teaching materials (*eg* de Lange and Verhage, 1992). Bishop (1988) sees it as indicative of the way values are assumed that the area of projects pointing to aspects of social life strongly influenced by mathematics, in particular aiming at criticising constructively the often implicit values embodied by these ideas, is not well served with in the reference literature. There is also a lack of appropriate documentation of mathematical applications in industry (Hunt and Neunzert, 1994).

The risk of disappointment has to be considered if critical analysis is performed on models developed by the students within project work, especially those related to the study of situations of socio-political relevance (*eg* environmental issues, equity issues) which are supposed to lead to some action (to inform other people, convince others). The perceptions of such a situation may be spurious and may even lead to the modeller trying to defend an unjustifiable position.

The fact that methodological and epistemological questions will not be asked by the students themselves (for example see Burrill, 1993: page 174; Graham, 1993: page 182; Bishop, 1988: page 173) matches the experiences of the author with students in seminars in mathematics education (for secondary and vocational teachers) many of whom are not likely to accept the plurality of models and show a disappointment about the lack of unambiguous solutions. This clearly shows a need for reflection.

REFERENCES

Bishop AJ, 1988, **Mathematical Enculturation**, Kluwer Academic Publishers, Dordrecht.

Blechmann I, Myškis A and Panovko A, 1984, **Angewandte Mathematik. Gegenstand, Logik, Besonderheiten**, Deutscher Verlag der Wissenschaften, Berlin.

Blum W, 1993, *Mathematical Modelling in Mathematics Education and Instruction*, in Breiteig T, Huntley I and Kaiser-Meßmer G, (eds.), **Teaching and Learning Mathematics in Context**, Ellis Horwood, Chichester, pp 3-14.

Booß-Bavnbek B and Pate G, 1989, *Expanding Risk in Technological Society Through Progress in Mathematical Modelling*, in Keitel C *et al.*, (eds.), **Mathematics, Education, and Society**, Science and Technology Education Document Series No. 35, UNESCO, Paris, pp 75-77.

Booß-Bavnbek B, 1990, *Rationalität und Scheinrationalität durch computergestzte mathematische Modellierung*, **proceedings, GI-20. Jahrestagung I. Informatik auf dem Weg zum Anwender**, Reuter A, (ed.), Springer, Berlin, pp 148-167.

Booß-Bavnbek B, Bohle-Carbonell M and Pate G, 1988, *Über die Risiken technologischer Lösungen im Grenzbereich unseres Wissens*, **Wissenschaftliche Welt, 32,** 2, pp 2-9.

Bunge M, 1985, **Treatise on Basic Philosophy,** Vol.7, Part II, Reidel, Dordrecht.

Burill G, 1993, *Daily-life Applications in the Maths Class*, in de Lange J, Huntley I, Keitel C and Niss M, (eds.), **Innovation in Maths Education by Modelling and Applications,** pp 165-176.

Davis PJ, 1989, *Applied Mathematics as a Social Contract*, in Keitel C et al., (eds.), **Mathematics, Education, and Society,** Science and Technology Education Document, Series No. 35, UNESCO, Paris, pp 24-27.

Feichtinger HG, Jablonka E, Winter H, 1989, **Mathematisches Know-How in Österreich: Angebot und Nachfrage**, unpublished research report, Österreichische Nationalbank, Wien 1989.

Finlay PN and King M, 1987, *Mathematical Modelling in Management*, in Berry JS *et al.*, (eds.), **Mathematical Modelling Courses**, Ellis Horwood, Chichester, pp 58-69.

Graham A, 1993, *Handling Data to a Purpose*, in de Lange J, Huntley I, Keitel C and Niss M, (eds.), **Innovation in Maths Education by Modelling and Applications**, pp 177-186.

Gross HE and Knauer U, 1989, *Mathematiker in der Wirtschaft*, in Maaß J and Schlöglmann W, (eds.), **Mathematik als Technologie? Wechselwirkungen zwischen Mathematik, Neuen Technologien, Aus- und Weiterbildung**, Deutscher Studien Verlag, Weinheim, pp 94-109.

Herget W, 1984, *Price Index: Mathematization Without a Happy Ending*, in Berry JS *et al.*, (eds.), **Teaching and Applying Mathematical Modelling**, Ellis Horwood, Chichester, pp 257-268.

Hunt J and Neunzert H, 1994, *Mathematics and Industry*, in Joseph A *et al.*, (eds.), **First European Congress of Mathematics, Vol. III, Round Tables**, Birkhäuser, Basel, pp 256-275.

Jablonka E, 1996, **Meta-Analyse von Zugängen zur mathematischen Modellbildung und Konsequenzen für den Unterricht**, Ph.D. Thesis, Technische Universität Berlin.

Keitel C, Kotzmann E and Skovsmose O, 1993, *Beyond the Tunnel-Vision: Analysing the Relationship Between Mathematics, Society and Technology*, in Keitel C and Ruthven K, (eds.), **Learning from Computers: Mathematics Education and Technology**, Springer, Berlin, pp 243-279.

de Lange J and Verhage HB, 1992, **Data Visualization**, Wings for Learning/Sunburst, Scotts Valley.

de Lange J, 1993, *Innovation in Mathematics Education using Applications: Progress and Problems*, in de Lange J, Huntley I, Keitel C and Niss M, (eds.), **Innovation in Maths Education by Modelling and Applications**, pp 3-18.

Lewis T and Gagel C, 1992, *Technological literacy: a critical analysis*, **Journal of Curriculum Studies**, **24**, 2, pp 117-138.

Medley DG, 1984, *Modelling 'Successful' and 'Unsuccessful'*, in Berry JS *et al.*, (eds.), **Teaching and Applying Mathematical Modelling**, Ellis Horwood, Chichester, pp 26-38.

Moscardini AO, 1989, *The Identification and Teaching of Mathematical Modelling Skills*, in Blum W, Niss M and Huntley I, (eds.), **Modelling, Applications and Applied Problem Solving. Teaching Mathematics in a Real Context**, Ellis Horwood, Chichester 36-42.

Niss M, 1989, *Aims and Scopes of Mathematical Modelling in Mathematics Curricula*, in Blum W *et al.*, (eds.), **Applications and Modelling in Learning and Teaching Mathematics**, Ellis Horwood, Chichester, pp 22-31.

Skovsmose O, 1994, **Towards a Philosophy of Critical Mathematics Education**, Kluwer Academic Publishers, Dordrecht.

4

A Case Study of Instruction and Assessment in Mathematical Modelling - 'the delivering problem'

T Ikeda

Yokohama National University, 156, Tokiwadai, Hodogaya-ku, Yokohama, 240, Japan.
e-mail: ikeda@ed.ynu.ac.jp

ABSTRACT

It is difficult for students in grades 7-9 to carry out a mathematisation activity by themselves. Social interaction plays a crucial role, with students interacting and exchanging mathematical thinking with one another in class. Initially we identify six components that will affect a mathematisation activity, then set up the teacher's role and the assessment viewpoints.

From the three experimental classroom settings of 'the delivering problem' at grade 7, we observe interaction between students, and infer that this interaction could promote modelling activity. We see from interviews and reports that some students realised the importance of using mathematical thinking and that they modified the conditions as they became aware of the issues.

From these results we expect that mathematical interaction between students will transfer to applying mathematical thinking within one's own mind. Therefore we consider that it is one of the aims of effective teaching and assessment for the teacher to encourage such interaction.

INTRODUCTION

We describe a mathematical modelling process with five states - real problem, classroom model, mathematical model, mathematical solution and real solution - and four stages - to mathematise (to clarify a real problem, to generate variables, to select variables, to set up conditions, to reflect), to solve mathematically, to interpret and to validate and modify. These five states and four stages are based on the view of Burghes and Borrie

(1978), Donald and Maki (1979), Treilibs, Burkhardt and Low (1980), and Blum and Niss (1989). Fig. 1 shows the five states and four stages.

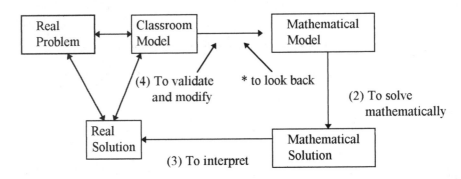

(1) To mathematise
     * to clarify a real problem
     * to generate variables
     * to select variable
     * to set up conditions

Figure 1: process of mathematical modelling

WHAT SHOULD THE TEACHER FOCUS ON IN A CLASSROOM SETTING ?

When we consider teaching mathematical modelling, we have to clarify the teaching aims and assessment. Initially, therefore, we analyse components which will affect the mathematisation activity.

Components which will affect the mathematisation activity

We focus on the mathematisation activity since it is the most important and the most difficult area of the mathematical modelling process. We identify six components that will affect such an activity (Ikeda, 1994).

1.   Mathematical knowledge and skill (enough to solve the classroom model mathematically),
2.   Knowledge about the real world situation,
3.   Interest in solving the classroom model willingly,
4.   Knowledge about the mathematical modelling process,
5.   Mathematical thinking that will promote the modelling process,
6.   Metacognitive skill about the mathematical modelling activity.

When we select and present a classroom model in a classroom setting, we have to pay attention to these six components. Furthermore, we consider that these six components relate directly to the teaching aims and assessment.

Teaching aims

We set up the teaching aims, concentrating on components 4, 5, 6 above, since they are concerned with the explorative, creative and problem solving abilities that it is important to foster in teaching mathematical modelling (Lesh *et al.*, 1986; Blum and Niss, 1989).

In component 5, we abstract three types of mathematical thinking:

    Type a Are there any vague conditions ?
              How can we clarify the vague conditions?
    Type b Does the variable affect the real solution ?
              To what degree does the variable affect the real solution ?
    Type c Is it easy to solve the mathematical problem under the present conditions ?
              How can we set up conditions in order to solve the mathematical problem easily?

It is important for the problem solver to acquire these three types of mathematical thinking and to apply them within their own mind as follows.

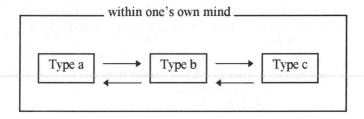

Furthermore, as for component 6, the process of applying the three types of thinking can be considered a metacognitive skill in a mathematisation activity, and we think it plays a crucial role in validating and modifying. Therefore fostering this ability can be considered a crucial aim in teaching mathematical modelling.

The teacher's role in a classroom setting

It is difficult for students in grades 7-9 to carry out a mathematisation activity individually, so interaction with other students allowing for the exchange of experiences and mathematical thinking plays a crucial role. Therefore the teacher needs to think about asking questions and focusing students' discussion on one issue in order that students can exchange ideas with one another as follows.

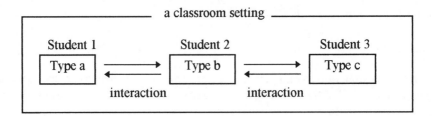

We can expect that, through such interaction, students will be able to transfer these ideas to their own thinking. From these considerations we propose the following basic framework of instruction - the teacher's role is to promote three types of mathematical thinking from students, and to focus the students' discussion on one issue to facilitate the exchange of ideas with one another.

Teaching styles for mathematical modelling

We suggest that there are four main styles of classroom teaching.

Teaching style 1: Whole modelling activity through the interaction between students and teacher in class.

Teaching style 2: Activity of 'mathematisation, solving mathematically, interpretation and validation' through the interaction between students and teacher in class. Students report on the rest of the modelling activity (individually or in groups).

Teaching style 3: Activity of mathematisation through the interaction between students and teacher in class. Students report on the rest of the modelling activity (individually or in groups).

Teaching style 4: Students report on the whole modelling activity (individually or in groups).

In styles 2, 3 and 4, the teacher should allow some students to present their reports in class so that all students can appreciate each other's thought processes. Teachers should select the style of classroom teaching according to their ability to execute the modelling activity (Ikeda and Yamazaki, 1992).

Assessment of classroom teaching and performance of students

We identify three main types of assessment.

  1. Assessment via students' discussion in teaching styles 1, 2, 3.

      (a) How many variables do students generate in the mathematisation ?
      (b) What kind of mathematical thinking interaction is there, and to what extent do students exchange mathematical thinking with one another in order to mathematise a real problem ?

2.   Assessment via interviews with students in teaching styles 1, 2, 3, 4.

   a) What kind of thought processes did the students use during the activity of
      mathematisation ?
   (b) What kind of thought processes did the students use during the activity of
      verification and modification ?

3.   Assessment via students' reports in teaching styles 2, 3, 4.

   (a) How do students mathematise the real problem (teaching style 4)?
   (b) How many times do students repeat the modelling process ?
   (c) How many times do students modify the conditions ?
   (d) Do students refer to the validity and limitation of a model ?

ABSTRACTION OF MATHEMATICAL THINKING WITH 'THE DELIVERING PROBLEM'

We analyse the three types of mathematical thinking at each stage of mathematisation by
considering 'the delivering problem' (Swets and Hartzer, 1991).   This analysis is
effective for the teacher when treating and evaluating students' activities in class.   When
we examine the mathematical thinking, we refer mainly to concepts of mathematical
thinking developed by Katagiri (1988).   Fig. 2 shows how the 'delivering problem' is
developed.

Figure 2: the Delivering Problem

Stage (1): To clarify the real problem

In this stage the method of delivering the
advertisements is considered. There are
two methods of delivering them.
Mathematical Thinking

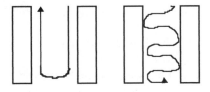

Type a : To find out the factors that may be essential to determine the easiest way of
delivering, for example, 'calories consumed' or 'walking distance'.

Type c : To consider which factors will be the simplest to solve. For example, to
determine which is simpler to solve mathematically: calories consumed or
walking distance.

Stage(2): To generate variables

If walking distance is an essential factor, then variables are generated to determine
which method will produce the shortest walking distance. Fig.3 shows an example of
generated variables.

Mathematical Thinking

Type a: To point out vague conditions that are not clear in a classroom model, for
example, whether cars use the street or not.
Type b: To find out several variables that influence the walking distance, for example,
'width of street', 'the number of shops', and so on.

Stage (3): To select variables

In this stage, necessary variables are selected from the generated variables. Fig. 3 shows
an example of variables that might be selected. We represent selecting a variable as ✓
and not selecting a variable as ✗.

Mathematical Thinking

Type b: To determine whether the variables found in stage (2) influence the walking
distance or not. For example, to determine whether the widths of Streets A and
B influence the walking distance or not.
    : To find out the relationship between the variables found in stage (2). For
example, the width of a shop is determined by the length of the street and the
number of shops.
Type c: To disregard some of the variables found in stage (2), because it is confusing to
consider too many variables at the same time. For example, to disregard
whether cars use the street or not, even though this influences the walking
distance a little.

Stage (4): To set up conditions

In this stage, conditions are set up with regard to selecting variables. Fig. 3 shows an example of conditions that might be generated.

(a) Number of shops ✓ 10 shops	(f) Width of Streets A and B ✓ same
(b) Size of each shop ✓ same	(g)Arrangement of shops on both sides ✓symmetry
(c) Width of street ✓ 5 m	(h) Whether cars pass on the street or not ✗
(d) Distance between shops ✗	(i) Number of people on the street ✗
(e) Length of street ✓ 100 m	(j) Amount of time spent waiting for ✗
	the traffic light to change

Figure 3: example of results of mathematisation

Mathematical Thinking

Type a: To set up conditions that are not clear in a classroom model - for instance to set up an arrangement having shops on both sides.

Type b: To examine how the walking distance will vary between the two ways of delivering when the width of the street, the number of shops, and so on, vary.

Type c: To set up special cases of the variables found in stage (3) to predict a real solution. For example determining that there are 10 shops on one side of the street is a special case.

    : To set up ideal cases of the variables found in stage (2), because it is confusing to consider too many variables at the same time. Making the arrangement of shops on both sides symmetrical would be an ideal case.

Stage(5): To reflect

Here all stages from (1) to (4) are validated. If some of the stages are not sufficiently developed to solve the problem mathematically, it is vital to point out the faults and modify them. For example, the following two additional variables might be generated:

(k)   location of each post - midpoint

(l)   zigzag method delivery - two ways

The mathematical problem might then be formulated.

EXPERIMENTAL CLASSROOM SETTING

We set up experiments in three 7th grade classes for two school units, which we denote as classes A, B and C. We selected teaching style 2. We analysed the students' discussions of mathematisation and students' reports based on the above mentioned assessment questions.

Analysis of classroom teachings

In classes B and C various opinions were obtained from students during the first mathematisation cycle. We could observe that the teacher was able to treat many varieties of mathematical thinking systematically by focusing the students' discussion on one issue. We could then observe interaction of mathematical thinking between students. From Fig. 4 we see that students could select variables by interacting with each other using mathematical thinking; one tried to clarify the vague conditions (type a) while the other tried to identify whether these conditions influenced the walking distance or not (type b). Furthermore, from Fig. 4, we see another case of interaction using mathematical thinking - one tried to investigate whether the conditions influence the walking distance or not (type b) while the other tried to simplify, idealise or specialise the conditions with the aim of getting a solution (type c). In class A only a few students offered opinions from stage (1) to stage (4). However we observed a lively discussion at stage (5). Furthermore, we could also observe interactions using mathematical thinking . We could see that students could modify the conditions based on previous interactions.

It can be said, therefore, that such mathematical interaction promotes the mathematical modelling activity in class.

Focusing on the mathematical thinking interaction between students is one of the most effective ways to assess such an activity in a classroom. Mathematisation activity depends on the ability of the students and the atmosphere in the class, so it is important for teachers to take these factors into account.

Analysis from interviews

From interviews with students, we could check that students realised that the more the consistency with the real world was emphasised, the more difficult it would be to solve the problem. On the contrary, the more the possibility to solve the problem mathematically was emphasised, the more the consistency with the real world would be broken. Students clearly realised the importance of interaction when using mathematical thinking to execute the modelling process well. However students cannot fully utilise concepts such as idealisation and specialisation to predict a real solution. This is a subject for future discussion.

Analysis of students' reports

We analyse how many and what kinds of conditions students changed at the validation and modification stage in reports. Fig. 5 shows the results of this analysis. We observed that they modified the conditions as they became aware of the issues. Some students changed one condition several times and derived interesting answers, but some changed

Thinking of clarification of conditions	Functional thinking	Thinking of simplification, idealisation, specialisation
1. What about the bus timetable?	2. No relation because we compare distance, not time.	
1. What about the width of the street?	2. As the width influences the walking distance, this information is necessary.	
1. What about the time spent waiting for the traffic signal to change?	2. This information is not necessary since time doesn't influence the walking distance.	
1. What about the number of closed shops?	2. As the number of closed shops will influence the zigzag way to deliver, it will influence the walking distance.	3. It is difficult to think about this in general terms, so it is better to disregard this factor.
1. What about the distance between the station and the bus stops?	2. This information is not necessary because the distance spent riding on the bus is not related to the walking distance.	
1. What about the width of streets A and B?	2. The walking distance will vary if the width of street A is different from that of street B.	3. Streets A and B should be regarded as having the same width.
1. What about the number of people on the street?	2. This information is necessary because if the street were crowded, the boy would have to walk further.	3. It is troublesome to think about this. 4. It would be good to think that the boy waits until the crowds go away.

Figure 4: protocol of class B

a few conditions at the same time and were unsure about how to arrive at a conclusion. We selected a few examples in which students could get an answer.

At first, one student who changed 'width of street' several times supposed that the number of shops was 10 and the length of the street 100m. After that, she denoted the width of the street by x and solved an inequality for x.

> delivering straight and back: $190 + x$
> delivering zigzag: $100 + 10x$
> $190 + x < 100 + 10x$ $x > 10$

She concluded as follows: 'If the width of the street were longer than 10m, it is better to deliver straight and back'.

In the following conditions, one student modified the size of each shop.

> 'number of shops: 10, location of post: midpoint, width of street: 10m, length of street : 100m'.

He concluded as follows : 'If the arrangement of shops on both sides were symmetrical, it is the same whether we deliver straight and back or zigzag'.

Changing one condition Total: 43		Changing two conditions Total: 16	
(1) Width of street	13	(1) + (4)	4
(2) Number of shops	11	(3) + (5)	3
(3) Location of each post	6	(5) + (7)	3
(4) Width of a shop	5	(1) + (5)	2
(5) Size of a shop	5	(2) + (6)	2
(6) Distance between shops	2	(2) + (7)	1
(7) Arrangement of shops on both sides	1	(3) + (6)	1
Changing three conditions Total: 9		**Changing four conditions Total: 6**	
(1) + (5) + (7)	3	(1) + (2) + (4) + (car)	3
(1) + (2) + (4)	2	(1) + (3) + (5) + (by two persons)	2
(2) + (4) + (5)	2	(1) + (3) + (6) + (7)	1
(5) + (6) + (7)	1		
(1) + (3) + (crowd of people)	1		

Figure 5: result of analyses of students' reports

CONCLUSIONS

It is difficult to execute a mathematical modelling process by applying mathematical thinking in isolation, but it is possible to execute a mathematical modelling process between students in a classroom setting through interaction using mathematical thinking. In fact, we observed interaction using three types of mathematical thinking between students where this interaction could promote a modelling activity in experimental classroom settings. We also observed, from interviews and reports, that some students realised the importance of interaction using mathematical thinking and that they modified the conditions as they became aware of the issues. From these results, we can expect that interaction using mathematical thinking between students will transfer to applying mathematical thinking by oneself. We consider that it is important for the teacher to pay attention to such interaction if the aims of effective teaching and assessment are to be achieved.

REFERENCES

Blum W and Niss M, 1989, **Mathematical Problem Solving, Modelling, Applications and Links to Other subjects - State, Trends and Issues in Mathematics Instruction**, Ellis Horwood, Chichester, pp 1-21.

Burghes DN and Borrie MS, 1978, **Mathematical Modelling**, ICSU Committee on the Teaching of Science.

Donald RK and Daniel M Jr, 1979, *Mathematical Models to Provide Applications in the Classroom*, **1979 yearbook, National Council of Teachers of Mathematics**, pp 1-7.

Ikeda T and Yamazaki K, 1992, *Study on Objectives in the Introduction of Mathematical Modelling in School Mathematics in Grade 7-9*, **Journal of Japan Society of Mathematical Education**, 75, 1, pp 26-32 (in Japanese).

Ikeda T, 1994, *Study on Mathematical Thinking and its Interaction in Mathematization*, **Paper presented at the 27th research meeting of Japan Society of Mathematical Education**, pp 317-322 (in Japanese).

Katagiri S, 1988, **Concretisation of Mathematical Thinking**, Meizi Tosyo (in Japanese).

Lesh R, Niss M and Lee D, 1986, *Application and Modelling*, **Proceeding of the Fifth International Congress on Mathematical Education**, Carss M, (ed.), pp 197-211.

Swets F and Hartzler JS, 1991, **Mathematical Modelling in the Secondary School Curriculum**, The National Council of Teaching of Mathematics, Reston, Virginia, pp 38-40.

Treilibs V, Burkhardt H and Low B, 1980, **Formulation Processes in Mathematical Modelling**, Shell Centre for Mathematical Education, University of Nottingham, Nottingham.

5

The Quest for Meaning in Students' Mathematical Modelling Activity

João Filipe Matos
Universidade de Lisboa
Campo Grande, C1-2° Piso
1700 Lisboa
Portugal
e-mail: ejfm@scosysv.fc.ul.pt

Susana Carreira
Universidade Nova de Lisboa
Quinta da Torre, 2825 Monte da Caparica
Portugal

ABSTRACT

The central claim that every mathematical model is based on a certain interpretation of reality motivates the search and examination of the meanings that support students' modelling activity. In this quest for meaning we analysed three episodes extracted from a modelling activity developed by a group of four tenth grade students, where the situation to be modelled involved the calibration of an hour-glass. Our results support the conclusion that students' models are mediated by their particular interpretations and discussion activities, and by their mathematical, technological and symbolic tools.

INTRODUCTION

The mathematical modelling of real world situations is often recommended for its potential in making mathematics meaningful to students. According to Blum (1991), the most important argument for the inclusion of applications and modelling in mathematics teaching and learning may be its contribution to the problem of the development of meaning.

Some of the recurrent questions students have in their minds when they do mathematics are related to the potential uses, purposes and relevance of the established curricular topics. There are people who hold a strong belief in the potential usefulness of mathematics and take that as an indisputable answer to the students' question: what is the point of this? Such an assumed applicability may even become a carrot to maintain students' expectations towards some relevant applications that might never emerge (Hobbs and Burghes, 1989).

Without denying that "mathematics has a vast, even oceanic, applicability to the real world" (Ormell, 1993:55), it is our conviction that the problem of meaning is not reducible to the utility or to the relevance of mathematics in solving real problems. The argument that relies on the utility of mathematics (the pragmatic argument) is just one among several others, including the psychological argument, to justify the importance of modelling and applications in mathematics learning (Blum and Niss, 1989; Blum, 1991).

There are many questions to be faced when the problem of meaning is raised in connection with the role of modelling and applications in mathematics learning. What can we say about the origins and nature of students' findings when they are involved in modelling activities? Where do their models come from? What elements are contributing to the sense they make of their mathematical thinking?

We have set ourselves the goal of searching for meaning and of investigating their origins in students' modelling activity. We shall do it by looking closely at student interactions, listening to their utterances, and by interpreting their actions and words.

THEORETICAL BACKGROUND: THE MAINSTREAM OF OUR APPROACH

Our approach to the problem of finding meaning in students' modelling activity draws from a set of relevant perspectives which have their basis in a socio-cultural view of the development of thought and learning.

Learning is viewed as a social and cultural activity. Meanings emerge as a result of a situated, shared and social activity. The setting where students develop this activity is not a neutral element and their actions are not immune to it.

It is also important to recognise that students' voices are not always the same (Wertsch, 1991). Sometimes they sound like interrogative voices; at other times they assume a more critical or, on the contrary, a more condescending tone. Students' voices may be more spontaneous or more conditioned and, eventually, they may sound like the voice of the teacher as in an act of ventriloquism (Wertsch, 1991).

Meanings are mediated by external signs. Some of the important external signs that mediate students' meanings are of a linguistic nature (Vygotsky, 1993). The words used by the teacher to describe a real situation that is chosen to be modelled by the class are important mediating elements in the images students make of the situation. The mathematical inputs or clues that are often included in such descriptions can have a pronounced affect on students' mathematical modelling.

Meanings also permeate students' discussion and their mediating tools. When one of those tools is a technological device, there are additional matters to take into account - like the logic of the tool's functioning and the logic of its use in a particular task (Skovsmose, 1994).

Students' own modelling processes are different from those of experts in crucial aspects. The first mark of distinction between students' modelling and experts' modelling is connected to the setting where the modelling activity takes place. The point is that students are confronted with a problem already formulated in terms of school mathematics - a problem about a real situation or phenomenon which, for several reasons, is found adequate to be presented to the class. The situation is somehow adapted for a certain group of students, and the problems are quite often simplified so that they remain within students' reach.

The reality that enters the classroom is a certain reality, a reality which differs from that of the research laboratory or of the real world exploration conducted by a team of scientific practitioners (Säljö and Wyndhamn, 1993).

THE EMPIRICAL SETTING OF THE STUDY

Our data result from an empirical study developed in a regular mathematics class over a period of four months (Matos and Carreira, 1994). The participants were the students and the teacher of a tenth-grade mathematics class.

During the first term the teacher brought simple modelling and application problems to the class and students were organized into small groups to work on them. The teacher undertook the supervision of students' work and she was seen by the groups as a consultant and as a guiding resource. By the end of the first term she introduced her students to the computer and taught them the basic features of the electronic spreadsheet. In the meantime, she and a team of researchers worked on the production of materials to support a weekly session on modelling problems during the second and third terms. From the curricular topics it was decided to explore the study of functions, including linear and other algebraic functions.

In each of the eleven modelling sessions three groups of students were observed, videotaped and audiotaped during their activity.

Each group had to come up with a collective report of each activity. The written reports were used as data sources in conjunction with the notes of the observers and the recorded tapes of the groups' activity.

DESCRIPTION OF EPISODES AND THE SEARCH FOR MEANINGS

In our analysis we shall focus on the activity of a group of students. They will be identified by the fictitious names Carla, Sofia and Roberto.

Carla was the student with the most successful background in mathematics. Her voice was heard by the group as one that deserved attention. She was not too fond of open-problem solving and showed her preference for structured exercises where she knew what had to be done and to be found.

Sofia seemed to enjoy working in real situations and was frequently in interaction with Carla. Together the girls made what could be called the kernel of the group. Sofia had some sharp ideas and intuitions and liked to put them forward.

Roberto was sometimes exuberant in his participation, but his performance in maths class had various ups and downs. He could be quite distant on some occasions as well as enthusiastic and active in others.

The real situation presented in this class involved the calibration of an hour-glass so that time could be measured in small intervals. It was decided to study a hypothetical hour-glass made up of two identical conic vessels united by their vertices (see Fig. 1).

Volume of each cone: 226 cm^3
Height of each cone: 6 cm
Draining Velocity: 0.5 cm^3/s

Figure 1: the diagram and information provided about the hour-glass functioning.

From the entire record of the activity we have selected three episodes found worthy of close attention.

EPISODE 1: LOCAL INVERSES AND FORMAL INVERSES

Sofia initiated the discussion about the hour-glass in operation. She described it in simple terms, saying that the flowing of the water would cause an increase of liquid in the lower container and a decrease in the upper container. She also mentioned that the water subtracted from the upper cone would be precisely the water added to the lower cone. Carla developed a mental image of the volumes changing, assuming that the upper cone would be initially full. She considered time changing in seconds:

Carla: "After the first second, it remained 225.5 [in the upper vessel]. That's 0.5 less. After two seconds, it's... 1 less. After three seconds it's 1.5 less, that is, 0.5 times 3. And in the lower one it increased. It became 0.5; 1; 1.5.

Roberto: "The amount that came into the lower one is equal to the amount that came out of the upper one. So, in the upper cone, the volume will be the total volume - which is 226 - minus 0.5 times the time value."

Carla: "And down there, the volume will be the total, which is 0, plus 0.5 times the time value."

Sofia wrote the corresponding formulae in her notebook:

$$V = 226 - 0.5t \quad \text{(upper cone)}$$
$$V = 0 + 0.5t \quad \text{(lower cone)}$$

At this point Roberto suggested a way of relating the volumes of the two cones but Carla immediately reacted to his conjecture.

Roberto: "I think that the lower cone must be the inverse of the upper cone".

Carla: "Not the inverse! How could that be, if we're about to see that the two volumes are directly proportional?"

She pointed to him that, after a second, the volumes would be 225.5 in the upper cone and 0.5 in the lower cone. Then she observed that the inverse (the reciprocal) of 225.5 was not 0.5 and she made the calculations to confirm it:

$$\frac{1}{225.5} = 0.0004 \ldots \neq 0.5$$

Sofia did not interfere in this dialogue and the discussion between Carla and Roberto was not prolonged. The group used the formulae to represent the two volumes as functions of time in their spreadsheet and graphed the two functions simultaneously (see Fig. 2).

In Portuguese, the word "inverso" is the mathematical term for the reciprocal of a number. The same word is used in common language to express the idea of inversion, reversion or contrary.

As they were interpreting the two straight lines obtained, the notion of inverse came back in a subtle way and, this time, in the comments of Carla.

Sofia: "This one that goes down is the upper cone."

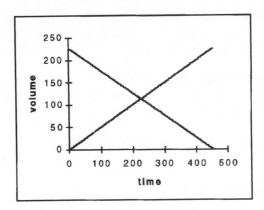

Figure 2: the graph of the two volumes against time obtained on the spreadsheet

Carla: "That's right. It starts with... and... it decreases. It loses to the other, and in the other it happens in reverse."

Meanings and their origins in episode 1

While Sofia was giving her description of the hour-glass behaviour, she used several opposite words. She mentioned increasing and decreasing, as well as adding and subtracting, as she spoke about the volume change in each of the cones. Carla began to express this variation in a more quantified manner and Roberto completed her reasoning by introducing new opposites: the water comes out of the upper cone and comes into the lower cone. In this way, an image of inversion was created connected with the recognition of a certain symmetry in the hour-glass behaviour. The word inversion became a natural term to represent the phenomenon - it translated what had been spontaneously captured from the pairs of opposites that filled the students' interpretations. We shall say, accordingly, that this type of discussion gave scope for the creation of a local model of the hour-glass functioning, which the idea of inversion is invoked to represent two opposite behaviours happening in the two vessels of the hour-glass.

However, when Roberto proposes this same notion to initiate a process of relating both volumes, Carla replies with another voice. She seems to back off from the local model used so far and switches to a formal model of inverse. She takes it from a scientific or scholarly position, and she disagrees with Roberto based on scientific concepts. She stresses her point of view by using mathematical arguments and she shows him that the inverse (the reciprocal) of 225.5 is not equal to 0.5. Everything gets a new formal aspect and the case is closed.

What is important to notice here is that the local model was intercepted by a formal and scientific model; the status of the latter proved to be strong enough to repress the more intuitive and tacit models, and discussion about the relationship between the two volumes did not continue. Nobody questioned the direct proportionality or tried to verify it, an attempt was made to graph the pairs of values in the two volume columns to see what would come up. This is quite in contrast with what is supposed to happen in an expert modelling situation, where conjectures would be submitted to test and confronted with the real situation.

The conjecture about a direct proportionality only meant a formal argument to reinforce a formal voice within the modelling process. This formal voice turned out to be an important constraint in the modelling process. It held back Roberto's suggestion even though it did not erase the former intuitive model. Obviously the position of the two straight lines in the graph invigorated the idea of inversion and Carla herself was caught up in the temptation of mentioning it.

So the idea of inversion was not totally dismissed but in the end it did not find a symbolic counterpart.

EPISODE 2: A GRAPH THAT IS DRAWN FROM RIGHT TO LEFT

Back at the beginning of the session the teacher made some comments to the whole class where she summarised the questions formulated.

Teacher "In question 1 you're going to explore how the volume changes with time. In 2, you'll see how the water level changes with the volume of water. In 3, what is finally required is that you look for the way the water level changes with time. So you may see a transitivity here..."

Once more it was Sofia who tried to figure out how the water level changed as the volume in the upper cone decreased. She drew the following diagram of the situation where she depicted the liquid remaining in the upper cone in successive instants (see Fig. 3).

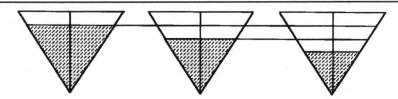

Figure 3: the scheme created by Sofia to represent the level variation in the upper cone

Roberto noted that the water level decreased with the volume of water. Sofia pointed out that constant decreases of level did not correspond to constant decreases of volume.

Sofia: "Only if they were cylinders would there be constant variations of volume for
         constant variations of level. But, as these are cones, the variations are not
         constant and so it's a different situation."

 As soon as the students got a formula for the height of the cone as a function of volume,
they implemented it in another column of their spreadsheet table. A first reading of the
values told them that the level of water started to decrease very slowly.

They wanted to graph the function level against volume on the spreadsheet, but first they
tried to imagine the type of graph that would appear on the screen. Roberto took the
initiative to draw a sketch on paper, and his representation was accepted by the two girls
(see Fig. 4).

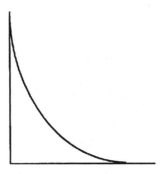

Figure 4: the graph Roberto sketched before plotting level against volume on the
spreadsheet.

He explained the general idea of his graph but he did not label the axes. Anyway it was
clear in Roberto's words and gestures that he was referring to the labels level and time
on the y-axis and x-axis, respectively.
Roberto: "The level begins with its maximum value and, as time passes, it decreases until
         it gets to the minimum value, which is zero, when all the water has come down
         to the lower container".

When they used the spreadsheet to plot the level against volume they found a different
representation from Roberto's (see Fig. 5).

At first they didn't understand what was wrong with their initial idea but after a while
Roberto came forward to clear up the matter.

Roberto: "Basically this is the graph that we're expecting with the only difference that it
         is reversed. Here it's as if time went backwards, as if time was running in the
         opposite way to ours".

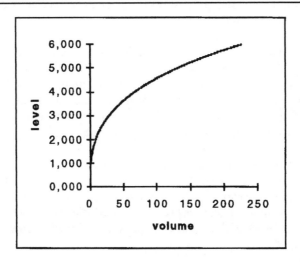

Figure 5: the graph of level against volume obtained on the spreadsheet

A few moments later Sofia had a remarkable observation about the graph obtained in the spreadsheet:

Sofia: "When the curve is drawn here on the computer, it begins from up there (pointing to the right edge of the curve) and it goes all the way to the axis line. So, it's as if the starting point was on the right and not on the left. And it really matches what we were thinking before".

Meanings and their origins in episode 2

The essential question treated in this episode was to understand how the water level would change with the water volume in the upper cone. In analysing how students dealt with this problem we can see two models of the situation. We will call them the dynamic model and the computer-mediated model.

The first model was developed through the pictorial representation made by Sofia. She depicted what she imagined to be the dynamic behaviour of the liquid flowing out of the upper cone. In her drawing the time variable was already an important feature of the model. It almost looked like successive shots in a film sequence of the upper vessel and so the conclusion was that the level decreases as the volume of water decreases. However, when Roberto proposed a graphical representation, he actually considered the variation of level with time. According to the teacher's remarks, in the beginning of the session the volume should have been taken as the new independent variable. However, this hint seemed to be ignored and instead the students' model grew out of a dynamic vision of the variation in level.

Even when the students obtained the true graph (level against volume) on the spreadsheet, they did not reject their dynamic interpretation. On the contrary, they searched for details that could support such an interpretation. For instance, they claimed that both graphs were compatible if one realised that time was inverted in the computer graph. The fact that the computer plotted the graph curve from right to left was used as a good argument to sustain the idea of time being inverted.

What has been described suggests the conclusion that the time variable became very central in the students' reasoning. Although having accepted the computer graph, they used a feature of the computer output (the graph is drawn from right to left) to reinforce their own perception of the relation between level and time. The meaningful idea of time as the independent variable seemed to win out over the apparent contradiction between Roberto's graph and the computer output.

EPISODE 3: WHY IS A CONE DIFFERENT FROM A CYLINDER?

The students finally plotted the graph level against time on the spreadsheet to get a better insight of how time should be marked on the hour-glass (see Fig. 6).

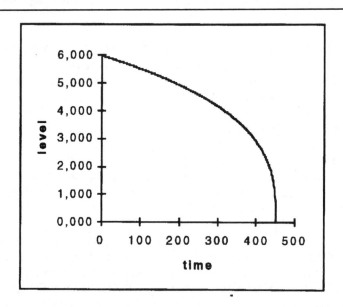

Figure 6: the graph of level against time obtained on the spreadsheet

They came to realise that Roberto's previous graph indicated a slow decrease towards the end of the emptying which did not match the real phenomenon. This way the students revised their graphical representation of the level-time function.

The new graph provoked some discussion between the students. They analysed the type of variation represented and discussed it.

Carla: "When there isn't much water left and the cone is almost empty, then it goes down faster, doesn't it"?

Sofia: "Yes, and the graph translates the water movement as it goes down in the upper cone".

Carla: "That's right. At the start, when there's still a lot of water, it has to be decided which is going down first... (smiling). But when there's only a little liquid, then it all goes down at once".

Sofia: "We got this curve because we're talking about a cone. If it was a cylinder it would be a linear model".

Roberto: "True, for in that case the water flow would always be the same... The width of the container would not change. But if it's a cone, and if it's quite full... the water flow is different. It makes a difference in the water flow if the cone is quite full or if it's almost empty".

Meanings and their origins in episode 3

Two ideas seem to cohabit in the students' reasoning. One is the idea that the shape of the vessel makes the level variation non-linear. The other is related to the notion of an increasing speed of water flow as the cone gets emptier.

Carla used a metaphor where she suggests that the liquid flow would not have a constant speed. Only Sofia kept an accurate idea about the influence of the vessel shape in the level function. Roberto aimed at combining both perspectives: in a way he followed Sofia's interpretation when he noticed that there was a variation in the width of the cone. At the same time he went along with Carla's image where the water did not flow at a constant rate - in the beginning it would be slower than at the end. It all seems to indicate that Roberto tried to accommodate both views of the problem. Thus there were three voices trying to explain the non-linearity. Sofia's voice pointed to the shape of the vessel, Carla's to the changeable speed of water flow and Roberto's showed a mixture of both. All of them brought different meanings to the situation but those meanings were not seen as conflicting. On the contrary, they were combined in a two-sided version of the phenomenon.

Once again we find aspects of students' modelling that are in contrast with the modelling behaviour of experts. For the experts, new conclusions or results within the same situation would have to be consistent with initial assumptions. In students' modelling there are persistent models and images that tend to impose themselves on the conceptualisation of the situation.

Finally, in experts' modelling there is a concern with the accuracy of descriptions involving variable identification. The linguistic aspect becomes of crucial importance as a means of expressing the object under study. In students' utterances certain linguistic and contextualised signs, like the expression "the water goes down faster", brought up a certain amount of ambiguity that mediated their interpretations and the meanings they took into the real situation.

CONCLUSIONS

To look for meaning in students' modelling activity implies the recognition that students' modelling behaviour is fundamentally tied to the setting where the modelling takes place. There are all sorts of mediating elements contributing to the emergence of meaning. We have identified and discussed some of those elements, namely students' discursive activity, the role and value of everyday language and its contextualisation in the analysis of a situation.

We have also looked at the way some students' models come from their intuitive views of the problem and how they can become rather persistent even when in the presence of contradictions. As Driver, Guesne and Tiberghien (1990: page 35) have stated, "students may ignore counter-evidence or interpret it in terms of their prior ideas". They can find their own processes for accommodating different interpretations and concepts, and of using their tools to reinforce such processes.

In many aspects there are notable differences between what students do in their mathematical modelling activity and what is supposed to be the performance of a modelling expert. This will not come as a surprise if one is willing to appreciate the reasons why students come up with their own sense of a certain real situation and its mathematical grounds. Students' criteria for coherence are not necessarily coincident with those of the scientists. To students, their own personal ideas and interpretations about natural events may appear to work quite well in practice (Driver *et al.*, 1990).

We want to conclude with the idea that mathematical modelling seems to provide a promising territory to explore the question of learning as an activity where students give meaning to ideas, problems, mathematical and non mathematical concepts.

REFERENCES

Bakhtin MM, 1981, **The Dialogic Imagination: Four essays by M. M. Bakhtin**, Holquist M, (ed.), trans. Emerson C and Holquist M, University of Texas Press, Austin, Texas.

Blum W, 1991, *Applications and Modelling in Mathematics Teaching - A review of arguments and instructional aspects*, in Niss M, Blum W and Huntley I, (eds.), **Teaching of Mathematical Modelling and Applications**, Ellis Horwood, Chichester, pp 10-29.

Blum W and Niss M, 1989, *Mathematical Problem Solving, Modelling, Applications and Links to Other Subjects State*, trends and issues in mathematics instruction, in Blum W, Niss M and Huntley I (eds.), **Modelling, Applications and Applied Problem Solving**, Ellis Horwood, Chichester, pp 1-21.

Driver R, Guesne E and Tiberghien A, 1990, *Children's Ideas and the Learning of Science*, in Lee V, (ed.), **Children's Learning in School,** Hodder and Stoughton, London, pp 33-40.

Hobbs D and Burghes D, 1989, *Enterprising Mathematics: A cross-curricular modular course for 14-16 year olds*, in Blum W, Berry JS, Biehler R, Huntley ID, Kaiser-Messmer G and Profke L, (eds.), **Applications and Modelling in Learning and Teaching Mathematics**, Ellis Horwood, Chichester, pp 159-165.

Matos JM and Carreira SP, 1994, **Modelação e Aplicações no Ensino da Matemática: Cinco estudos de caso**. Projecto MEM, Departamento de Educação da Faculdade de Ciências de Lisboa, Lisboa.

Ormell CP, 1993, *A Pedagogy Based on Projective Modelling*, in Lange J, Keitel C, Huntley I and Niss M (eds.), **Innovation in Maths Education by Modelling and Applications**, Ellis Horwood, Chichester, pp 53-62.

Säljö R and Wyndhamn J, 1993, *Solving Everyday Problems in the Formal Setting: An empirical study of the school as context for thought*, in Chaiklin S and Lave J, (eds.), **Understanding Practice - Perspectives on activity and context**, Cambridge University Press, Cambridge, pp 327-341.

Skovsmose O, 1994, **Towards a Philosophy of Critical Mathematics Education**, Kluwer, Dordrecht.

Vygotsky LS, 1993, **Thought and Language**, (Portuguese Translation), Martins Fontes, São Paulo.

Wertsch, JV, 1991, **Voices of the Mind**, Harvester Wheatsheaf, London.

6

Some Mathematical Characteristics of Students entering Applied Mathematics Degree Courses

P L Galbraith
University of Queensland
Brisbane, Queensland 4072
Australia
e-mail: p.galbraith@mailbox.uq.oz.au

C R Haines
City University
Northampton Square
London EC1V 0HB
e-mail: c.r.haines@city.ac.uk

ABSTRACT

Several studies indicate that first year students have inadequate conceptual knowledge, flawed procedural knowledge and lack the coordinating abilities necessary to remedy broken knowledge (Galbraith, 1982; Gray, 1975; Tall et al., 1993). These factors are of signal importance in degree courses that include applications modelling since students cannot invoke mathematical knowledge and skills over which they have inadequate mastery. There are a variety of strategies to meet these issues, some of which place a heavy reliance on computer software.

From this standpoint, the research identifies gaps in knowledge and demonstrates the competences of students by investigating in detail their mechanical, interpretive and constructive skills. The methodology focuses on those skills identified as central to teaching programmes which attempt to use computers effectively. The construction and the development of the test instrument, including individual items, is discussed and we report on the results from its application to 244 students. The analysis draws comparisons between the mechanical, interpretive and constructive skills and the difference between students in mathematics and in engineering.

INTRODUCTION

In considering challenges involved in the teaching of mathematical modelling and applications the natural tendency is to focus on difficulties encountered in bridging the real world to the world of mathematics. The modelling process and suggested related meanings for the term 'applications' have been set in context by Blum (1993). Carr

(1993) refers to "the famous Open University seven - block diagram" which in one form or other continues to provide guidance for the activity of modelling - both teaching and doing (Table 1). Again the emphasis in such visual and verbal mnemonic aids is focused on real situation - mathematics links, for example in moving from assumptions to formulation, or from mathematical results to interpretation, or from interpretation to evaluation. We could say that the emphasis tends to be on the 'intervals' rather than the 'notes'. Indeed we may be tempted to assume the 'notes' are correct - for instance mathematics knowledge exists as a usable package if only we can bring it into direct contact with an application context.

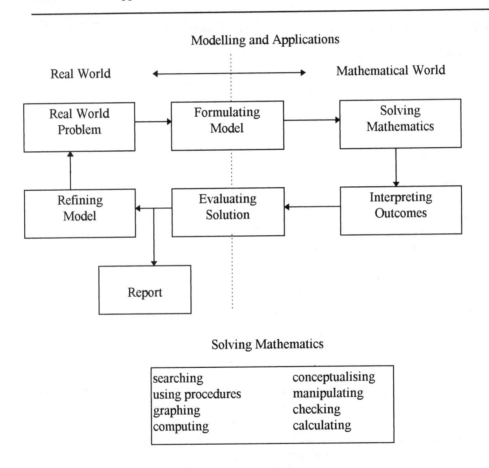

Table 1. A representation of the modelling process in seven stages, highlighting the role of the *solving mathematics* stage and its importance for students to have adequate conceptual and procedural knowledge.

However, here we need to examine this assumption - that the hardest part of the job is done if we can build bridging links between the real context and the mathematical world of concepts and procedures. Studies addressing the understanding of basic concepts and procedures demonstrated by tertiary mathematics students have been conducted regularly over time (Buckland, 1969; Gray, 1975; Clement, Lochead, and Soloway, 1980; Galbraith, 1982; Tall and Razali, 1993).

All these studies show that, in spite of years of extensive teaching and learning experiences, many students continue to have inadequate conceptual knowledge, flawed procedural knowledge and lack the coordinating abilities necessary to mend broken knowledge, or even to relate their mathematics sufficiently to identify where broken knowledge exists. Furthermore the conceptual and procedural difficulties have strong roots in pre-tertiary mathematics. Tall and Razali (1993) note that weaker students suffer from the continued "misinterpretation that algebra is a menagerie of disconnected rules to do with different contexts". Particular problems were noted in the coordination of graphical and algebraic representations when slight variations were introduced into the context.

A common thread in these studies is the powerful negative influence of fragmented learning, and the apparent absence of cognitive strategies to coordinate conceptual and procedural knowledge, as for example in the sense of Hiebert and Lefevre (1986). So we need to probe within mathematics itself, as well as within the linkages between context and manipulation, in order to understand better the difficulties so many students have in successfully applying mathematics generally and in particular in mathematical modelling.

An analogy serves to clarify this interest further. Consider the problem in which a semi-infinite rectangular plate (of width a and unbounded length b) has one edge kept at a constant temperature T and the other edges maintained at zero temperature. The temperature distribution $u(x,y)$ is sought.

The problem involves solving $\qquad \dfrac{\partial^2 u}{\partial x^2} + \dfrac{\partial^2 u}{\partial y^2} = 0$

subject to the four boundary conditions: $u(0,y) = 0$, $u(x,y)$ finite, $u(a,y) = 0$ and $u(x,0) = T$.

The first three conditions lead to

$$u(x, y) = \sum_{n=1}^{\infty} B_n \exp\left(\frac{n\pi y}{a}\right) \sin \frac{n\pi x}{a}$$

Applying the fourth condition yields

$$T = \sum_{n=1}^{\infty} B_n \sin \frac{n\pi x}{a}$$

and the solution is completed by fitting a Fourier Series to find the constants B_n.

This essential phase of the solution means delving into Fourier Series until a satisfactory solution can be imported back to the problem context. The effectiveness of the solution is only as good as the quality with which the problem solver can access the necessary mathematics, which can vary from completeness through various degrees of fragmentation. It is the accessibility and use made of such supporting knowledge that is our interest here. The applied mathematics problems used in structuring the course (to be described) are dependent for successful solution on a well developed network of supporting concepts and procedures. The way in which students are able to invoke, coordinate, and apply their basic mathematical knowledge, is central to the success of making application contexts work.

The context of our study is a first-year programme in mathematics at City University, London. The students are enrolled on a variety of degree courses, such as Actuarial Science and Engineering, each of which has a particular interest in applying mathematics.

THE TEACHING PROGRAMME

A key feature of the teaching programme is a sequence of computational mathematics workshops in which a graph plotting package, a symbolic algebra package and spreadsheets are used to structure student learning. Concepts and procedures are introduced through practical contexts within which the students learn, and from which they are expected to anchor and generalise their learning. Hence the programme involves not only linkages between application contexts and mathematical manipulations, but between computer and pen-and-paper representations of knowledge. A method of using experimental workshops in this way is discussed by Atkins, Haines and Orman (1996), but the reader can sense the flavour of the program from the following illustration. The detailed introductions provided for the use of Graphic Calculus (1986) and Derive (1990) are curtailed.

Sample workshop

An engineering test rig is to be set up in a laboratory which is 12m × 5m. The test rig takes up half the area of the laboratory, is centrally placed, and the strip around the test rig is a safety zone. The problem is to find the width of the safety zone.

(1)   Draw a diagram of the situation -let the width of the safety zone be x metres and show that x must satisfy the equation $4x^2 - 34x + 30 = 0$. Can you find the width of the strip?

(2)   If the laboratory had been 15m × 4m, find the quadratic equation which must be satisfied. Can you find the width of its safety zone in this case?

Generalise the problem for a laboratory a metres by b metres, for integers (a,b).

(3)   If the area of the laboratory is 60m^2, how many different sized laboratories are possible? List the corresponding values of a and b.

(4)   From your diagram show that x must satisfy the equation
$8x^2 - 4(a+b)x + ab = 0 (*)$

[Advice is provided at this point on using *Graphic Calculus*, and students are then asked to use the package to...]

(5)   Draw graphs for the left hand side of (*) for all possible values of a and b.

(6)   Estimate the strip width in each case.

(7)   List and solve by an appropriate method all equations defined by (*).

[Advice is then given on using *Derive* to solve the equations in (7) and the procedure is applied as a check]

(8)   Return to the tabulated list of solutions to (7) and identify those of the solutions that are also solutions to the practical problem.

(9) Generalise to new conditions - for instance, laboratories having a total area of 24m^2.

Several other application contexts are used to structure these workshops - the longest ladder that can be carried around a selection of corners, the largest hospital trolley that can be wheeled around a variety of corners, shortest path problems, models of growth and recession, investment, resisted motion under gravity, optimisation of volumes and surface areas, illumination, cost matrices and allocation problems.

We now pose the central pedagogical question.

What are key mathematical attributes that are required consistently for students to successfully solve structural problems and then generalise their solutions to new variations of the context?

From a context analysis of the learning tasks for this course we have identified two attributes which we believe are constant in their presence and varied in the different disguises that they wear:
1. parameterisation,
2. coordination of graphical and algebraic representations.

We believe that these two key attributes are distinctive but linked. Schoenfeld (1986) has drawn attention to the special challenge involved in coordinating knowledge between different symbol systems (as occurs with the algebraic characterisation of graphical forms). He points out that many students seem to make virtually no connections between reference domains and symbol systems that appear closely related. He has raised the possibility that computer-learning environments, through the very ease with which examples can be generated might promote empiricism rather than logical deduction in connecting theory and form.

Consequently, in respect of (2), the computer-based training context adds another dimension. We need to be interested in whether insights generated through the use of software are quickly forgotten or whether they are internalised as learning that can be reconstituted through other means such as verbal dialogue and pen-and-paper argument.

It is relevant then to assess what understandings students have in respect of these two attributes on entering the course, and we concentrate on this aspect. However it is also appropriate to assess the extent to which these entering characteristics have a measurable impact on performance in the course. A further important question is the extent to which the teaching programme produces specific gains in the identified attributes.

ITEM CONSTRUCTION AND PURPOSE

We take the view that knowledge may be usefully classified in terms of concepts and procedures (Anderson, 1990; Hiebert and Lefevre, 1986), and that retrieval of mathematical knowledge from memory storage may be understood in similar terms. Conceptual knowledge is assumed built by the construction of new relationships between existing information, or through linking existing knowledge to some new information - as for example the successively generalised concept of multiplicative inverse developed from real numbers, through matrices, to functions. Conceptual knowledge is assumed stored as a linked network of individual units, where the more elaborate the network the more points of access there are for activation to occur. Inadequate conceptual knowledge means that a needed piece of information will not be retrieved when required (blank response), or that some incomplete or inaccurate version will be acted on (for instance, 1/A written as the inverse of matrix A). The connection to processes in the modelling activity will be immediately apparent.

Procedural knowledge is acquired by practice that involves the execution of a routine (production rule) in response to an activating condition, for example

$$\text{If } ax_1 = ax_2 \text{ and } a \neq 0 \text{ (condition) then } x_1 = x_2 \text{ (production rule)}$$

Flawed procedural knowledge may be a consequence of either a mis-applied or mis-remembered condition such as losing a solution from $x(x - 1) = x(2x - 3)$ through cancelling x, or a mistaken rule, for instance, $(x + a)^2 = x^2 + a^2$.

The investigations into student understanding cited in above identified many misconceptions and flawed procedures that can be interpreted in these terms.

In doing mathematics, learners may merely be required to complete a *mechanical* routine. At a higher level they may be required to *interpret* information in order to reach a conceptually-based conclusion. At a higher level still is the requirement to *construct* a solution. Typically this will involve the creation of new links and interplay between concepts and procedures that must be generated as part of the solution process.

We believe that existing knowledge, and that to be acquired, can usefully be considered in terms of these three descriptions:

1. mechanical,
2. interpretive,
3. constructive.

The capacity of learners to invoke mathematics in new situations is strongly dependent on their ability to operate at interpretive and constructive levels, particularly the latter. We therefore assess the knowledge of the students in this project in terms of these three forms.

Consistent with our earlier remarks, our interest here is in the mathematical requirements and performance of students during those phases of application problems where the focus is on mathematical processing. The items embody concepts and procedures identified as central to the two attribute areas identified previously. It is further noted that the knowledge required for the items, although central to successful functioning, is grounded in secondary school mathematics.

The test items serve several roles in this research.

- As a measure of entering mathematical competence - that is of the extent to which students have a grasp of basic concepts and procedures required for successful completion of the application problems presented in the course.

In doing so they serve a *diagnostic* role in which the pattern of performance pin-points areas of strength or weakness and a *predictive* role in which the final performance in the course is related (for example by regression) to the entering performance level.

- As pre-and-post measures to assess the impact of the teaching program in deepening the understanding and capacities of students in concepts and procedures that are central to it.

Following upon this, they serve to assess the extent to which the computer-based learning sequences were effective in producing enhanced learning.

The content of the items was selected to reflect learning specifically targeted by the technology.

EXAMPLES OF TEST ITEMS

Mechanical

The items in this group require the students to perform some standard procedure that is cued in the wording of the question.

Item 1M $x^2 - ax + 12 = 0$ represents a family of equations. Four members of the family are obtained by giving a the values 5, 6, 7 and 8. For what values of a can the equations be solved by factorising the left hand side?

A. 5 only   B. 7 and 8   C. 6 and 7   D. 8 only   E. none

The solution involves either factorising the left hand side for the given values of a or in evaluating the discriminant and selecting those values of a for which it is a perfect square.

In terms of cognitive operations a direct cue is provided (factorise) which should activate procedural knowledge to perform the manipulations. That is, the activating condition does not have to be inferred from the mathematical context, and the students are tested simply on the proficiency with which the nominated procedure can be used. It will be noted that the idea of parameterisation is present but not an essential element for solution.

Interpretive

The items in this group require the retrieval of conceptual knowledge and its application to identify a correct alternative. These items do not involve mathematical procedures.

Item 2I Which of the following could be the equation of the graph shown?

A. $y = (x - 2)^2(1 - x)$ B. $y = (2 - x)^2(1 - x)$
C. $y = (x - 2)^2(x - 1)$ D. $y = (x - 1)^2(x - 2)$ E. none of these

Reasoning such as the following is required. Since a double root occurs at $x = 2$ and a single root at $x = 1$ the equation is either $y = (x - 2)^2 (1 - x)$ or $y = (x - 2)^2 (x - 1)$, noting the equivalence of alternatives A and B. Since for large x the graph behaves like $y = x^3$, alternative C is correct.

In terms of cognitive operations the student needs to draw upon knowledge about the behaviour of polynomial graphs for large x, about the algebraic meaning of single and repeated roots shown graphically, and about the equivalence of perfect squares. They must then coordinate the knowledge to eliminate alternatives systematically. This requires access to a well-developed network of linked conceptual knowledge, but no actual mathematical procedure needs to be completed. Coordination of algebraic and graphical forms is essential for success on this item.

Constructive

These items involve the use of both conceptual and procedural knowledge in which necessary procedures have to be introduced by the student. Responses involve the construction of a solution rather than the selection of an alternative.

Item 2C The equations of two graphs are $y = 3/x$ and $y = x^2 - 4$. Obtain a cubic equation whose solution gives the x-coordinate of the point(s) of intersection of these two graphs. How many positive roots does this equation have?

The solution involves recalling that the required equation is obtained by equating $3/x$ and $x^2 - 4$ (concept), simplifying $3/x = x^2 - 4$ to provide a cubic equation in some form (procedure) and sketching a graph such as Fig 1 (procedure), and recognising that one intersection to the right of 0 means one positive root (concept).

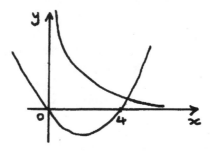

Figure1: graphs for Item 2C

In terms of cognitive operations the solution involves an interplay of conceptual and procedural knowledge. The recognition of relevant concepts activates the formal conduct of procedures where successful use involves further knowledge of algebraic simplification rules, the shape of the respective graphs, and the interpretation of the intersection in terms of the solution to the question. It is the internal activation of procedures through the exercise of conceptual knowledge that causes constructive items to differ from mechanical items (where procedures are externally cued) and from interpretive items (which involve concepts but not procedures). The term *constructive* is used to indicate that the student must act to set up a solution framework rather than respond in a singular way to a precisely targeted question. A consequence of this is that an extended response is required in contrast to multiple-choice responses that are appropriate for mechanical and interpretive items. (The term *constructive* is used in the sense above and has no definitive connection to constructivism as used in the academic literature.)

PRELIMINARY RESULTS

The observed average responses from 244 mathematics and engineering students are given in Table 2. Examples of the test items 1 (mechanical 1M), 2 (interpretive 2I) and 3 (constructive 2C) were given above. From Table 2, it can be seen that in the whole group 59% of the students got item 1M correct, that only 30% were correct on item 2I and that just 14% could achieve correct answers on item 2C. The mechanical and the interpretive items, as multiple choice items, were scored 0 (correct) or 1 (incorrect). The range of expected responses on the constructive items was wide. These items were marked on various scales from 0, 1, 2 for the shorter questions to 0, 1, 2, 3, 4, 5 for those with more complex structures. In the preliminary analysis of this paper the scores on the constructive items have been rescaled to the same scale (0,1) as for the other items.

Consider first the overall performance of the students. On the six items in each of the groups, the mean (0.52) for mechanical skills suggests that there is only a 52% chance of students being successful on these straightforward tasks - tasks are firmly embedded in the secondary curriculum and which are essential to the successful completion of many modelling activities. Similarly the mean (0.42) for interpretive skills shows that retrieval of conceptual knowledge is not as well advanced as that of using mechanical skills. Further, the interaction of both conceptual and procedural knowledge where procedures have to be introduced by the student is the least developed amongst the 244 students (mean 0.34). On an item-by-item basis, this falling off in performance from mechanical through interpretive to constructive skills is generally evident from Table 2, (Item 2M mean 0.41, Item 2I mean 0.30, Item 2C mean 0.14).

The standard deviations on mechanical skills (0.22) and on constructive skills (0.19) draw attention to the wide variations in the data. The variations in interpretive skills are a good deal narrower (s.d. 0.08). Item 3M does not fit this pattern since only 17% of the students could carry out a mechanical task involving the 'addition' of two given graphs. The corresponding graphical interpretive (3I) and constructive (3C) tasks were achieved

	mechanical skills			Observed Average interpretive skills				constructive skills			
item	all groups 244 students	all groups female 45 students	all groups male 199 students	Act Sc 23 students	all groups 244 students	all groups female 45 students	all groups male 199 students	Act Sci 23 students	all groups 244 students	all groups female 45 students	all groups male 199 students	Act Sci 23 students
1	0.59	0.53	0.60	0.77	0.41	0.58	0.62	0.52	0.59	0.51	0.65	
2	0.41	0.51	0.39	0.35	0.25	0.31	0.74	0.14	0.18	0.13	0.40	
3	0.17	0.16	0.18	0.30	0.41	0.50	0.60	0.36	0.30	0.37	0.44	
4	0.74	0.65	0.77	0.86	0.34	0.35	0.34	0.37	0.14	0.18	0.14	0.22
5	0.44	0.47	0.43	0.50	0.38	0.42	0.37	0.53	0.67	0.75	0.65	0.89
6	0.77	0.82	0.76	0.95	0.45	0.47	0.44	0.67	0.22	0.18	0.24	0.53
mean	0.52	0.52	0.52	0.62	0.42	0.39	0.42	0.59	0.34	0.36	0.34	0.52
s.d.	0.22	0.20	0.21	0.25	0.08	0.07	0.09	0.12	0.20	0.23	0.19	0.21

Table 2. The table shows the observed group averages for the six items in each category: mechanical, interpretive and constructive skills. For example 47% of the female students are likely to be successful on item 6I.

by 48% and 36% of the students respectively. This suggests that, for some graphical tasks, conceptual knowledge amongst this group of students is better developed than procedural knowledge, but the story is more complicated. Items 5 were also graphical tasks, but item 5C involved the interaction between these two kinds of knowledge together with student initiative in introducing their own procedure (constructive skills). It was achieved by 67% of the group, strongly contrasting with 44% and 38% for items 5M and 5I.

Amongst the sample of 244 students, there were 45 female and 199 male students. Not unexpectedly, Table 2 does not show a strong overall gender effect as all the students in the sample were undergraduates on courses which require strong mathematical skills. However, the data do show substantial differences on individual items. For example, on the mechanical item 2M above, 51% of females were correct compared to 39% of males, whilst on item 4M 65% of females were correct compared to 77% of males. Other minor differences in mechanical skills are also apparent. Similar comments can be made about interpretive skills on item 1I for which the performance of the females (41% correct) did not match that of the males (58% correct).

Given the diverse student group, it is useful to compare the performance of the actuarial science students whose formal entry qualifications are generally high, with that of the complete sample. Four of the 45 female students were in this group. Table 2 shows that these students' skills are equally developed across all areas. The means, across all items, for their mechanical, interpretive and constructive skills are 0.59, 0.51 and 0.54 respectively, whereas the means for the whole sample are uneven and fall off considerably. Notice that if an actuarial student attempted item 6M, then they were likely to get it right. Note too that the standard deviations of the actuarial science students are generally high. Here too there are wide variations in performance. This professional group of students is well-equipped to undertake modelling tasks which in subsequent parts of their course are primarily of a statistical nature.

The summarised items and their rank order performance are shown in Table 3. Item 2M was attempted by most students, with a 41% success rate. The second most popular item was 4M, on which students achieved a 74% success rate. As far as mechanical skills are concerned, it could be said that the students were not confident, since the rank order of frequency of attempt does not correspond closely with the performance outcome. Alternatively, students may have been confident of their success at a particular mechanical item but the results show that in fact they could not do it. This is shown by the fact that item 1I was the most frequently attempted with a 55% success rate but, for example, item 2I was the least popular. Perhaps in this case the knowledge required and its application was closer to that of their recent experience prior to entry to the courses. Results for constructive skills are similar to those of the mechanical skills. Item 1C was attempted by most students, with a 52% success rate, but the least popular item 5C had a better success rate (67%). Perhaps in this case, the item was perceived to be too difficult for some students, who therefore did not attempt to answer it.

These item summaries are to give the reader a picture of the scope of the items. They are not worded in the mechanical, interpretive and constructive styles presented to the students. For such details refer to the three examples of test items given above and to Galbraith and Haines (1995).

CONCLUSIONS

These preliminary investigations into students' mathematical skills have helped us to understand the difficulties faced by students on courses where applying mathematics is central to the curriculum. We have shown that, where mathematical modelling requires strong mechanical skills even at a routine level, there is a strong probability that students may not succeed unless adequate support mechanisms are included. Our study is not diagnostic in the sense of highlighting a need for so-called remedial action by top-up short courses provided by whatever means. However it does highlight the need for careful planning and execution of modelling activities, encouraging and motivating students by recognising their capabilities and limitations.

Further, the modelling activity and associated situations where we apply mathematics need well-developed interpretive and constructive skills. We have demonstrated that entry skills to a wide range of courses are such that students may find it difficult to interpret information and modelling results and that their constructive skills will need careful nurturing. On entry, students are unlikely to be successful on tasks which depend on these skills.

ACKNOWLEDGEMENT

The authors acknowledge the assistance of the British Council in supporting the collaborative research reported in this paper

items in rank order by performance	% correct	summarised item	rank order of item by frequency of attempt
mechanical			
6	0.77	$y=2x^2-bx^3$ cuts the x-axis at $x=4$: find b	6
4	0.74	use $y=mx+c$ to find equation to a line	2
1	0.59	factorise $x^2-ax+12$ for $a=5,6,7,8$	3
5	0.44	find a local maximum of $y=x(x-3)^2$	5
2	0.41	identify equivalent forms of $(x-2)^2(2+x)^2$	1
3	0.17	given graphs y_1 and y_2 to find minimum (y_1+y_2)	4
interpretive			
1	0.55	$y=x^2-ax+6$: position of vertex as given by a	1
3	0.48	graph of $y=af(x)+b$: effect of parameter a	3
6	0.45	for graphs y_1 and y_2 identify properties of (y_1-y_2)	6
5	0.38	identify reflected and translated graph	5
4	0.34	interpreting a in $y=a(x-3)^2+2$ as steepness	4
2	0.30	matching $y=(x-2)^2(x-1)$ to a given graph	2
constructive			
5	0.67	given $y=(x-a)(x-2a)(x-3a)$, draw graph for $a=-2$	6
1	0.52	draw $y=(x-1)^2-2$ by translating given graph	1
3	0.36	given $1/\cos x, 1/\sin x$ draw $y=2/\cos x+2/\sin x$	4
6	0.22	draw $y=3^x-2x^2-1$ using values at $x=0,1,2$ etc.	5
4	0.14	given $(2,-2)$ on $y=-ax^2$, draw it translated $(-3,1)$	2
2	0.14	given $y=3/x$, $y=x^2-4x$, solve and interpret cubic	3

Table 3. This table summarises the mechanical, interpretive and constructive items and gives the performance of the students in terms of proportion correct and their rank order by frequency of attempt.

REFERENCES

Anderson JR, 1990, **Cognitive psychology and its implications**, 3rd edition, WH Freeman and Company, New York.

Atkins NG, Haines CR and Orman BA,1996, *Teaching and Learning: Experimental Workshop,*. in Haines CR and Dunthorne S, (eds.), **Mathematics Learning and Assessment - Sharing Innovative Practices,** Edward Arnold, London.

Blum, 1993, *Mathematical modelling in mathematics education and instruction*, in Breitag T, Huntley I, and Kaiser-Messmer G, (eds.), **Teaching and Learning Mathematics in Context**, pp 3-14, Ellis Horwood, Chichester.

Buckland PR, 1969, *The mathematical background of teachers in training*, **The Mathematical Gazette**, **53**, 386, pp 357-362.

Carr A, 1993, *Problems and projects for teaching modelling*, in Breitag T, Huntley I, and Kaiser-Messmer G, (eds.), **Teaching and Learning Mathematics in Context** ,pp 15-25, Ellis Horwood, Chichester.

Clement J, Lochead J and Soloway E, 1980, *Positive effects of computer programming on the students' understanding of variables and equations*, **Cognitive Development Project**, Dept of Physics and Astronomy, University of Massachusetts.

Derive: **A Mathematical Assistant for your Personal Computer**, 1990, Soft Warehouse Inc., Honolulu, Hawaii.

Galbraith P.L, 1982, *The mathematical vitality of secondary mathematics graduates and prospective teachers*, **Educational Studies in Mathematics**, **13**, pp 89-112.

Galbraith PL and Haines CR, 1995, *Students' Mathematical Characteristics: Some Developmental Skills for Undergraduate Programmes*, **Bulletin IMA**, **31**, pp 175-179.

Graphic Calculus: *A graphic approach to the calculus*, 1986, Rivendale Software, Kenilworth.

Gray JD, 1975, *Criticism in the mathematics class*, **Educational Studies in Mathematics**, **6**, pp 77-86.

Hiebert J and Lefevre P, 1986, *Conceptual and procedural knowledge in mathematics: an introductory analysis*, in Hiebert J, (ed.), **Conceptual and Procedural Knowledge: the case of Mathematics**, pp 1-27, Lawrence Erlbaum Associates, Hillsdale, NJ.

Schoenfeld AH, 1986, *On having and using geometric knowledge*, in Hiebert J, (ed.), **Conceptual and Procedural Knowledge: the case of Mathematics**, pp 225-264, Lawrence Erlbaum Associates, Hillsdale, NJ.

Tall D and Razali MR, 1993, *Diagnosing students' difficulties in learning mathematics*, **International Journal for Mathematical Education in Science and Technology**, **24**, 2, pp 209-222.

Section B

Assessment at Tertiary Level

7

The Assessment Factor -
by Whom, for Whom, When and Why

Leone Burton
University of Birmingham, Edgbaston, Birmingham B15 2TT, UK.
e-mail: l.burton@bham.ac.uk

ABSTRACT

In constructing a course and deciding upon how its effectiveness with students is to be judged, the argument is made that much more needs to be considered than is apparently the case when a test is set. An appeal is made to consider the mathematics and the learners as heterogeneous and to build curricula, the package of syllabus, pedagogy and assessment, on differences rather than on sameness. The purposes of assessment are discussed and attention is drawn to their formative, informational uses as being as important, if not more so, than their summative function. The chapter concludes with a review and discussion of a number of different strategies for assessing mathematics.

INTRODUCTION

Assessment, especially in higher education and particularly in Mathematics Departments, has, until recently, been considered unproblematic from the point of view of those setting and marking assignments. The assessment scene has been dominated by unseen paper and pencil examination papers, as a way of testing what knowledge and skills students have acquired on the course being assessed. Such tests are, presumably, widely accepted by practitioners as both legitimate and valuable, otherwise how can one explain their prevalence? Indeed Niss (1993a: page 4) pointed to ..."an increasing mismatch and tension between the state of mathematics education and current assessment practice" (see also Niss, 1993a). The students, when asked, are less unequivocal. When asked to compare the writing of an essay for assessment with the taking of an examination, one student said:

> I feel I know and understand it a lot more than I would after revision for an exam because, well, people who have got this far know how to do exams, and

you can sometimes do exams by a minimal amount of work. And that doesn't necessarily mean you understand it. Whereas an essay-based subject implies more understanding of the subject.

One explanation for the acceptability of examinations is located in the assumed legitimacy of testing the reproducibility of knowledge. Behind this lies a belief in a mathematics which is fixed, given, absolute. The teacher's role then is seen as the transference of that given knowledge into the heads of the students. Meaning, within this paradigm, lies in the knowledge. The epistemology, the theory of how people come to know, which is driving the teaching and assessing is based on reproduction and is consistent with a behaviourist stance on the psychology of learning. Another explanation is located in the accessibility of tests to so-called 'objective' marking, an accessibility which is itself tested on many occasions when an academic attempts to make sense of a student's papers. Even were we to agree about the 'objectivity' of marking of mathematics tests, we would still have to consider what we are sacrificing in using this assessment style. In the words of Chris Rust, speaking to the University of Birmingham Teaching Forum, "Most of the assessments done in HE do not test the objectives of the course they purport to examine". Dr Rust was reporting on some recent findings which showed that successful students often use very sophisticated ways to spot the cues that signal likely assessment areas and choose to learn what they know will be assessed.

In other words, by working within this paradigm, one is ensuring that students adopt the same learning assumptions:

- the syllabus content is not questioned,
- learning happens by telling/being told, that is the learner is dependent upon the teacher/text,
- students must be receptive, usually implying passivity in class,
- good learning is individual and frequently aided by competition,
- testing is the best (the only?) way to find out what students know.

Even amongst those working within educational testing who accept the problematic nature of test outcomes, there is not an equivalent acceptance that such testing, itself, is problematic and the search continues to make so-called 'objective' testing more 'objective'. It is not, of course, the first time that some people have spent large parts of their lives searching for the Holy Grail but, in my view, it continues to be a considerable waste of energy. More useful, it seems to me, would be for those constructing mathematics curricula to pay attention to the reasons why they make this particular choice of content, for whom it is intended, how the target students might engage with it and by what means the efficacy of the choices are to be evaluated. That is, we must acknowledge the demand made by Dr Rust and others that assessment addresses the purposes of the course and, I would add, respects the constituency of students taking it. Further, we need to accept that every course is a result, probably unique, of a combination of choices, preferences and demands and that it is better to make these clear in the course planning so that the course is an obvious response to cultural conditions rather than to accident or prejudice. As Luke Hodgkin pointed out in 1976, there are:

> ... important ways in which scientific knowledge changes character, even locally, with changing culture. This determining character of social institutions on the knowledge acquired is well-known among scientists: X was at Princeton, at Grenoble, at Warwick and so she sees a differential equation in a particular way, even has a particular notion of 'proof'. But this knowledge is not spread outside the restricted community of scientists, so that outsiders still believe in a universal agreement on an absolute 'rightness' which is something they themselves have been taught to believe in - at school, and in mathematics in particular.

Thus, in constructing a course, we not only need to identify its content as, conventionally, we have been accustomed to do, but to interrogate that content with respect to the choices and possibilities it offers rather than working on the assumption that it is fixed. However, even if we have done that, we have only begun. What is to be taught does not indicate how; that is, what are the pedagogical styles through which the students will experience the content? Not independent of those choices is the range of evaluation strategies which will be used at two levels; first to provide evidence of what the students are learning on the course and second to provide evidence that the course itself is meeting the objectives which we had for it. I wish to suggest that the basis for all these decisions is that heterogeneity in the mathematics classroom is the norm, not the exception.

(HOMOGENEITY) AND HETEROGENEITY

Why is heterogeneity a necessary assumption? I am asking that we move from the position that the knowledge which is to be transmitted is the controlling aspect of the mathematics classroom. Despite the loss of certainty due to Gödel and the shift from a mechanistic to the dynamic and interconnected world of quantum mechanics, much mathematics continues to be presented to students statically, uncritically and uncreatively. Referring to physics students, Kim Thomas (1990: pages 172, 173) wrote that they:

> ... held a view of science as 'objective' and 'value-free': physics in particular was seen as a 'fundamental' subject, able to reveal universal and immutable truths about the nature of the physical world ... This rigidity in the view of their subjects was mirrored by a rigidity in teaching methods ... they were not, generally, required to discover for themselves; the point of experiments, for example, was to illustrate a received 'truth' rather than to allow students to make findings. Dissatisfaction with this method of teaching came largely from the very high achievers and the very low achievers.

I do not think mathematics is very different from physics in these respects. Indeed, Ole Skovsmose (1993: page 170) asserts that:

> Mathematics is conceived as a single homogeneous system of knowledge. And, naturally, the aim of teaching must be to inform the students as well as possible. The teacher has to introduce the students to the subject, and ... the focus is on the relationship between the individual student and that parcel of the curriculum that is presented by the textbook and explained by the teacher. In that way homogeneity of knowledge introduces authority into the classroom.

It is impossible, however, for us to predict what any one learner will take away from any particular learning experience, as demonstrated in the important research done by Brenda Denvir (1986a,1986b). Learning is a process of meaning making by learners not of being handed meaning by their tutors. Accept this, and we shift the focus from the perceived homogeneity of the mathematics which we are intending to convey, to the heterogeneity with which any new piece of information is encountered and understood by the members of a class. This shift is an extremely challenging one as it strikes at the heart of the authority of mathematics.

Under these circumstances, the best we can do is respect the differences in knowing which we take into the classroom and the differences in knowing which result from what has happened in the classroom. Our learners are heterogeneous. They have different styles of knowing, they know different amounts and make different connections between their knowledge. They take different things away from the classroom. A clip from an Open University video of a secondary classroom (EM236 Excerpts 1-11), where the teacher's agenda is about sequences and, particularly, their nth terms, emphasises this heterogeneity. The teacher asks the class to consider a sequence represented by three dot patterns consisting of two, six and twelve dots in a rectangular array. He invites the pupils, first, to indicate how they 'see' these arrays. The first response is as a rectangle, 2 x 1, 3 x 2 and 4 x 3. The second response is as a square plus a column, (1 x 1) +1, (2 x 2) + 2 and (3 x 3) + 3. The third response is 'completing the square' and then taking away a row, thus, (2 x 2) - 2, (3 x 3) - 3 and (4 x 4) - 4. The pupils are then asked to supply an expression for the nth term of the sequence, from each visualisation of the pattern, and the expressions $n(n+1)$, n^2+n and $(n+1)^2 - (n+1)$ are obtained. The three expressions are shown to be the same and the point is made, by the teacher, that everyone thinks that their way of 'seeing' is the only way and that what is important is to look at the many different ways there are of imaging a mathematical idea and the impact each has on the movement from image to representation. The teacher emphasises that it is the pupil's job to look for similarities and differences between their way of seeing and recording and those of the other pupils in their group - and of the textbook author! It might be argued that mathematics is happening throughout this sequence whereas I believe that it is possible to argue that the mathematics starts to happen for these students when they observe the generality, that is that their own preferred particular expression *is* the same as the other particular expressions. I am suggesting that simply because numerical and algebraic expressions can be used is not evidence that the students can and are thinking mathematically but this is the kind of evidence on which we, as teachers, often rely. For an expansion on mathematical thinking, see Mason *et al.* (1982).

Research on the undergraduate teaching of engineering has also demonstrated that if the teacher's and the student's thinking styles are similar communication can take place, but differences in thinking style can often be an explanation for failure in communication and consequently, in learning (Cowan, 1975). Every classroom can demonstrate heterogeneity of the kind described, and many other kinds, if the students are given the opportunities to display and consider it. In every classroom our way of 'seeing' some mathematics is likely to be only one of many possible ways and our choice lies in celebrating these differences and, out of them, building mathematical strengths, or ignoring them and merely pretending a homogeneity exists.

PLANNING A CURRICULUM

With heterogeneity in mind return to our curriculum planning agenda we are setting out to consider why this content, for whom it is intended, and how they will engage with it and by what means the learning and teaching are to be evaluated. These questions are inter-dependent in that the what of the content cannot be dissociated from the how or from the means used to evaluate the learning. Unlike a classroom which is predicated on transmission and homogeneity, the assumption of heterogeneity leads to the responsibility of ensuring that every student has some opportunities to utilise their preferred learning, assessing and thinking styles. This is especially the case for those styles which are different from our own. Further, the valuing of heterogeneity in the classroom makes possible the encouragement of effective learners who can make use of many different styles, choosing appropriately between them. It also encourages the tutor to state their own preferred style as well as looking to identify and distinguish those which are different. What are the purposes of assessment in all of this,?

WHY ASSESS?

The first, and in my view the most important purpose of assessment is to provide information to learners that enables them, in the words of Hatch and Gardner (1990: page 416), to

> ... build on their own intellectual capacities and take optimal advantage of the educational resources around them.

That is, assessment informs learners about their own learning Such information is being sought, and found, at all times in the classroom whether the formality of assessment is happening or not. Learners engage persistently in cycles of self- and peer-evaluation as a necessary part of their learning. Such formative assessment can have either an encouraging or a debilitating affect on a students' propensity to continue learning. Rather than let informal assessment happen in an *ad hoc* way, possibly allowing misperceptions and misunderstandings to develop, I believe that it is essential to integrate evaluative strategies into the learning, teaching and thinking styles upon which the course is drawing.

The second purpose of assessment is to enable the teacher to build up a generative picture of the learner, identifying strengths and diagnosing weaknesses so that further appropriate learning choices can be made and a sense gained of how effective the course is proving to be. Tutors need, in Robert Glaser's words (Glaser,1990: page 482) to

> become especially adept at developing instructional situations in which significant aspects of students' thinking and problem solving can be observed, assessed and constructively guided.

A third purpose of assessment is to provide a mechanism for a sharing, listening, responsive critique between learners upon which, in my view, effective learning is dependent. This form of critical evaluation depends upon the growth of a learning community which itself requires honesty, trust and respect between learners themselves, as well as between learners and tutors. A critical community creates its own culture within which are expressed and unexpressed rules, obligations and requirements. The transmission classroom subsumes a mathematics culture which usually remains uninterrogated despite having its own particular constraints and clear behavioural norms. A learning community overtly addresses the nature of the learning process and recognises its 'cultural' embeddedness and complexity. One result of this process is a sensitive awareness on the part of tutors and learners to the learning which is taking place and, consequently, to the effectiveness of the course itself. The tutor in such a classroom requires that interpersonal behaviours are institutionalised as a necessary part of the learning process and this, in turn, depends upon students accepting the value of communicating with each other for a mutually beneficial result. It is essential that this style of working is negotiated with, and understood by, the students who are likely to have had many years of individualistic, competitive, transmissive learning experiences. Those who are at the university to obtain a piece of paper which is a passport to better living are likely to be strongly influenced by competition and resistant to collaborative strategies, at least until they find that these strategies function well in terms of their learning. Involving people from industry, as those who have done so know, can be a very effective way of finding new voices to emphasise the need for team work and collaboration in preference to those behaviours which have featured more traditionally.

This leads to a further purpose of assessment. The most frequently cited reason for assessing is summative. We do not, however, have to think of summative assessment in a constrained way. Elements of assessment which are team-based, are highly attractive to industry because they are more connected to expected behaviours in the workplace than is sitting an examination paper for three hours and adequately reproducing known knowledge. Indeed I think that we are foolish to disregard the amount of support for innovation which is potentially within the world of work and, equally, to disregard the increasing amount of experience which school students are having with alternative styles for reporting upon their mathematics work. Whether we test to extinction or we involve our students in a range of innovative strategies for assessing their learning, they will make judgements about what they are learning and the teaching, learning and assessing styles being employed to facilitate that learning. Integrating assessing into their course experience requires that we make careful choices between a range of assessment styles, choosing appropriately and differently, depending upon the course and the students.

ASSESSMENT STRATEGIES

Reflection

Many years ago Piaget claimed that to know was to act. A lot of water has passed under the bridge of knowledge since then, but most people would accept that to know and be unable to act is not to have very valuable knowledge. However, I would go further and say that developing knowing in the absence of activity is extremely difficult for the learner. Learning mathematics demands active engagement with some ideas as well as with the multiple possibilities of interpretation of those ideas held by others in the learning community. One reason, I believe, why computers have been efficient learning tools in some settings is that ways of using them to promote engagement have been developed, and engagement stimulates the formulation of questions which open the door to learning activity. However what is missing from learning through activity is how the learner, and the tutor, surface the learning. How do they become aware of what is now known which was not known before, and of what is still not known? The crucial connector, in my view, is reflection. Rather than persistently rushing on to the next topic, or to learning the next valuable nugget of knowledge, time is better spent in reflecting on the process which has just been experienced. As Confrey (1990: page 109) puts it:

> In mathematics the reflective process, wherein a construct becomes the *object* of scrutiny itself, is essential. This is not because, as so many people claim, mathematics is not built from sensory data but from human activity (mathematics is a language of human action) ... to create such a language we must reflect on that activity, learning to carry it out in our imaginations and to name and represent it in symbols and images.

Opportunities for reflection are therefore, I would claim, a necessary part of every mathematics course and need to be created, protected and used to provide the students with feedback on their progress as well as to ensure that the tutors are in touch with the learning which is taking place. Such reflection can be incorporated into the formal assessment of the course, if required, in the form of a learning journal or log. (For examples see Borasi and Rose, 1989; Borasi and Siegel, 1990; Powell and Ramnauth, 1992). It can also be built into the learning strategies. For example, instead of 'delivering' a lecture, hand it out in advance on paper; require the students to work together in small groups, going through the text and identifying their own questions about the text. Ask them to prioritise these questions on the assumption that many questions will be shared by many groups and in this way they will listen to the questions of others maximising the benefits for the learning community. Use the lecture time to answer the questions of the groups, allowing each group one question in turn. So far the students have invested their own working time in this class; the tutor has not been required to spend any more time than is currently spent. If students have outstanding questions at the end of the class, they could hand them in for the tutor to prepare answers for the following class. Alternatively the group questions could be handed in before the class and the answers prepared by the tutor to utilise the class time. This loses the immediacy and the personal aspects of the question-asking but provides tutors with advance warning (good for decreasing feelings of vulnerability). However it also loses the benefit of encouraging students to pay attention to others' questions in case they lead to a response which is relevant to one of their own. Students could be asked to write a reflective piece on their questions and the ways in which they would now deal with them, perhaps as part of a learning log. If, and how, you use the suggestion will depend upon your own local circumstances. We let a student have the last word:

> I always wonder at the way a lecturer stands there and writes it all on the board and you copy it all down because, to be honest, it's a waste of time - you may as well just have printed notes in the first place and discuss them in class instead of just sitting there copying stuff off the blackboard.

Project work

The incorporation of courses or modules on mathematical modelling into many mathematics degrees has had an interesting effect on assessment. Clearly an effective evaluative strategy for a project which has required a group to devise a mathematical model and investigate its implications is unlikely to be a three-hour unseen examination. However, even where there is an institutional requirement for a summative examination, and it is less and less the case nowadays that the requirement is institutional, there is every opportunity to try to ensure that the assessment matches the style and content of the material being assessed. For example in a teacher education course with which I was closely connected, students were required to sit a two-hour, unseen written examination at the end of their second year, and this affected their choice of options in the next two years. The course developers were anxious to ensure that all aspects of the course, which had been based on the use of 'enquiry' methods of learning mathematics, were reflected in the assessments. We also wanted to ensure that the assessments addressed the learning which was central to becoming an effective teacher. So the examination had to be contiguous with the course and appropriate to an intending teacher, not an easy challenge for an examination to meet. How did we do it? In the term during which the summative examination was set, all students met and worked at some mathematical enquiries which, unknown to them, were also given to pupils. The students were invited to bring their work to the examination, at which the work of the pupils on one of the enquiries was used to provide the vehicle for the students to reflect on what they had done, on what the pupils had done and to apply these reflections to their professional concerns with respect to teacher decision-making in the classroom. Since a major part of teacher decision-making is interactive we felt that it was not too distant from the classroom to ask for reflection on work that had just been provided. One of the students' greatest surprises was in the depth and extent of the pupil work compared with their own. I do not want to claim too much for this examination; but it is just one example of how we can make the constraints work towards consistency. We should discard the excuse that something is not of our choosing and we would prefer to do it differently, but we are forced to conform with the demands of 'the system'.

Alongside the use of project work there has been the growth of a range of assessment strategies through which such projects are evaluated. The work of the Assessment Research Group (Houston et al., 1994; Burton and Izard, 1995) has been at the centre of these developments in the UK and a range of publications provide public documentation of the difficulties encountered and the strategies developed. Many Mathematics Departments are working with industry and commerce to ensure that students are constructing mathematical models which have to undergo the stringency of reality testing rather than pursuing modelling of an imaginary, or even pseudo-real, kind. I have already referred to the positive impact of such collaboration on many aspects of courses, as well as on the particulars of the modelling course itself. Interestingly, once the door is opened to some mathematical modelling, tutors begin to see the relevance of those strategies to other parts of the mathematics degree course. I think that is in part, an explanation for the growth of so-called innovative strategies in teaching and assessing mathematics, reflected for example in the forthcoming publication of a Mathematics Learning and Assessment Pack (Haines and Dunthorne, 1996) This pack includes examples of project work in pure mathematics and statistics, as well as the more familiar modelling projects and also looks at mathematical modelling in non-mathematical disciplines. The assessment techniques which are discussed include poster presentations, oral communications, written reports, peer and group assessments and the use of learning diaries and comprehension exercises. They are certainly innovative when compared with the range of assessments on many mathematics degree courses, but are by no means as radical as they could be if we were to begin to take seriously the implications of the inferences that I have been drawing between the teaching of absolute mathematics and the recognition of the socio-cultural role of the learning community in the construction of understanding.

Stephen Brown (1986: page 208-9) describes a unit on number theory that he taught jointly with Marion Walter. With the intention of deriving a formula to generate primitive Pythagorean triples, they asked the students to propose some answers to $x^2 + y^2 = z^2$. As he points out, the equation :

> ... is not even a question. How can one come up with answers? Yet the students dutifully did come up with answers, because they carried along a host of assumptions that we in fact have trained (implicitly) them to accept ... we encouraged students to ask such new questions as:

For what *rational* numbers x, y and z, is it true that $x^2 + y^2 = z^2$?
For what natural numbers is it true that $x^2 + y^2 = z^2$ is 'almost' true?

Realising that we had implicitly assumed that the question was algebraic, the students began to ask a host of geometric questions that derived from connotations of the algebraic form. What followed immediately was one of the most intellectually stimulating units that either of us had previously experienced with our students ... we began to appreciate that such deviations from standard curriculum are not mere frills. That is, in exploring such questions as the 'almost' primitive Pythagorean triplet question, all of us gained a much clearer understanding of what the actual primitive Pythagorean triplet question was in fact about ... we began to realise that an implicit part of the common language that we share with students is one which focuses upon and points so strongly towards the search for solutions and answers, that we continue to search for answers even when no question is asked at all!

Were I wanting to assess some work of this kind, I would ask the students to justify their own questions in terms of what they learnt by pursuing them and, possibly, distribute these individual papers so that group summaries could be prepared which synthesised the class work.

Profiling

We hear persistently about the need to involve students in their own learning. Reasons given include motivation, trust and self-knowledge and certainly there is evidence of just how effective it can be. One student says:

You have so many exams here that, by the time you get to the third or fourth, you haven't a clue what was in the first two courses you've just been examined for. But, in the essays, you do so much research actually remember the information and some of the details.

The development of profiling is now recognised as a powerful means of promoting learner self-responsibility and providing opportunities for tutors to have an enlightening dialogue with their students. School students are becoming more and more familiar with the use of profiles and many universities (including my own) have introduced a student profile into the assessment pattern of certain Faculties. To be effective, a profile should contain learning objectives as well as learning outcomes and these must be reviewed regularly as a way of charting progress.

However, the other side of involving students in their own learning is consultation and participation in that learning. Asking students what they feel about their courses can be quite shattering, but it is clear to me that the process of evaluating and assessing is one which requires close collaboration between students and tutors so that everybody learns to recognise and "distinguish quality work in themselves as well as in others" (Burton and Izard, 1995: page 3). That is to say - everybody - tutors and students, needs to learn:

*   what constitutes quality work,
*   how to recognise when, and how, it is produced,
*   what is required to be a critical member of the learning community.

I am suggesting that what quality work is *not* is the correct reproduction of given text. If we are sure of that, we have an obligation to our students to determine what it *is*. This is not a trivial exercise for those, whether tutor or student, who have been accustomed to think of quality work as consisting of model answers to predetermined questions. Open the door to student interpretation and consequent justification, and new criteria are necessary for judging 'quality'. Are we sure that we recognise quality when we see it, or only when it looks like what we expected to see? Again, how do we ensure that we are challenged and astonished by our students' work, celebrating its diversity rather than its uniformity? Finally, does the learning environment of our classroom nurture a community, representative of all its members and their diverse styles? In what ways do we understand 'community' and how do we ensure that our understanding is shared with the students? Do we build 'community' into our learning, teaching and assessing strategies? How do we demonstrate that we value and expect critique? Do we, ourselves, offer it? Are we happy with one student's reaction to their Mathematics and Theoretical Physics degree:

> Tell them that they can be really smug around non-mathematicians because everyone else in College thinks we are all super intelligent. "That is the one subject I could not have done", they say.

A tutor comment was:

> Hopefully this was said in humour, but the last point is valid. You need a particular ability, not given to everyone, to be successful at mathematics.

It seems to me that both the student and the tutor are demonstrating that kind of elitism which makes mathematics unattractive to many students.

In summary, in looking for effective evaluative strategies, I think it is helpful to use certain rules of thumb.

*   Is the assessment appropriate to what is being assessed?
*   Does the assessment provide the student with an opportunity to demonstrate positive achievement?
*   Are the criteria for successful performance clear to everybody?
*   Is the assessment appropriate to all of those who are being assessed?
*   Does the assessment contribute to the learning, particularly by promoting reflection?

- Across the course, is there an adequate range of different assessment styles so that every student has the possibility of using their preferences and strengths?

CONCLUSION

I started out asking Assessment - by whom, for whom, when and why? I hope that it is now clear that, in my view, assessment is taking place all the time, not only informally, because making judgements is something that we all do about ourselves and about others, or formally by tutors but also as self, peer and group assessments. Assessment does not have to be understood as something that is always done to students by tutors; it does not have to constitute a threat. If we could construct a situation in which assessment is seen to be an integral part of the learning environment, and practised in many different ways, examinations (as presently understood) would be recognised for the very limited information they offer. The assessment factor would then be just one of the contributory influences ensuring constructive learning experiences for students on undergraduate mathematics degrees. I may not believe in searching for the Holy Grail - but I can dream, can't I?

REFERENCES

Borasi R and Rose B, 1989 *Journal writing and mathematics instruction,* **Educational Studies in Mathematics**, **20**, 4, pp 347-365.

Borasi R and Siegel M, 1990, *Reading to learn mathematics: new connections, new questions, new challenges,* **For the Learning of Mathematics**, **10**, 3, pp 9-16.

Brown S I, 1986, *The Logic of Problem Generation: from Morality and Solving to Deposing and Rebellion,* in Burton L, (ed.), **Girls Into Maths Can Go**, Holt, Rinehart and Winston, London.

Burton L and Izard J, 1995, **Using FACETS to Investigate Innovative Mathematics Assessments**, School of Education, University of Birmingham, Birmingham.

Confrey J, 1990, *What Constructivism Implies for Teaching,* in Davis R *et al.,* **Constructivist Views on the Teaching and Learning of Mathematics**, NCTM, Reston, Virginia.

Denvir B and Brown M, 1986a, *Understanding of Number Concepts in Low Attaining 7-9 year olds. Part 1: Development of a Descriptive Framework and Diagnostic Instrument,* **Educational Studies in Mathematics**, **17**, pp 15-36

Denvir B and Brown M, 1986b, *Understanding of Number Concepts in Low Attaining 7-9 year olds. Part II: The Teaching Studies,* **Educational Studies in Mathematics**, **17**, pp 143-164.

Glaser R, 1990, *Toward New Models for Assessment*, **International Journal for Educational Research**, **14**, 5, pp 475-483.

Haines CR and Dunthorne S, (eds.), 1996, **Mathematics Learning and Assessment : Sharing Innovative Practices**, Edward Arnold, London.

Haines CR and Izard JF, 1995, *Assessment in Context for Mathematical Modelling*, in Sloyer C, Blum W and Huntley I, (eds.), **Advances and Perspectives in the Teaching of Mathematical Modelling and Applications**, Water Street Mathematics, Yorklyn, Delaware.

Hatch T and Gardner H, 1990, *If Binet had looked beyond the classroom: the assessment of multiple intelligences*, **International Journal of Educational Research**, **14**, 5, pp 415-429.

Hodgkin L, 1976, *Politics and Physical Sciences*, **Radical Science Journal**, **4**, pp 29-59.

Houston SK, Haines CR and Kitchen A, 1994, **Developing Rating Scales for Undergraduate Mathematics Projects,** University of Ulster, Coleraine.

Mason J, Burton L and Stacey K, 1982, **Thinking Mathematically**, Addison Wesley, Wokingham.

Niss M, (ed.), 1993a, **Investigations into Assessment in Mathematics Education: An ICMI Study**, Kluwer Academic Publishers, Dordrecht.

Niss M, (ed.), 1993b, **Cases of Assessment in Mathematics Education: An ICMI Study**, Kluwer Academic Publishers, Dordrecht.

Powell AB and Ramnauth M, 1992, *Beyond Questions and Answers: Prompting Reflections and Deepening Understandings of Mathematics Using Multiple-Entry Logs,* **For the Learning of Mathematics**, **12**, 2, pp 12-18.

Skovsmose O, 1993, *The Dialogical Nature of Reflective Knowledge*, in Restivo S, van Bendegem P and Fischer R, (eds.), **Math Worlds**, State University of New York Press, Albany, NY.

Thomas K, 1990, **Gender and Subject in Higher Education**, Open University Press, Buckingham.

8

Assessment of Complex Behaviour as Expected in Mathematical Projects and Investigations

John Izard
Australian Council for Educational Research, Hawthorne, Victoria, Australia
e-mail: izard@acer.edu.au

ABSTRACT

No single assessment method is capable of providing evidence about the full range of achievement. This paper reviews the problems faced in devising better assessments to monitor learning, and provides practical suggestions for meeting these problems.

The design of sound assessments, the development of assessment skills, and methods of describing progress are essential requirements for both teacher/lecturer and students. The assessments need to have curriculum relevance, be practical and fair, and provide useful information for further learning. The assessment strategies and the approaches to analysis of assessment data presented in this paper are applicable to traditional examinations, project and investigation reports, presentations and posters, judgements of performance and constructed products, observations of participation, collaborative group work and ingenuity. The paper concludes with advice on monitoring the quality of the assessment process, illustrated with examples from recent investigations.

INTRODUCTION

One valid approach to assessment is to make observations about every part of student learning. While this approach may be feasible in a limited, well-funded research study, in most situations there are insufficient resources and time to make such exhaustive observations possible. Instead, one has to select valid samples of evidence from the skills and achievements of interest.

Since one of the targets of learning is to provide students with the capability to complete tasks successfully, one way of assessing their learning is give a number of these tasks to be done under specified conditions. Some tasks (such as traditional essays and multiple-choice test items) are relatively easy to score, but lecturers and teachers make assumptions about the extent to which these items are consistent with each other. Often the differential difficulty of such items is ignored; easy items give an inflated estimate of achievement and difficult items do the opposite (unless there are benchmarks to give candidates appropriate credit for their skills and achievement). Assessment evidence can be obtained from traditional open-ended examinations, from multiple-choice tests, or from assessments of performance such as a project or investigation report, from a written or oral presentation of a topic, or from ratings of a construction or other product. In some cases, assessment evidence from several sources is combined (Izard, 1991). When assessment evidence is gathered for a topic, some way has to be found to monitor its quality and internal consistency, and a way has to be found to locate the candidates on a meaningful achievement continuum.

Multiple-choice tests are often subjected to detailed analysis to check that the items are internally consistent (that is, are measuring the same type of skill or knowledge). This is a form of quality control; items which do not distinguish between those students who are able and those who lack knowledge may be deleted from the test. The same type of analysis can also be carried out for items worth more than one mark, like traditional mathematics problems and essays. Such an analysis may be called a partial-credit analysis. Those who have full knowledge receive full credit, those with partial knowledge receive partial credit, and those without knowledge receive no credit. As with the multiple-choice test analyses, items which do not distinguish between able students and less able students are queried. Many who assess performance fail to investigate whether each of the separate performance assessments are distinguishing between students who are able and those who are not, as judged by the overall performance test. Since performance tests often involve transient evidence, where a record is not or cannot be kept for independent checking, it is important that the assessment process involves a quality control procedure.

Assessment by observations about every part of student learning may be feasible in a limited, well-funded research study, but in most situations there are insufficient resources and time to make such exhaustive observations possible.

Projects and investigations are often promoted as more comprehensive, valid (authentic) and meaningful tasks, and therefore better approaches to assessment of learning, because candidate skills have to be integrated in a meaningful whole. Projects and investigations require valid assessment strategies just as much as traditional indicators of learning, but the complexity of the tasks, the fact that different candidates often do different projects, and the diversity of examiners, lead to potentially unreliable (and therefore invalid) assessments. Capricious assessment practices fail to give candidates due credit for their efforts, make it difficult for students to be consistent in evaluating their own work, and discredit the institution issuing such assessments.

MEASUREMENT INDICATORS

The assessment of complex behaviour, as exhibited in mathematical performance on projects and investigations, requires an assessment strategy that takes account of all the factors considered important in describing quality performance on that complex behaviour. For example, Le Masurier (see Haines, Izard and Le Masurier, 1993) had video recordings made of the oral presentations by students who had solved a mathematical modelling task. The factors considered important were expressed as descriptors of behaviour.

1. *Establishes a rapport with the audience.*
   Engages and keeps the interest of the audience, eye contact, enthusiasm, personal qualities, *etc.*

2. *Makes an effective delivery.*
   Uses correct speed, pace, tone, pauses, intonation, rhythm delivery is well rehearsed and spoken, not read.

3. *Has a good command of spoken English.*
   Uses sentence structure correctly, good vocabulary and grammar, appropriate use of appropriate language.

4. *Gets over the point using a good structure.*
   Explains well, shows evidence of structure, gives appropriate reference to mathematics, shows understanding.

5. *Gives a clear explanation of the problem and its outcome.*
   Gives a clear explanation of the problem, and an effective presentation for the audience.

6. *Plans and organises overall delivery well.*
   Shows evidence of a plan, organisation (at group level), work by a team, group coheres well.

7. *Makes appropriate use of visual and other aids.*
   Chooses the right medium for communication and uses it well, aids present when needed, explains aids well.

8. *Establishes a rapport with the audience.*
   Sets out OHP transparencies, posters, *etc* well, labelling is good, produces a high quality of appropriate aids.

Each descriptor was rated on a scale from 0 to 4. The 0 reflected absence of evidence of the behaviour, a 1 indicated evidence at a low level, and a 4 indicated that the behaviour had occurred at the highest level.

When judging such complex behaviour, it is important to realise that there are several aspects (or facets) where variation may occur even when the judges are in agreement that the list of descriptors (1 to 8) does summarise the essential elements to be judged. If each candidate makes more than one presentation then it is possible that the candidate's performance will vary. If more than one judge is involved in the assessment process, the judges may have different understandings of the meanings expressed in the descriptors. Further, it is possible that the descriptors themselves describe behaviour that is differentially difficult or that some descriptors are inconsistent with others.

Crouch and Haines (1995) used the same rating scale for oral communication in a study of self, peer and tutor assessment of group oral presentations following the completion of some mathematical modelling tasks. Each descriptor was rated on a scale from 0 to 2. The 0 reflected absence of evidence of the behaviour, a 1 indicated partial evidence, and a 2 indicated clear evidence of the behaviour. Crouch and Haines provide an appendix which summarises the design of the group data collection, identifying the academic, self and peer judges. They found that the self and peer assessments often showed discrepancies, probably because these students had not had explicit training in recognising quality work. Even students who had achieved high marks had difficulty in recognising quality work from their peers.

Can it be said that a student has sound mathematical modelling skills if they cannot detect different quality in modelling consistently?

Houston (1995) used rating scales with 12 descriptors to assess group posters reporting work on a mathematical modelling module. Six of the descriptors were concerned with content, five with presentation and appearance, and one with student performance in defence of their poster during discussion. Peers as judges were generally consistent with the lecturer, but varied considerably in leniency-stringency. Peer judges also had difficulty in using the rating scales; Houston suggested that if they were involved in the development of the descriptors they would be better able to use them.

Le Masurier and Dunthorne (1995) used a different rating scale for assessing posters describing a delivery modelling problem. They also found that the students had difficulty in making consistent assessments of the posters.

Goldfinch and Goodall (1995) described the analysis of assessment data obtained for a mathematics module as part of a first-year engineering course. Their criterion-referenced marking scheme addressed six aspects of communication, nine aspects of execution, and five aspects of interpretation in assessing some extended mathematical modelling tasks. They found that the assessments were inconsistent. There tended to be some agreement between assessors but the judges differed in stringency - the same performance was given differential credit by different judges. Note that this inconsistency only became evident because they used more than one examiner for a relatively small sample of papers. (This issue relates to the design of the data to be collected. The issue of designing data collection to illuminate the assessment procedures is taken up in more detail later in this paper.) They concluded that use of a checklist with sub-tasks may clarify the meaning of these sub-tasks, but training in the use of rating scales by examiners was probably necessary to achieve reliability.

In another study, Goodall, Hall and Walker (1995) investigated the influence of choice of module on final results. The modules varied in the mode of assessment - traditional examination and, in some cases, coursework as well. Their study included investigating the contribution of the major project. They concluded that there appeared to be a considerable lack of equity in the assessments, and that the assumed degree of agreement between examiners on a 'tightly-marked' mathematics examination was not reflected in the actual results.

These studies share a number of features. All five studies identified difficulties due to the inexperience of judges (whether student or lecturer) in recognising and consistently rewarding quality work. The researchers recognised that the use of existing data (produced from routine examination records) did not permit a full exploration of the assessment context or provide many leads on how to advise both teachers and students. In each case, the researchers were able to speculate about plausible explanations but could not decide between such explanations without further data collection. The discussion below describes some ways to improve the data collection process by ensuring that the issues of importance can be addressed.

Another feature shared by these studies is the diversity of the concerns that real-life assessment must meet, such as the necessity for:

- group activities to be assessed as well as individual activities,
- ensuring that choice of subject does not affect the chance of success,
- the development of student skills in recognising quality work from others,
- the development of the capacity to recognise one's own quality work.

Clearly, assessment procedures should cope with the real-life complexity. There are two essential requirements: the assessments must give due credit for quality work, and the assessments must distinguish between the performances to the extent that they differ in quality. While the fact that some assessment areas are more controversial than others suggests that all judges will not reach perfect unanimity, equity requirements do imply that there should be some reasonable measure of agreement between judges.

Simplifying the assessment procedures has a (superficial) appeal. Making the assessment tasks more mechanical and less complex eases the cognitive load of the judges in reaching a decision. However, the simplification process makes the assessment task less like a sample of the real world and authenticity (and hence validity) suffers. Further, so-called objective testing only refers to the scoring process - the choice of the items to use in the 'objective' test is subjective. Clearly leave valid assessment must be reliable and consistent.

There are a number of other potential pitfalls in interpreting assessment data. The candidate sample will not be a random sample of the whole population of secondary or tertiary students; the candidates are atypical in that they will have higher general abilities, will have received a higher than average duration of schooling and, because some of this (recent) teaching has been common to this candidate group, intraclass correlation will be higher than expected with a random sample. Accordingly, traditional inferential statistics will need cautious interpretation and effect-size measures are required to interpret the data.

In real life, assessments often combine results on multiple-choice items, traditional mathematical problems, and projects and investigations. When evidence on several topics is combined, some way has to be found to check that the intended weightings are reflected in the actual weightings (Izard, 1991). If this is not done the allocation of marks reported to the students (and external examiners) will not be those actually used. Misleading examination candidates by using a different allocation of marks to the intended and published allocation could lead to legal challenges! Carrying out such analyses provides both quality control information, data on the topics that students find difficult, and also information on the quality of the assessment procedure. This information about what students find difficult is vital for the teacher and lecturer. In contrast, competitive assessments place an emphasis on relative position in a group, sometimes to the exclusion of meaningful feedback for teaching/learning purposes.

A test analysis software package called QUEST (Adams and Khoo, 1993) can be used to conduct such analyses. The QUEST software is an application of Item Response Modelling (sometimes known as Item Response Theory, IRT, or Latent Trait Theory, LTT) and implements an extension of an analysis approach first derived by a Danish statistician, Georg Rasch (Rasch, 1960). QUEST can be used to partition the contributions of both student achievement and item complexity to an assessment score.

QUEST can be used for traditional examinations, whether essay or worked problems, on rating scales for projects, investigations, reports, presentations and posters, judgements of performance, and also for questionnaires. The scoring and analysis of questionnaire items provides essential information about the views of those who completed the questionnaire, identifies the extent of agreement with particular views, and shows which items are consistent with the other items (that is, assessing the same or similar ideas). Results of the analyses of test data include graphs showing how the assessment device or scale had performed, and how the candidates had performed.

The use of the analyses for quality control purposes in identifying discrepancies and inconsistencies can provide a better approach to informing candidates how the different aspects of complex behaviour are assessed. The analyses require the appropriate data to be collected. The discussion now addresses features of design which can, if met, lead to more efficient and economic analyses.

DESIGN REQUIREMENTS FOR TEST ANALYSIS

For an analysis to provide useful information, the data for the analysis must include the appropriate indicators in order to address the appropriate issues or aspects. For example, if wishing to compare scores obtained by female candidates with those obtained by male candidates one has to know which are male and which are female.

Such an analysis is impossible unless the data for each candidate include this information. Similarly, an analysis of the contribution of each test item requires the data for each candidate to include each response to the items. This information cannot be retrieved from total scores for candidates.

If there are four items and seven candidates the largest possible number of item-candidate pairs will be 28. The matrix as shown in Fig. 1 has a bullet (•) for each interaction between an item and a candidate.

				Candidate			
Item	1	2	3	4	5	6	7
1	•	•	•	•	•	•	•
2	•	•	•	•	•	•	•
3	•	•	•	•	•	•	•
4	•	•	•	•	•	•	•

Figure 1: item-candidate pairs for four items and seven candidates

An analysis is possible even if some items are not attempted by all candidates. Often we can assume that items not attempted provide evidence that the work was not known. Or we may have to refer to such missed items as missing data where such an assumption is inappropriate. For example, some examination papers may have had faults; it would not be fair to penalise the candidate for errors made by the examiners. In the matrix shown in Fig. 2 reasonably accurate estimates for those who received the faulty items (shown with a ?) are possible because the gaps in the evidence are limited.

				Candidate			
Item	1	2	3	4	5	6	7
1	•	•	•	?	•	•	•
2	•	•	•	•	•	?	•
3	•	?	•	•	•	•	•
4	•	•	?	•	•	•	•

Figure 2: incomplete item-candidate pairs for four items and seven candidates

An analysis is possible if many items and candidates are not paired, but if there is not some form of connectedness in the data the analyses may lead to ambiguous or misleading results. For example, the matrix shown in Fig. 3 will still given reasonably accurate estimates for subsets but, because the gaps in the evidence are substantial, the performance on items 1 and 2 with candidates 4 to 7 cannot be related to the performance of items 3 and 4 with candidates 1 to 3. Items and candidates cannot be put on the same continuum through limitations in the data collection design.

			Candidate				
Item	1	2	3	4	5	6	7
1				•	•	•	•
2				•	•	•	•
3	•	•	•				
4	•	•	•				

Figure 3: unconnected item-candidate pairs for four items and seven candidates

Assessing performance with multiple judges

When assessing projects and investigations, or mathematical communication skills, the data are often transient. The judgements have to be made when the performance occurs. Such judgements are susceptible to systematic bias which may not give candidates a mark which is consistent with their performance. Sometimes additional examiners are used to assure fairness and consistency but such assessments are rarely analysed to ensure that this intention has been realised. Unfortunately, expert mathematicians and statisticians are rarely conversant with current assessment theory and practice. The discussion which follows looks at some strategies for exercising quality control in such situations.

To distinguish between the examiner assigned to the candidate under current assessment procedures and the additional examiners, the term judge has been used instead of examiner for the latter. For convenience, the term judges means the group of judges and the examiner.

Work of a similar standard may be judged differently by different judges, even though the judges are in agreement about the key elements to be identified in student performance (perhaps expressed as a similar rank order). These differences would be a consequence of variations in stringency and leniency, or may be a consequence of one judge seeing differences in black-and-white terms while another judge may see more shades of grey when making distinctions in the value of the student performance. Also, performance may be judged as similar in standard as a whole although different judges may vary in their emphasis of the components they considered in reaching their decision about the quality of each performance. It should be noted that it is not the intention of this activity to achieve complete consensus between each of the judges. Such consensus might neglect key issues - more valid approaches will represent the true diversity of the real world, rather than considering issues where there is agreement and ignoring issues where experts differ. However there should be some consensus between expert mathematicians about the non-controversial topics, and the statistical analyses are intended to reveal those topics where there was consensus and those where there was less agreement.

An assessment data collection strategy can be designed to gather information about how the judges varied, how the candidates varied and how the items (sometimes called descriptors) were internally consistent. For each item, several ratings should be possible. These ratings should be ordered; the top rating should be judged to reflect the highest band of performance and the bottom rating the lowest. To investigate how judges vary it is necessary to have more than one judge assess each person, and to have an overlap in the judges of the persons assessed.

Design requirements for multiple judges

For an analysis to provide useful information, the data for the analysis must include the relevant indicators in order to address the appropriate issues or aspects. For example, if wishing to compare scores awarded by female judges with those awarded by male judges one has to know which judges are male and which are female. Such an analysis is impossible unless the appropriate data are recorded.

If there are four judges and seven candidates the largest possible number of judge-candidate pairs will be 28. This can be represented as a matrix as shown in Fig. 4, where each bullet (•) indicates one such pair.

				Candidate			
Judge	1	2	3	4	5	6	7
1	•	•	•	•	•	•	•
2	•	•	•	•	•	•	•
3	•	•	•	•	•	•	•
4	•	•	•	•	•	•	•

Figure 4: judge-candidate pairs for four judges and seven candidates

An analysis is possible even if some judges and candidates are not paired. For example, the matrix shown in   Fig. 5 will still give reasonably accurate estimates because the gaps in the evidence are limited.

				Candidate			
Judge	1	2	3	4	5	6	7
1	•	•	•		•	•	•
2	•	•	•	•	•		•
3	•		•	•	•	•	•
4	•	•		•	•	•	•

Figure 5: incomplete judge-candidate pairs for four judges and seven candidates

An analysis is possible if many judges and candidates are not paired, but if there is not some form of connectedness in the data the analysis may lead to ambiguous or misleading results. For example, the matrix shown in Fig. 6 will still give reasonably accurate estimates for subsets, but because the gaps in the evidence are substantial, the performance of judges 1 and 2 with candidates 4 to 7 cannot be related to the performance of judges 3 and 4 with candidates 1 to 3. Judges and candidates cannot be put on the same continuum through limitations in the design.

				Candidate			
Judge	1	2	3	4	5	6	7
1				•	•	•	•
2				•	•	•	•
3	•	•	•				
4	•	•	•				

Figure 6: unconnected judge-candidate pair for four judges and seven candidates

Similar arguments can be applied to the set of descriptors. Some gaps can be tolerated, but judges who systematically omit certain items on a rating scale (or who systematically mark a particular item in a different way) or who fail to provide ratings on sufficient numbers of items can distort the assessment process.

				Judge				
Candidate	A1	A2	A3	A4	A5	A6	X	Y
1	•						•	•
2	•						•	•
3	•						•	•
4		•					•	•
5		•					•	•
6		•					•	•
7			•				•	•
8			•				•	•
9			•				•	•
10				•			•	•
11				•			•	•
12				•			•	•
13					•		•	•
14					•		•	•
15					•		•	•
16						•	•	•
17						•	•	•
18						•	•	•

Figure 7: designed Sample for Data Collection

Fig. 7 shows a designed sample plan for collection of assessment data on a mathematical modelling task. Note that the plan does not require large numbers of candidates, here 18 candidates are assessed by their usual examiner/judge. For example, judge A1 assesses three candidates (1, 2, and 3). Judges X and Y assess the work of all 18 candidates and serve as a link between the usual judges.

These ratings may be analysed initially using the FACETS program which partitions effects due to judges (those doing the ratings), effects due to candidates (those who provided the samples of student performance being assessed), and effects due to descriptors (the statements considered to reflect the key elements that should be taken into account by the judges). This type of analysis is a rating scale equivalent to conventional test item analysis which identifies inconsistent items and assigns the remaining items to a difficulty or complexity scale. In a FACETS analysis, judges are placed on a scale from lenient to stringent, students are placed on a scale (in the same metric as the judge scale) from low achievement to high achievement, and the descriptors are placed on a scale (also in the same metric) from low demand to high demand. Inconsistent ratings are identified separately, and each judge, student, and descriptor is considered on the basis of all the ratings to identify judges, students or descriptors which differ from others in that category.

This type of analysis shows the extent to which the results from a performance assessment make sense and can indicate where there is room for the examiners to work on attaining shared meanings for the assessment terms they use, at least as judged from their assessment behaviour using real candidates.

GLOBAL MEASURES OR DETAILED MARKING PLANS?

Rating scales similar to those developed by the Assessment Research Group (Burton and Izard, 1995) have been shown to have greater reliability than global measures which some examiners like to use. The major problem found with use of such global measures is that substantial discrepancies occur between examiners. When this is recognised as inappropriate the examiners are unable to reduce the discrepancies (Izard, 1991). Without reducing such discrepancies the judgements made are unreliable (inconsistent) and candidates generally feel such assessments are capricious. There is no evidence that identifying factors separately destroys the unity of the total assessment - indeed there is considerable evidence that identifying such factors contributes to the unity by ensuring that each examiner is consistent with others, and that a single examiner remains consistent from performance to performance.

A way forward

More sophisticated test analysis software like QUEST (Adams and Khoo, 1993) and FACETS (Linacre, 1990; Wright and Linacre, 1994) is required for investigating the quality of the assessment processes, particularly where there are items giving partial credit or where tests are multiple-marked (that is, each test is scored by several examiners). A number of studies have used such software to explore the ways in which the examiners agree and differ. For example, members of the Assessment Research Group have developed sets of descriptors of 'content' for modelling, pure mathematical and statistical investigations on a common basis. These, together with associated descriptors for both oral and written communication skills, form rating scales which have been shown to be useful for assessment (Haines, Izard, Berry et al., 1993).

Application of the procedures in field trials and the use of video recordings shows that the oral descriptors perform well in widely differing situations (Haines, Izard and Le Masurier, 1993). Other test analyses and commentaries are provided in Izard (1992), Izard (1994), Haines and Izard (1994), and Izard and Haines (1994).

Involving students in the assessment process

The involvement of the student in the assessment procedures has a formative effect on the education process. The advice which is given to the student in expressing the intended outcome clearly will affect the way in which the task is carried out. It is important to the student's learning process that they know the critical points which will be assessed, and to be aware of the actual rating form used by the judges. Some modern assessment schemes include the possibility of the student negotiating an individual assessment mode which deals with examinations, coursework and projects. It may also be possible for the student to choose whether to work in a group or independently.

Self assessment is also important, as it helps the student to understand and evaluate the task which has been undertaken. Clearly, students who cannot recognise high-quality work produced by their peers (or by themselves) have little claim to soundly-based knowledge. Being able to distinguish between work of high quality and work of lesser quality is an indication of knowledge. The inclusion of any student self-assessed mark in the overall rating given to a project is a matter for debate, as is the way in which that assessment is carried out. A mixture of peer and self assessment is a feature of the procedure for some group projects in universities. There is no doubt that some students perform better when they understand and are involved in the assessment of their own work. Studies by the Assessment Research Group, carried out in the United Kingdom, have shown that many undergraduate mathematics students cannot make such distinctions when assessing their own work and the work of peers (Burton and Izard, 1995).

A view of assessment

In this paper it has been asserted that assessment of student learning provides evidence to enable educational decisions to be made. This evidence should help us evaluate (or judge the merit of) teaching programmes, or we may use the evidence to make statements about student competence or to make policy decisions about the development of teaching expertise. Clearly the quality of the evidence is a critical factor in making sensible decisions. To improve this quality, the tasks chosen should not be limited to the usual pencil-and-paper type of task (as used in national examinations). The use of carefully selected, relatively small samples of student work provides an opportunity to gather a wider range of evidence than is usual in selection testing. Remember that evidence of what has been achieved is required, as well as what needs to be achieved. It also helps to gather evidence of the progress being made towards achieving goals and demonstrates the relative value of particular educational policy initiatives.

Many educational systems include a component of school-based assessments to access a wider range of information than measured by traditional examinations, and to allow such information to be gathered over a period of time. An insight into the place of assessment in the creative learning environment usually present in project and investigational work is provided by Burton (1992). Some such school-based assessments include solving complex problems similar to real-life problems. Assessments like these are presumed to be more accurate descriptions of student achievement because of the more diverse, and multiple, sampling of performance and because of the authentic nature of the tasks. However, if such measures are not consistent, validity claims may be spurious. At the IAEA 1993 conference in Mauritius, Izard (1994) reviewed assessment experience from three different settings (system-wide assessment at the end of secondary school in one Australian State and some universities in Australia and the United Kingdom).

CONCLUSION

Difficulties in achieving consistent assessments and useful strategies for overcoming such difficulties have been discussed in this paper. Developing a marking schedule for examiners to use is not a sufficient requirement to ensure that the scoring will be consistent - such schedules need to be validated in context. If the judges do not use the schedule correctly or at all, then the assessments almost certainly will be flawed. The procedures for test construction and analysis described in this paper have been developed over many years of practical work in the field. Some of the advice has arisen from research into test analysis, and some has been derived from the practical experience of large numbers of research and development staff working at various agencies around the world. Improving the quality of the evidence is not an easy task. Reading a book about the procedures will not suffice - improving one's skills as a test constructor will require working as a test constructor in concert with others, and analysing the results with the help of those experienced in the development of tests and in interpretation of test data.

REFERENCES

Adams RJ and Khoo ST, 1993. **QUEST: The Interactive Test Analysis System.** (Computer program), Australian Council for Educational Research, Hawthorn, Victoria.

Burton L,1992. *Who Assesses Whom and to What Purpose?* In Stephens M and Izard J, (eds.), **Reshaping Assessment Practices: Assessment in the Mathematical Sciences Under Challenge,** Australian Council for Educational Research, Hawthorn, Victoria, pp1-18.

Burton L and Izard J, (eds.), 1995, **Using FACETS to Investigate Innovative Mathematics Assessments**: A report of on-going research by the Assessment Research Group (ARG), University of Birmingham, Birmingham.

Crouch R and Haines C, 1995, *Understanding Self, Peer and Tutor Marking in Mathematical Tasks: Analysing Oral Communications*. In Burton L and Izard J, (eds.), **Using FACETS to Investigate Innovative Mathematics Assessments:** A report of on-going research by the Assessment Research Group (ARG), University of Birmingham, Birmingham, pp 14-28

Goldfinch, J and Goodall, G (1995). *A Study of the Reliability of a Marking Scheme for Mathematical Modelling Assignments*. In Burton L and Izard J, (eds.), **Using FACETS to Investigate Innovative Mathematics Assessments:** A report of on-going research by the Assessment Research Group (ARG), University of Birmingham, Birmingham, pp 4-13.

Goodall, G, Hall, P and Walker, C (1995). *Equity of Assessment in a Module-Choice Situation*. In Burton L and Izard J, (eds.), **Using FACETS to Investigate Innovative Mathematics Assessments:** A report of on-going research by the Assessment Research Group (ARG), University of Birmingham, Birmingham, pp 29-37.

Haines CR, Izard JF, Berry JS, *et al.*, 1993, **Rewarding Student Achievement in Mathematics Projects, Research Memorandum 1/93**, City University, London.

Haines CR, Izard JF and Le Masurier D, 1993, *Modelling Intentions Realised: Assessing the Full Range of Developed Skills*, In Breiteig T, Huntley I and Kaiser-Messmer G, (eds.), **Teaching and Learning Mathematics in Context**, Ellis Horwood, London, pp 200-212.

Haines CR and Izard JF, 1994, *Assessing Mathematical Communications About Projects and Investigations*. **Educational Studies in Mathematics**, 27, pp 373-386.

Houston SK, 1995, *Using Rating Scales for Tutor and Peer Assessment of Mathematical Modelling Posters*. In Burton L and Izard J, (eds.), **Using FACETS to Investigate Innovative Mathematics Assessments:** A report of on-going research by the Assessment Research Group (ARG), University of Birmingham, Birmingham, pp 45-52.

Izard JF, 1991, *Issues in the Assessment of Non-Objective and Objective Tasks,* in Luitjen AJM, (ed.), **Issues in Public Examinations** (Proceedings of the 16th IAEA conference, Maastricht, The Netherlands, 18-22 June 1990.) Lemma, BV Utrecht, The Netherlands, pp 73-83.

Izard JF, 1992, *Patterns of Development with Probability Concepts*. In Stephens M and Izard J, (eds.), **Reshaping Assessment Practices: Assessment in the Mathematical Sciences Under Challenge,** Australian Council for Educational Research, Hawthorn, Victoria.

Izard JF, 1994, *Strategies for Assessing Projects and Investigations: Experience in Australia and United Kingdom*. In Mauritius Examinations Syndicate (eds.), **School-based and external assessments**, Mauritius Examinations Syndicate, Mauritius, pp 214-222

Izard J and Haines C, 1994, *Validating the Assessment of Projects and Investigations*. **Acta Didacta Universitatis Comenianae (ADUC) - Mathematics**, Issue **3**, pp 83-96.

Le Masurier D and Dunthorne S, 1995, *Use of FACETS to Analyse the Use of Criterion-Based Rating Scales to Assess Students' Posters*. In Burton L and Izard J, (eds.), **Using FACETS to Investigate Innovative Mathematics Assessments:** A report of on-going research by the Assessment Research Group (ARG), University of Birmingham, Birmingham, pp 38-44.

Linacre, JM, 1990, *Modelling Rating Scales*, Paper presented at the Annual Meeting of the American Educational Research Association, Boston, MA, USA, 16-20 July 1990, [ED318 803].

Rasch G, 1960, **Probabilistic Models for Some Intelligence and Attainment Tests**, Danmarks Paedagogiske Institut, Copenhagen.

Wright BD and Linacre JM, 1994, **FACETS** (computer program) (Version 2.80), MESA Press, Chicago, Illinois.

Software

The QUEST computer program (Version 2) was written by Raymond J Adams and Siek-Toon Khoo and published by The Australian Council for Educational Research Limited (ACER) in 1993. QUEST is Copyright © The Australian Council for Educational Research. Information can be obtained from ACER, 19 Prospect Hill Road, Camberwell, Melbourne, Victoria 3124, Australia [e-mail: quest@acer.edu.au].

The FACETS computer program (Version 2.80) was written by Benjamin D Wright and John M.Linacre and published by MESA Press in 1994. FACETS is Copyright © John Michael Linacre, 1987-1994. Information can be obtained from MESA Press, 5835 S. Kimbark Avenue, Chicago, Illinois 60637, United States of America. [e-mail: MESA@uchicago.edu]

9

Mathematical Proficiency on Entry to Undergraduate Courses

Peter Edwards
Bournemouth University, 12 Christchurch Road, Bournemouth BH1 3NA.
e-mail: pedwards@bournemouth.ac.uk

ABSTRACT

Several reports have been published recently in the UK indicating a decline not only in the number of students offering A-level Mathematics on entry to undergraduate courses but also in general mathematical proficiency. How do universities cope with this in engineering and science, for example, where the ability to apply mathematics to solve real-world problems is paramount? One solution widely used is diagnostic testing on entry to a course, with a subsequent programme of extra mathematics tuition if needed. Some of the results from one such diagnostic test, given to a cohort of first-year engineering undergraduates in their induction week, are presented here. These and other indicators certainly seem to show that the mathematical competence of current UK undergraduates is a cause for concern, with many gaps in background knowledge and an abundance of mathematical misconceptions. Primarily this paper relates opinions and reactions from within the United Kingdom, though there are indications that this problem is not restricted to the United Kingdom.

INTRODUCTION

Amongst the activities undertaken in mathematical modelling (about which many books have been written - see, for example, Townend *et al.*, 1995) are Formulate the real-world problem, Formulate the mathematical problem and Solve the mathematical problem. In order to solve a modelling problem, therefore, students need a number of skills, amongst which are the ability to *abstract* a mathematical problem from a real-world situation, *identify* the appropriate mathematical tools needed to solve the problem, and *use* the appropriate mathematical tools to solve the problem.

Over the past decade, with experience in teaching undergraduate engineers and as a part-time tutor for the Open University's Modelling with Mathematics course, the author has detected an improvement in students' ability to abstract, but a decline in students' ability to identify and use appropriate mathematical tools. During 1995, general awareness of this decline was raised by several reports and newspaper articles. Amongst these three reports stand out - not only for their content, but also for their commissioning bodies. The first (Sutherland and Pozzi, 1995) was commissioned by the Engineering Council. The second (James *et al.*, 1995), reported the findings of a working group set up under the auspices of the major UK engineering and mathematics institutions (including accrediting bodies for engineering degrees). The third (Howson *et al.*, 1995), was commissioned by the London Mathematical Society, the Institute of Mathematics and its Applications and the Royal Statistical Society. When such august bodies are anxious about declining standards, this has to be a cause for concern - especially in a country where wealth emanates, to a large extent, from engineering, manufacture and commerce.

All of the reports expressed particular concern over the dilution of pre-A-level mathematics (up to the age of 16), especially in the areas of algebraic manipulation and graphical work. The subsequent affect at university level was highlighted by all the reports including, in the James report, concern over the maintenance of standards in mathematics on engineering degree courses in order to preserve the differentiation between Chartered Engineer status and that of Incorporated Engineer. The Howson report offered one possible cause of the current decline: "In recent years, English school mathematics has seen a marked shift of emphasis, introducing a number of time-consuming activities (investigations, problem-solving, data surveys, *etc.*) at the expense of 'core' techniques". For the mathematical modelling process this is not too detrimental a step since an investigative approach to mathematics enhances abstraction skills and qualitative problem-solving abilities. On the other hand, if a student is neither capable of recognising the need for a particular mathematical method for solving a problem nor capable of using such a method, then any advantages of such a shift are lost.

During this time even the British press began to lament the mathematical performance of students. Although such articles were confined mainly to the Education pages, the *main headline* in The Times of December 7 1995 exclaimed *Calculators banned in exams after maths standards fall*. An example quoted later in the article indicated that only a quarter of the 600,000 school pupils who took the GCSE Mathematics examination could calculate 15 per cent of 80 correctly.

Against this background, many lecturers have felt the need to gauge for themselves the mathematical capability of their new students.

ASCERTAINING STUDENTS' MATHEMATICAL ABILITY

In order to accommodate those students with an insufficient grounding in mathematics, it is vital to identify those who will need extra mathematical support. Relying on the results of end-of-school examinations alone is no longer sufficient. A recent study (Angel and Edwards, 1995) has shown that a pass in A-level Mathematics does not guarantee success in the more analytical undergraduate engineering subjects. The same paper, however, investigated the results obtained from a particular mathematics diagnostic test used at Bournemouth University. A significant correlation was found between the test results of over 200 students and their performance in end-of-first-year Technology examinations (but with 'value added' during the year). In identifying students who need extra support - and as a predictor of student performance - the diagnostic test proved more useful than end-of-school examination results.

Interesting though such a test's predictive properties may be, it is the analysis of students' incorrect answers that can provide a measurable insight into particular mathematical deficiencies. For areas such as the mathematical modelling of engineering applications, these deficiencies can result in an inability to cope with the mathematics encountered at university level.

REDUCING UNDERGRADUATE MATHEMATICAL CONTENT

The Product Design and Manufacture Department at Bournemouth University evolved from an engineering department offering traditional mechanical and electronic courses into one in which holistic product design was a major goal. As a consequence, the 'traditional' mathematics was trimmed to support the engineering design applications in the new courses. Even from the inception of such courses, the mathematical ability of incoming students was seen to be deficient. Since much of the material normally taught in the latter years of school, such as calculus, was to be included in the undergraduate programme it was decided, therefore, that student ability on entry should be determined by a diagnostic test containing mathematics only at GCSE level (up to 16 years old). It was thought that students who were diagnosed sound at this level would have the knowledge and capability to learn successfully the mathematics taught during their course. This has proved to be a general feeling amongst lecturers who teach non-specialist mathematicians (Edwards, 1996).

ANALYSIS OF A SELECTION OF DIAGNOSTIC TEST QUESTIONS AND ANSWERS

Some details of diagnostic tests used at five British universities can be found in an Open Learning Foundation report (Edwards, 1996). An important feature common to all the tests, most of which used multiple choice questions, was the capability to

(a)   determine the prevalence of a particular misunderstanding, or misconception (in multiple choice tests by the use of 'meaningful' distracter answers, eg $9 - 3 \times 2 = 12$),

(b)   identify a forgotten, or never learned, process (by encouraging students to answer the "I don't know" option if an answer is unknown). Some example questions from the test given to over 200 Bournemouth students, are shown below, with an indication of the success rate. The correct answers have their response percentage underlined.

Example 1.

The value of $\sqrt{6^2 + 8^2}$ is:
[A] 14 [B] 10 [C] 70 [D] 7 [E] Don't Know

Responses:
[A] 8.7% [B] <u>76.4%</u> [C] 1.4% [D] 6.7% [E] 6.7%

Feedback, in a subsequent session with the students, showed that although most students were able to answer this numerical problem correctly in the test, when asked in the feedback session to find the expansion of the related problem, $(a+b)^2$, over half answered $a^2 + b^2$.

Example 2.

The value of 1/2 + 2/3 is:
[A] 3/5 [B] 7/6 [C] 3/6 [D] 8/6 [E] Don't Know

Responses:
[A] 24.5% [B] <u>61.1%</u> [C] 5.8% [D] 6.3% [E] 2.4%

Feedback revealed students have become too dependent on calculators - and, in this test, calculators were not allowed. Decimal answers *may* be acceptable when combining numerical fractions in real-world problems, but a later question in the test required the addition of algebraic fractions - with a much lower percentage success.

Example 3.

Think of a number, treble it, add fifteen and divide the result by three. If x is the number, the *simplest* algebraic expression for this is:
[A] (3x + 15)/3 [B] 3x + 5 [C] x + 15 [D] x + 5 [E] Don't Know

Responses:
[A] 60.6% [B] 6.3% [C] 5.3% [D] <u>26.4%</u> [E] 1.4%

Comment: The ability to be able to abstract a mathematical expression from a statement in English is essential in mathematical modelling. In this question, most students abstracted the mathematics successfully - the pitfall was the word 'simplest'.

Feedback: Excellent example of Read the Question (in the test, the word 'simplest' is in bold italics *and* underlined) and Read all the Options (too many saw the first answer and stopped there). Nevertheless, there were still 13% whose answers were otherwise wrong or 'I don't know'. An interesting point raised here was that of mathematical vocabulary. Nearly half of those answering [A] argued that $(3x+15)/3$ was *already* the 'simplest' answer and attached no further significance to the word.

Example 4.

Make a the subject of $v = u + at$:

[A] $(v - u)/t$ [B] $(v - u)t$ [C] $v/(ut)$ [D] $v/u - t$ [E] Don't Know

Responses:

[A] 62.0% [B] 8.2% [C] 6.7% [D] 6.7% [E] 16.3% (!)

Feedback revealed students too frequently use 'cross-multiply' and 'change side, change sign' and, then, incorrectly. Many who had not studied A-level Mathematics did not recognise this equation as a standard model from kinematics. Nevertheless, most were aware of the requirements of rearranging equations (GCSE mathematics very often uses equations from technology or science outside the GCSE syllabus).

Example 5.

The slope of the straight line joining two points is given by the difference between the y co-ordinates divided by the difference between the x co-ordinates. For the line joining the points with co-ordinates (-2,6) and (6,2) the slope is :

[A] -2 [B] 2 [C] -0.5 [D] 0.5 [E] Don't Know

Responses:

[A] 10.6% [B] 23.1% [C] 24.5% [D] 21.2% [E] 20.7% (!)

Comment: Again, a definition in English requiring mathematical abstraction.

Feedback: Most confusion was due to "Which is x and which is y?", together with problems such as how to handle 6-(-2). Most of those answering [E] indicated that they were unable to translate the English into Mathematics.

Further examples and comments from the same Diagnostic Test can be found in a separate paper (Edwards, 1995a).

SUBSEQUENT ASSISTANCE FOR WEAKER STUDENTS

Most of the extra support for weaker students at Bournemouth uses class contact. In recent years this has varied from a three hours-per-week, six-week course reviewing and extending the mathematics of the diagnostic test, to one and two hour-per-week, full-session basic mathematics courses. For some topics this is supplemented by the use of CAL material specifically written in-house for these students (Edwards, 1995b). Students are grouped according to their results into three separate levels. Those with a score in the lowest level are advised that they should attend all extra mathematics classes (for which a register of attendance is taken), those in the middle level are told that it may be helpful for them if they were to attend, and those who do well in the test are told that it is not necessary to attend. However, the subject of the following week's extra mathematics lecture is posted on a notice board so that the more able students can decide whether attendance could be to their benefit. Results for one cohort of students (Edwards and Angel, 1995) have shown that the extra classes rarely turn a weak student into an outstanding student, but that they can add sufficient value to a student's underlying mathematical capability in order to ensure success in analytical subjects.

INTERNATIONAL COMPARISONS

The results from the above diagnostic test and the three previously mentioned reports commissioned by the various institutions indicate a problem within the UK. However, is it an international problem?

In an article in the Sunday Times (14 May 1995) entitled *Britain gets a minus in maths*, Professor David Burghes of Exeter University's Centre for Innovation in Mathematics Teaching drew national comparisons from a mathematics test taken by 14 year-olds in, amongst others, England, Germany and Hungary. When asked, for example, to determine the value of $\dfrac{64 \times 0.3}{0.32}$, only 5% of UK pupils obtained the correct answer, while the figures for Germany and Hungary were 28% and 51%, respectively. For the problem, 'Factorise $x^2 - 3x$', the figures were 8%, 19% and 34%, respectively. Although these are just two examples from a more comprehensive study, they do give an indication of the differences between these countries and reflect approaches to learning. Professor Burghes points out that in Hungary, for example, the researchers " ... hardly saw a pupil who was not trying hard." Further salient points made in the article were that both Hungary and Germany restrict the use of calculators until pupils reach secondary school, and all students in Germany are expected to continue mathematics until the age of 18 (whereas in Britain most drop the subject at 16).

There seems also to be an unfavourable comparison at A-level. With respect to the standing of A-level Mathematics, the James report points out that "A supportive indicator of the reduced standing of A-level mathematics is that in Hong Kong the UK GCE A-level attainments are factored by 0.6 when compared with HK A-level attainments."

Making international comparisons can lead to generalisations such as 'The mathematical ability of students in country X is better than that in country Y'. Studies into the mathematical ability of 12-15 year-olds in England, Scotland and Germany (Blum *et al.*,1993; Burghes and Blum, 1995) indicate that, overall, mathematical ability of German pupils *is* possibly better than that in England - but *not* in all areas. In a test given to pupils in all three countries, although the German 12-15 year-olds scored consistently the highest in *Number* and *Algebra* questions, they scored consistently the lowest in *Shape and Space* (graphs and geometry). Various reasons for these anomalies are suggested within the study.

Although the evidence presented so far indicates that the decline in mathematical ability seems to be more marked within the UK, other countries should not be complacent. For example, in *relative* terms, the 8%, 19%, 34% correct answers to 'Factorise $x^2 - 3x$' show UK students to be the weakest when answering this question, but in *absolute* terms neither the 19%, nor even the 34%, are particularly good success rates. Perhaps mathematical weakness is not solely restricted to the UK.

Concern over the mathematical proficiency of college students in the USA prompted the Mathematical Association of America (MAA) to establish a subcommittee on Quantitative Literacy Requirements *in 1989*. A report from this subcommittee, available from the MAA's World Wide Web pages (MAA, 1996), indicates that general mathematical knowledge in America is "in a sorry state" and recommends amongst its conclusions that "Colleges and Universities should *expect* every college graduate to be able to apply simple mathematical methods to the solution of real-world problems." The expectation mentioned here needs action - but what are the best ways of trying to ensure success for those with a weaker mathematical background?

CAN UNIVERSITIES ACCOMMODATE MATHEMATICALLY WEAKER STUDENTS?

Compounded by the widening of access to UK undergraduate courses due in part to the rapid expansion of UK university provision in the early 1990s, there has been an increase in the number of undergraduates who

(a) have an inadequate mathematical knowledge-base to study, for example, engineering applications of mathematics,

(b) are accepted on courses without a mathematics qualification that may have been previously required (in order, perhaps, to ensure that universities meet their student targets).

British universities should not see such students as being *unable* to undertake undergraduate mathematical study - often they can absorb the mathematics at *that* level as well as anyone. Rather, they are sometimes *incapable* of successful learning due to the lack of the background knowledge so important in underpinning the acquisition of knowledge at the higher level.

In order to accommodate mathematically less able students, some universities are changing the emphasis of their undergraduate programmes. These and other universities are also, however, researching new teaching and learning methods, including the introduction of foundation years, access courses, peer tutoring, supplemental instruction, student-centred learning and computer-aided learning. In order to direct such support, widespread use is made, especially during the students' induction week, of mathematics diagnostic testing.

CONCLUSIONS

Many involved with mathematical modelling have welcomed the introduction of investigative mathematics into the UK school syllabus, as this gives pupils an introduction to abstracting mathematics from real-world problems. However, this does seem to have been at the expense of the accumulation of a firm mathematical knowledge base and the development of mathematical skills - and this can be a problem on entry to, say, engineering degree courses. Evidence of a lack of knowledge and the preponderance of mathematical misconceptions has been given here using, amongst others, some of the questions from a diagnostic test written and used by the author.

A brief international comparison shows that even though the problem is particularly marked within the UK, non-UK countries cannot afford to be complacent. Non-UK academics should monitor carefully the experiences of their UK colleagues in order to lessen the likelihood of the problem occurring in their own countries.

Whilst the problem continues, it is the responsibility of universities to accommodate mathematically weaker students. However it is felt that universities should also be more responsible for input into discussions on the content of mathematics taught in schools since, unless major initiatives are forthcoming, mathematical weakness on entry to undergraduate courses will become the norm.

REFERENCES

MAA, 1995, *Quantitative Reasoning for College Graduates: A Complement to the Standards*, Mathematical Association of America. This document can be found on the MAA World Wide Web page at http://www.maa.org/ql/ql_toc.html

Angel M and Edwards P, 1995, *Factors predicting the Academic Performance of Students in University Product Design Departments*, **Proceedings of the 2nd National Product Design Education Conference**, Coventry University, Coventry.

Blum W, Burghes DN, Green N and Kaiser-Messmer G, 1993, *British/German Comparative Project: Some Preliminary Results*, **Teaching Mathematics and its Applications**, **12**, 1, pp 13-21.

Burghes, DN and Blum, W, 1995, *The Exeter-Kassel Comparative Project. A Review of Year 1 and Year 2 Result,*. **Proceedings of a Seminar on Mathematics Education**, The Gatsby Charitable Foundation, London.

Edwards P, 1995a, *Some Mathematical Misconceptions on Entry to Higher Education*, **Teaching Mathematics and its Applications**, **14**, 1, pp 23-27

Edwards P, 1995b, *A Visual Approach to Understanding Mathematics*, **CTI Maths and Stats Newsletter**, **6**, 3, pp 2-7

Edwards P, 1996, *A Survey of Mathematics Diagnostic Testing on non-Specialist Mathematics Courses*, The Open Learning Foundation, London.

Howson AG, Barnard AD, Crighton DG, Davies N, Gardiner AD, Jagger JM, Morris D, Robson JC and Steele NC, 1995, **Tackling the Mathematics Problem**, London Mathematical Society. London.

James DJG, Lepper AM, Carroll JE, Cornish A, Croll JGA, Maudsley D, Swanson SAV and Young NJ, 1995, **Mathematics Matters in Engineering**, The Institute of Mathematics and its Applications, Southend-on -Sea.

Sutherland R and Pozzi S, 1995, **The Changing Mathematical Background of Undergraduate Engineers - A Review of the Issue***s*, The Engineering Council, London.

Townend MS, Edwards P, Goldfinch J, Hamson M, Norcliffe A and Stone J, (eds.), 1995, **Mathematical Modelling Handbook - A Tutor Guide**, Pavic Publications, Sheffield Hallam University, Sheffield.

10

Evaluating Rating Scales for the Assessment of Posters

Ken Houston
University of Ulster, Jordanstown, Northern Ireland, BT37 0QB.
e-mail: sk.houston@ulst.ac.uk

ABSTRACT

This paper deals with the use of posters by university students as a means of communication and assessment. There is a summary of a literature review and a rationale for getting students to create posters. It is claimed that this activity is enjoyable and beneficial for students.

The assessment of posters is an important issue, and the main purpose of this paper is to describe the development and evaluation of assessment criteria and rating scales. Early trials indicate that students have difficulty with some aspects of poster creation, in particular including a proper statement of the problem they are working on. The findings reported in the paper give direction for teaching and classroom management.

INTRODUCTION

There is ample evidence in the literature supporting the view that students of many subjects enjoy and benefit academically from participating in poster sessions. This evidence is reviewed by Berry and Houston (1995) and is collected from a wide variety of sources. Berry and Houston introduced poster sessions to their courses on mathematical modelling in 1993 and have continued to use them. Their students prepare posters to communicate their project work to both peers and tutors who subsequently assess them. This paper describes the evolution of the rating scales which are used in this assessment process, and which can also be used by students to carry out a self assessment of their own work before submitting it to the scrutiny of others.

The paper begins with a brief review of the rationale for using posters for communication and assessment, and then describes recent experiences when implementing poster use. The development of poster assessment criteria is described and their use evaluated.

Evaluation data are available from four experiments:

- a workshop in 1993 at the 6th International Conference on the Teaching of Mathematical Modelling and Applications (ICTMA-6),
- two cohorts of students at the University of Ulster,
- one cohort of students at the University of Brighton.

The data are analysed using FACETS, a program that carries out a many-facet Rasch analysis, (Linacre, 1990; Linacre and Wright, 1994), the results are discussed, and pedagogical implications are presented.

STUDENTS USING POSTERS

Lecturers in higher education in a variety of subject areas have used posters successfully with their students, but the only reference in the literature before Berry and Houston (1995) to the use of posters in the mathematics classroom is a report of a poster contest organised by the American Statistical Association for students in grades K-12 (Denson, 1992).

The literature review suggests that poster sessions for students:

- are an excellent alternative medium for developing communication skills,
- encourage the thorough investigation and concise reporting of a topic,
- provide opportunities for self and peer assessment and for peer learning,
- promote a positive attitude in students.

Berry and Houston also argue that the use of poster sessions:

- introduces students to another aspect of professional practice,
- encourages deep learning by requiring students to think about the pedagogical purposes of their poster and also through the critical reading of their peers' posters,
- helps expose and confront misconceptions in student understanding,
- puts emphasis on concept as well as procedure in mathematics.

CONTEXT OF THIS STUDY

The students at the University of Ulster who participated in this study were enrolled on a one-semester module on Mathematical Modelling and were either in the second (final) year of a Higher National Diploma (HND) course or the first year of a Bachelor of Science (BSc) degree course. The module deals with Mathematical Methods, Mathematical Models and Mathematical Modelling. The Modelling was conducted in small groups of preferred size four (but occasionally of three or less). Students carried out two mini-projects, each of about four weeks' duration. They wrote a report and, for one project, gave a seminar presentation; for the other they presented a poster at a poster session.

In 1993, while the method and its assessment were being developed, students were not summatively assessed on their poster work. However, some of their posters were exhibited at a workshop at ICTMA-6 in Delaware and the draft assessment criteria tested. Workshop participants - all experienced modelling lecturers - used the draft criteria to assess the posters. These data were analysed using FACETS and reported by Haines and Izard (1995) in the post ICTMA-6 book edited by Sloyer *et al.* (1995).

After the Delaware meeting, Houston, in collaboration with Lamon, developed the draft assessment criteria in the light of the outcome of this analysis and wrote a commentary. These criteria were tested during 1994 with a cohort of students at Ulster and were used for both peer and tutor assessment. They were published in Houston *et al.* (1994) and their testing was reported by Houston (1995). Furthermore, assessment criteria based on these, but developed by Le Masurier and Dunthorn (1995), were used for peer and tutor assessment at the University of Brighton during 1993-94.

Finally, in 1995, students at the University of Ulster used posters as an *additional* means of communicating information to their peers, along with a seminar and written notes on a particular Mathematical Model which they had researched but not created.

Much of the work outlined above is discussed later in the paper.

ASSESSMENT CRITERIA

The assessment criteria in Table 1 were developed by Houston in discussion with his students in 1993. They are grouped into two sections, which deal with the *content* and the *presentation* of the poster respectively.

At a workshop at ICTMA-6 in Delaware in 1993, 14 participants used these criteria to assess 10 posters created by students at Ulster. The judges had no previous experience in assessing posters, had not contributed to the discussion leading to the development of the

Content		Difficulty for student (rank)
PS1	States the problem	1
PS2	Outlines the solution	5
PS3	Reports the results	3
Presentation		
PS4	Uses bold headings	6
PS5	Uses good layout design	4
PS6	Use illustrations	7
PS7	Aesthetically pleasing	2

Table 1. Assessment criteria (1993) and rank order of difficulty for students.

criteria and were not trained in their use. It is recommended that at least some of these conditions should be met before attempting serious summative judgments (*cf* Haines and Izard, 1995). Nevertheless, the 14 judges agreed fairly closely on the rank ordering of the posters, if not on the absolute scores awarded.

The FACETS analysis showed which criteria were more difficult to meet and which were easier. The rank order of difficulty is given in Table 1. Clearly students did not in general 'State the problem' very well in their posters, but they 'Used illustrations' to good effect.

These draft criteria were further developed by Berry and Houston (1995).

After ICTMA-6 Houston, in correspondence with Lamon, developed these criteria into those given in Appendix 1 and Table 3 and published in Houston *et al.* (1994). They also wrote an explanatory commentary following the patterns established by the Assessment Research Group (Houston *et al.*, 1994). A detailed analysis of these criteria is presented in the next section.

Le Masurier and Dunthorn (1995) also developed the criteria and wrote a commentary which is very similar to the one in Appendix 1 They carried out a FACETS analysis of the use of their criteria by 2 tutor-judges and 13 peer-judges of 5 posters created by 14 students working in groups of 2 or 3. They found that 'States the problem' was the easiest criterion to satisfy and 'Producing an aesthetically pleasing poster' the most difficult. They found that "the students' assessments correlated resoundingly well with [those of] the tutors", although they expressed concern that some of the student judges were not judging in the same way as the others. These authors concluded that "criteria should be issued and discussed before students produce their poster".

EVALUATION OF THE 1994 ASSESSMENT CRITERIA

During the second semester of the 1993-94 session, the assessment criteria in Appendix 1 were extensively tested at the University of Ulster. The class consisted of 23 BSc students and 22 HND students. All students took part in the preparation of posters and 40 took part in the peer assessment exercise at the poster sessions. There were 13 posters, 7 from the HND group and 6 from the BSc group. Students worked in groups of 3 or 4, except for one mature student (number 20) who worked on his own and who did not take part in the judging. Of the 40 student judges, 21 judged 4 (or 5) groups, 4 judged 3 groups and the rest judged only 2 groups because they were absent from one of the sessions. (The poster sessions were held on consecutive days). Students were assigned to judge 2 posters from their own class (BSc or HND) and 2 from the other class. This information is summarised in Appendix 2. The groups are labeled 7 to 19, with 7 to 13 being HND groups and 14 to 19 being BSc groups. Students 1 to 22 were HND students and students 31 to 54 were BSc students. Student 40 withdrew early in the semester and students 1, 2, 5, 31 were absent on both days and so did not take any part in the judging. Number 25 is the tutor, who judged all the posters. An X in the table indicates group membership, while a dot indicates that the particular student judged that group.

The data were analysed using FACETS. A tick in a box labeled 'High' scored 4, the next box 3, the next 2 and a 'Low' box scored 1. 'Not shown' (N.S.) scored 0, and if a judge deemed the descriptor 'Not Applicable' (N.A.), or if they did not tick any box, then that descriptor was omitted from the analysis.

The rank order of the posters, as determined by all the judges, is compared with the rank order as determined by the lecturer alone in Table 2.

Groups	7	8	9	10	11	12	13	14	15	16	17	18	19
All judges	6	5	7	3	9.5	1	11	8	3	3	9.5	12	13
Lecturer alone	2	4	8.5	10	8.5	7	11	6	2	2	5	12	13

Table 2. Rank order of groups as determined by
(a) all the judges together, (b) the lecturer alone

Where two or more posters have the same score, they are each given their mean rank order. The Spearman Rank Order Correlation Coefficient was 0.64, which indicates that there was some disagreement between the "expert" judge and the whole group.

It must be noted that the 12 descriptors are equally weighted in this analysis. It would not necessarily be like this in a summative assessment. For example, it may be desirable to give greater weight to *Content* than to *Presentation*.

Columns 1 and 2 of Table 3 give the rank order of difficulty of the criteria as determined by the whole group of judges and by the lecturer alone in the 1994 exercise.

Students were clearly having difficulty writing a satisfactory statement of the mathematical problem that arises in the modelling, and in reporting overall conclusions.

The FACETS analysis also suggested that some of the student judges did not give the task sufficient attention or consideration. This may have been due to negligence or, more likely, they may have been unable to make better judgments given their level of understanding at that time. These students had not been involved in the discussion leading to the development of the criteria, and they had not been trained in their use. The criteria were merely handed to the students by the lecturer, with very little explanation or interpretation. Consequently they did not have ownership of them, and failed to use them properly. Considerable work needs to be done in this respect before students can be relied upon to function as accurate peer assessors.

The same assessment criteria were used in 1995, but this time for a slightly different purpose. In this session, one of the four-week modelling projects was omitted. Instead students were asked to prepare a seminar describing a mathematical model which they had read about in the literature, and were given a list of topics and associated references. They were asked to prepare a class hand-out on this topic, to give an oral presentation and to prepare a poster on their topic. This was intended to be a peer teaching activity wherein the students would learn, first by independent research, and then from one another. The work of the presenting group was assessed using the criteria described in Houston *et al.* (1994) for written modelling reports, for oral presentations and for poster presentations (*ie* the criteria in Appendix 1), with the tutor as the only assessor. The whole class was also assessed on their understanding of the models by written examination at the end of the semester. The results of this will be published elsewhere (Houston and Lazenbatt, 1997).

The rank order of difficulty of scoring on the poster descriptors is given in column 3 of Table 3 and compared with the tutor rank order from 1994.

	Content	Rank (whole group) 1994	Rank (tutor alone) 1994	Rank (tutor alone) 1995
PO1	States the problem clearly	6	5	1
PO2	States the problem succinctly	4	8	3
PO3	Outlines the solution			
	3.1 Describes the model	5	2	6
	3.2 States the mathematical problem	1	1	2
	3.3 Reports on the mathematical solution	2	6	4
PO4	Reports conclusions	3	3	5
	Presentation			
PO5	Designs the poster logically	11	9	7
PO6	Uses different fonts effectively	9	12	8
PO7	Uses illustrations effectively	10	10	9
PO8	Produces an aesthetically pleasing poster	7	7	10
PO9	Text is concise	8	4	11
PO10	Demonstrates understanding of project through discussion	12	11	12

Table 3. Assessment criteria, 1994, and rank order of difficulty as judged by
(1) Whole group (1994), (2) Tutor alone (1994), (3) Tutor alone (1995).

Again, 'Stating the mathematical problem' was very difficult to score well on, and this time students did not 'State the problem clearly' in their posters.

It is possible that these criteria, developed for the assessment of posters for modelling projects, are not entirely suitable for the assessment of posters prepared as a visual aid to a seminar. It was intended by the tutor that they should be free standing, but this may not have been understood by the students. A particular anomaly came to light which illustrates this. The student seminars were held at weekly intervals. At the end of each seminar, the tutor would judge each handout, poster and oral presentation. The assessment tick sheets were used by the tutor to help inform the choice of grades. The groups presented in numerical order, with group 1 going first. The data in Table 4 were available at the end of the

GROUP RANK ORDER (1995)

Group	Average. Score	Grade
6	3.2	β+
5	2.8	β-
2	2.7	β-
7	2.6	β-
1	2.5	β+
4	2.1	γ+
3	1.3	δ

$\alpha = 4.0$ $\qquad$ $\beta = 3.0$ $\qquad$ $\gamma = 2.0$ $\qquad$ $\delta = 1.0$

Table 4. Group assessment scores, 1995.

semester, where the 'grade' is the grade awarded by the tutor immediately after the presentation and the 'average score' is the FACETS score for that poster calculated at the end of the semester when all the data were available. (It was deemed necessary to give rapid feedback to students rather than have them wait to the end of the semester).

The anomaly arises with group 1 which was awarded a grade of β+ while its eventual rank position in the class suggests that β- would have been more appropriate. Was the lecturer over generous to start with, becoming more stringent as the semester proceeded, or was he being inconsistent when using the tick sheets to determine grades? A closer inspection of the scores for groups 1, 6 (the best group) and 7 (the group closest to group 1) revealed the data in Table 5.

Group 1 had scored lots of 3s and 4s and this influenced the tutor to award β+ even though some descriptors had scored 0. Group 1 had not included anything on the mathematical problem, but yet this 'experienced' judge did not mark them down for this. At this stage although he had expected the poster to include such things, he wavered, wondering if the criteria were appropriate after all. The later judgments of the other groups suggested that they were all appropriate, but on judging the first seminar there was some doubt in his mind. This demonstrates quite clearly that, when any of the parameters in an assessment are changed, the whole process must be thought through very carefully before consistent, reliable assessments can be made.

SCORES ON ASSESSMENT SHEET

	Group 1	Group 6	Group 7
PO1	4	3	2
PO2	4	3	2
PO3.1	3	3	2
PO3.2	0	2	3
PO3.3	0	2	3
PO4	0	1	2
PO5	3	4	3
PO6	3	4	3
PO7	3	4	3
PO8	3	4	3
PO9	3	4	3
PO10	4	4	2
Avg.	2.5	3.2	2.6
Grade	β+	β+	β-

Table 5. Poster assessment scores for groups 1, 6, 7 in 1995.

CONCLUSIONS

This paper has outlined the rationale for requiring students to use posters as a means of communication and assessment. It has described how the method has been used in practice,

and it has reviewed and commented on the development and use of criteria for assessing posters.

It was discovered that some students had difficulty applying the criteria in a suitable way, and that their judgments were not consistent with the consensus view. It is recommended that students be involved in developing the criteria themselves, or at least be trained in their use, so that they have a greater measure of 'ownership' of them. This would be particularly important if it is intended to use students' peer assessment for summative purposes.

It was also discovered that, in their posters, students at Ulster were not very good at stating problems, either the original modelling problem or the corresponding mathematical problem. It is recommended that this particular aspect of poster creation be emphasised in teaching. On the other hand, students at Brighton found this the easiest criterion to score on. Perhaps this indicates some difference in the teaching approaches at these two universities and is an avenue for further investigation. Students at both Brighton and Ulster found it difficult to score well on reporting their conclusions and their evaluation of their project. Again, this is something that should be emphasised in teaching.

Finally, based on the experience of the author, it is recommended that very careful consideration be given to the criteria before using them for a purpose for which they were not created, no matter how similar the purposes may seem!

REFERENCES

Berry J and Houston K, 1995, *Students Using Posters as a Means of Communication and Assessment,* **Educational Studies in Mathematics, 29**, pp 21-27.

Denson PS, 1992, *Preparing Posters Promotes Learning*, **Mathematics Teacher, 85**, pp 723-724.

Haines CR and Izard JF, 1995, *Assessment in Context for Mathematical Modelling,* in Sloyer C, Blum W and Huntley I (eds.), **Advances and Perspectives in the Teaching of Mathematical Modelling and Applications,** Water Street Mathematics, Yorklyn, Delaware, pp 131-149.

Houston SK, 1995, *Using Rating Scales for Tutor and Peer Assessment of Mathematical Modelling Posters,* in Burton L and Izard J (eds.), **Using FACETS to Investigate Innovative Mathematics Assessment**, University of Birmingham, Birmingham, pp 45-52.

Houston SK, Haines CR and Kitchen A, 1994, **Developing Rating Scales for Undergraduate Mathematics Projects**, University of Ulster, Coleraine.

Houston K and Lazenbatt A, 1997, *Peer Tutoring in a Modelling Course,* submitted for publication.

Le Masurier D and Dunthorn S, 1995, *Analysing Criteria-Based Rating Scales for Assessing Students' Posters,* in Burton L and Izard J (eds.), **Using FACETS to Investigate Innovative Mathematics Assessments**, University of Birmingham, Birmingham, pp 38-44.

Linacre JM, 1990, *Modelling Rating Scales,* Paper presented at the Annual Meeting of the American Educational Research Association, Boston, Massachusetts, 16-20 July 1990, [ED318 803].

Linacre JM and Wright BD, 1994, **FACETS: Many-Facet Rasch Analysis**, MESA Press, Chicago, Illinois.

Sloyer C, Blum W and Huntley I, 1995, (eds.), **Advances and Perspectives in the Teaching of Mathematical Modelling and Applications,** Water Street Mathematics, Yorklyn, Delaware.

APPENDIX 1

Poster Presentation

Content	High			Low	N.S.	N.A.
PO1 **States the problem clearly**	☐	☐	☐	☐	☐	☐
PO2 **States the problem succinctly (or concisely)** Student has summarised the essence of the problem, rather than merely restating it.	☐	☐	☐	☐	☐	☐
PO3 **Outlines the solution**						
3.1 Describes the model Student has considered all relevant facts and information. Student has explained critical assumptions and relationships. Student has made sensible and appropriate use of personal knowledge and experience.	☐	☐	☐	☐	☐	☐
3.2 States the mathematical problem Appropriate mathematics, tools and resources have been brought to bear on the problem.	☐	☐	☐	☐	☐	☐
3.3 Reports on mathematical solution Mathematics used is correct. Mathematical terminology is correct. Description of mathematical methods is succinct.	☐	☐	☐	☐	☐	☐
PO4 **Reports conclusions** Conclusions are related to the modelling assumptions. More than an answer is produced . Model is used to describe, explain or make predictions about the phenomenon under consideration. Student has investigated a necessary and sufficient number of cases.	☐	☐	☐	☐	☐	☐

	Presentation	High			Low	N.S.	N.A.
PO5	**Designs the poster logically** Layout is logical and easy to follow.	☐	☐	☐	☐	☐	☐
PO6	**Uses different fonts effectively** Headings are bold. Highlighting is used when appropriate.	☐	☐	☐	☐	☐	☐
PO7	**Uses illustrations effectively** Illustrations are necessary ant sufficient to aid understanding of the text.	☐	☐	☐	☐	☐	☐
PO8	**Produces an aesthetically pleasing poster**	☐	☐	☐	☐	☐	☐
PO9	**Text is concise** Overall presentation is of agreed size. Uses English language correctly.	☐	☐	☐	☐	☐	☐
PO10	**Demonstrates understanding of project through discussion** Optional - depends on the primary purpose of the poster and the nature of the poster session.	☐	☐	☐	☐	☐	☐

APPENDIX 2

Students   Groups

	7	8	9	10	11	12	13	14	15	16	17	18	19
1		X											
2	X												
3		X	•	•						•	•		
4		X	•	•						•	•		
5		X											
6				X	•	•	•						
7	•	•				X							
8	X	•	•							•	•		
9			X	•	•						•	•	
10	•	•				X		•	•				
11				X	•	•	•					•	•
12	X	•	•							•	•		
13	•	•				X		•	•				
14				X	•	•						•	•
15			X								•	•	
16	•				X	•			•		•		
17	X		•						•	•			
18			X	•	•						•	•	
19					X	•				•			•
20							X						
21	•				X	•				•			•
22				X	•	•						•	•
25	•	•	•	•	•	•	•	•	•	•	•	•	•
31											X		
32											X	•	•
33			•	•					X	•	•		
34								•				X	•
35				•	•					X	•	•	
36								•				X	•
37				•	•					X	•		
38			•	•					X	•	•		
39								•				X	•
41	•	•											X
42		•	•					X					
43											X	•	•
44								X	•	•			
45											X	•	•
46		•						X		•			
47			•	•					X	•	•		
48	•	•						•	•				X
49		•	•					X					
50	•	•											X
51								•				X	•
52				•	•					X	•	•	
53				•	•					X	•	•	
54	•	•						•	•				X

11

Deriving Learning Outcomes for Mathematical Modelling Units within an Undergraduate Programme

Andrew Battye and Maggie Challis
Sheffield Hallam University, City Campus, Sheffield S1 1WB, England
e-mail: a.r.battye@shu.ac.uk **or** *m.challis@shu.ac.uk*

ABSTRACT

The move in British universities towards modularisation of the curriculum brings with it the risk of fragmentation of learning in such a way that it is difficult to monitor learning and progression. One way to ensure that units of study can be seen to fit into a coherent progression route for individual learners is to describe them in terms of anticipated learning outcomes. In this way it is possible for both staff and students to have a clear picture of any unit, in terms of its content, delivery and assessment methodologies.

This paper addresses a definition of mathematical modelling, the aims of mathematical modelling and the anticipated learning outcomes of mathematical modelling that may be derived from these aims. It ends with the challenge of identifying anticipated learning outcomes appropriate to level.

INTRODUCTION

"Nobody seemed to know where they came from, but there they were in the Forest" (Milne 1926: chapter 7.)

We have long been impressed by the deep philosophy embedded within Milne's Winnie the Pooh books (Battye, 1991). Although masquerading as simple children's stories, time and time again we come across those telling little phrases that show how much broader their interpretation can be (Williams, 1995).

Like Kanga in the Milne books, learning outcomes have appeared from nowhere in our educational Forest. Trying to find out where they came from is difficult, although the reasons why they are important have been well rehearsed. Describing programmes (a collection of courses with some commonality), courses (such as a named degree), units or modules (the smallest portion of study that results in a specified number of credit accumulation learning points - hereafter called modules), in terms of learning outcomes has many advantages:

- students are able to see what is expected of them and develop strategies for achieving the target outcomes,

- students can make better, more informed choices of which modules to take,

- flexibility and therefore access are improved, especially in the context of credit accumulation and transfer,

- outcomes and assessment methodologies may be negotiated with students, thus increasing student involvement in their educational process,

- a learning outcomes approach enables a wider range of achievements to be overtly recognised and given credit - for example core skills such as problem solving or communication skills,

- specifying learning outcomes can act as a first step in reviewing teaching, learning and assessment methods.

There are, of course, counter arguments to the use of learning outcomes. These relate to issues such as who should set the outcomes and whose values they represent; whether readily assessable outcomes constitute evidence of learning or mere behaviour adaptation; the difficulty in getting the level of specificity right to avoid either broad generalisations or lapsing into triviality; the risk that the flexibility afforded by describing courses in learning outcomes may lead to fragmentation and lack of coherence in the overall learning experience.

However despite these reservations and not in any way to belittle them, there is a strong trend in the UK, particularly in those universities that are ex-polytechnics, to modularise the curriculum and describe each module in terms of learning outcomes.

The shift from curricula described in terms of syllabus content to those described in learning outcomes has happened quickly. In 1992, the Council for National Academic Awards was advocating in its guide to good practice in framing regulations (CNAA, 1992) the use of aims and objectives: "A programme of study should have stated *aims* and *objectives* which the curriculum, structure, teaching methods and forms of assessment are designed to fulfil. .. The *aims* should include the development to the level required for the award of a body of knowledge and skills appropriate to the field of study and reflecting academic developments in the field... The statement of *objectives* should show how the programme of study will fulfil the aims." Nowhere were learning outcomes mentioned.

However, the Higher Education Funding Council in England (HEFCE, 1994) stipulates that its assessors, when inspecting institutions of higher education, should be looking for objectives that are ... "normally expressed in terms of the expected learning outcomes of the academic programme and relate to the acquisition of knowledge, the development of understanding and other general intellectual abilities, the development of conceptual, intellectual and subject-specific skills, the development of generic or transferable skills, or the development of values, of motivation or attitudes to learning."

So learning outcomes are now inextricably tied in with a teaching and learning assessment process upon which future funding of higher education institutions depends. In this light, it is perhaps not surprising that they are being adopted so readily.

WHAT DOES A LEARNING OUTCOME LOOK LIKE?

The definition that has been adopted at Sheffield Hallam University (1993) is that ... "learning outcomes reflect the changes which have taken place in the individual as a result of going through a learning process". Within a module description learning outcomes would normally be introduced by a phrase such as 'The student will be able to......'. The outcomes are usually seen as including:

- subject based outcomes, knowledge and comprehension, the ability to apply knowledge in different situations and processing skills acquired through the use and application of knowledge,

- personal outcomes, including interpersonal skills like teamwork and negotiation, and intrapersonal qualities like motivation, initiative and critical self-reflection.

In addition to these broad outcomes, it is usual to consider:

- the behaviour looked for in the learner,
- the degree of independence or autonomy expected,
- the complexity and/or significance of the situation in which the learner is expected to demonstrate learning.

By specifying these aspects through learning outcomes, it has been argued that level, or progression, can be addressed. This can be supported by Bloom's (1956) well-known taxonomy of cognitive learning, which indicates that the use of knowledge becomes more complex as it moves through the stages of recall, comprehension, application, analysis, synthesis and evaluation. This approach to defining level is one that has been used at Sheffield Hallam University within its undergraduate awards framework, although there are debates within the higher education sector about generic definitions of level that could be applied across the whole sector, or even whether there is any need to break undergraduate learning into defined levels at all (Winter, 1994).

If there is to be some form of level descriptor, however, this must also be related to the context within which learners are expected to demonstrate their learning. The activities required may move from 'basic' to 'complex' or are intended to be carried out 'under supervision' or 'independently'.

Thus, as an example, a first year module may require that learners:

- recognise, with minimal prompting, engineering problems in which the application of the basic laws of fluid mechanics and thermodynamics may be appropriate

while a third year unit may require that they:

- evaluate alternative solutions to energy management problems using both technical and socio-economic criteria.

It can be seen from these examples that the structure and phrasing of learning outcomes is designed to ensure that it is possible to identify whether the student has achieved the intended learning. This is made easier by ensuring that the introductory verb is one that lends itself to assessment. Verbs such as 'understand' or 'appreciate' are difficult to assess, while 'describe', 'evaluate' and 'apply' are more amenable to measurement. Where *understanding* is intended to be a key feature of the learning process, it may be better incorporated into the aim of the unit or programme, rather than built into specific learning outcomes.

It is important that learning outcomes are consistent with the teaching and learning strategy for the programme, and support the assessment methodology that will be adopted. For example, if specific skills are to be built in, it would be difficult to see how these could be taught by lectures or assessed through a written examination.

Equally, the number of learning outcomes and their presentation should not become a 'wish list' - what an ideal student might achieve - but rather the behaviour which might reasonably be expected of a student at the point of summative assessment of the module (or programme). They may also be seen as anticipated outcomes, rather than a set of strictly defined competences, all of which must be assessed. This enables a degree of flexibility to be built into the process so that contingencies can be met, formative assessment can be incorporated, and negotiation between students and staff may form part of the final outcomes that will be summatively assessed.

MATHEMATICAL MODELLING AT SHEFFIELD HALLAM UNIVERSITY

In the new undergraduate provision at Sheffield Hallam, mathematical modelling no longer appears explicitly as a named module in the first year of either the degree or HND in Computing Mathematics. Thus the challenge we face is how to ensure that this important aspect of mathematical thinking is incorporated into the overall structure, through other named mathematical modules (where modelling is an intended component). In order to achieve this it was important to agree, amongst the members of the mathematics teaching team, exactly what the distinguishing features of mathematical modelling are, how they could best be described in terms of learning outcomes and how they could be incorporated into the whole programme at appropriate stages and levels.

The first two stages of this process were addressed at an internal mathematics curriculum development workshop. Through an open discussion on the features of mathematical modelling, we attempted to reach an agreed definition that could be used as a basis for the next stage of our discussions.

DEFINITION OF MATHEMATICAL MODELLING

The definition of mathematical modelling that emerged from the workshop was:

"a process that involves responding to a real situation, abstracting a problem using some simplification and assumption, establishing a response to the problem (which may involve the use of mathematical visualisations and symbols) and evaluating and communicating that response to self and others".

This definition is more detailed than those *implied* by Burghes (1986)

"using mathematics to solve *real* problems"

or Huntley and James (1990)

"the real challenge (of mathematical modelling) is to find relevant *questions* and *answers* from a (real) situation that may look initially chaotic".

However, despite the absence of any universally accepted definition of modelling, the initial discussion provided a framework within which to reach agreement on the aims of teaching mathematical modelling within the undergraduate curriculum.

AIM OF MATHEMATICAL MODELLING

The aim of mathematical modelling was defined by the group as the enabling of students to develop and acquire:

i) strategies for problem solving through the skills of
        problem specification - including the use of imagery
        problem verification - including the internal consistency of the problem
        problem validation - including the need to view each problem in its real world context,
ii) an appreciation of aspects of conditionality that are inherent in problem solving,

iii) an ability to engage in the critical usage of modelling,

iv) participation and communication skills.

This list may be compared with other defined aims of mathematical modelling, for example that offered by Usiskin (1991):

i)   to foster creative and problem solving attitudes, activities and competences,

ii) to generate a critical potential towards the use and misuse of mathematics in applied contexts,

iii) to provide the opportunity for students to practise applying mathematics that they would need as individuals, citizens or professionals,

iv) to contribute to a balanced picture of mathematics,

v) to assist in acquiring and understanding mathematical concepts.

LEARNING OUTCOMES FOR MATHEMATICAL MODELLING

Our newly defined aims were further refined into anticipated learning outcomes that might be applicable to the overall undergraduate curriculum in mathematical modelling.

The student should be able to:

i)   describe how modelling forms a part of a broader scheme of mathematical, personal and interpersonal communication systems,

ii) specify the domain of interest of a model,

iii) identify, select and collect data that are relevant to the situation to be modelled, taking into account assumptions made by self and others about the problem,

iv) abstract and represent context-relevant aspects of the process/system to be modelled,

v) express the known limitations and conditions of applicability of a model, and relate these to the real world,

vi) identify the mathematical nature of a model, and advocate an appropriate action plan to operate on it to explore the situation,

vii)interpret a model solution in the context of the problem requirements,

viii) communicate effectively in a range of group contexts, including peer and client, expert and non-expert groups,

ix) represent situations in a variety of ways - for instance diagrammatic, symbolic, textual, verbal.

THE CHALLENGE

The problem that remains is to determine how these learning outcomes could be further refined in order to indicate achievement at different levels within an undergraduate curriculum. Is it possible to write learning outcomes at levels 1, 2 and 3 (that is years 1, 2 and 3 of a first degree), or are there some cognitive skills involved in the modelling process that are only applicable at particular levels? Are Bloom's definitions of cognitive progression appropriate to our context of mathematical modelling? Should we seek an alternative taxonomy within which to describe achievement and progression - for example an 'experiential taxonomy' (Steinaker and Bell, 1979) or an 'affective domain taxonomy' (Krathwohl, Bloom *et al.*,1964)? Should we abandon the idea of measurement of staged progression and aim instead for a defined end point?

CONCLUSIONS

Having begun this attempt to identify anticipated learning outcomes in mathematical modelling we realised that there was not even a universally accepted definition of modelling. This presents us with real problems if we want to articulate our perceptions of what students should know and be able to do in mathematical modelling at various stages of their undergraduate careers. However we remain to be convinced that there is a very real need to identify such outcomes by level. It is certain that the use of learning outcomes in the course definitions we work to can make these documents more useful and transparent than were the traditional syllabi of yesteryear. They can thus act as a positive force in this era of increased student choice. There remain in our minds, however, questions about how these defined learning outcomes are derived, and whether all aspects of the undergraduate curriculum can be neatly fitted into levels, the appropriateness of whose descriptors themselves may be called into doubt.

REFERENCES

A Guide to Writing Learning Outcomes, 1993, Learning and Teaching Institute, Sheffield Hallam University, Sheffield.

Battye AR, 1991, *Modelling Air Resistance in the Classroom*, **Teaching Mathematical Modelling and its Applications**, **10**, 1, pp 32-24.

Bloom B, (ed.), 1956, **Taxonomy of Educational Objectives**, **Book 1, Cognitive Domain**, Longman, London.

Burghes DN, 1986, in Berry JS *et al.*,(eds.), **Mathematical Modelling Methodology, Models and Micros**, Ellis Horwood, Chichester.

CNAA 1992, Council for National Academic Awards **Academic Quality in Higher Education: A Guide to Good Practice in Framing Regulations**, CNAA, London.

HEFCE 1994, Higher Education Funding Council for England, **The Quality Assessment Method, from April 1995**, Circular 39/94, HECFE, London.

Huntley ID and James DJG, 1990, **Mathematical Modelling - A Source Book of Case Studies,** Oxford University Press, Oxford.

Krathwohl D, Bloom B *et al.*, 1964, **Taxonomy of Educational Objectives : Affective Domain**, David Mackay Company, New York.

Milne AA, 1926, **Winnie the Pooh**, Methuen and Co Ltd, London.

Steinaker N and Bell M, 1979, *The Experiential Taxonomy - A New Approach to Teaching and Learning*, **Academic Press**, New York.

Usiskin Z, 1991, in Niss M *et al.*, (eds.), **Teaching Mathematical Modelling and Applications,** Ellis Horwood, Chichester.

Williams JT, 1995, **Pooh and the Philosophers**, Methuen London.

Winter R, 1994, *The Problem of Educational Levels, Part II: A New Framework for Credit Accumulation in Higher Education*, **Journal of Further and Higher Education**, **18**, 1, pp 92-106.

Section C

Secondary Courses and Case Studies

12

An Operative Approach to Formal Reasoning

Paola Forcheri and Maria Teresa Molfino
Istituto per la Matematica Applicata del Consiglio Nazionale delle Ricerche, Genova, Italy.
e-mail: forcheri@ima.ge.cnr.it **or** *molfino@ima.ge.cnr.it*

ABSTRACT

A knowledge-based learning environment, able to assist students performing the deductions needed to solve mathematical problems of a symbolic nature, was designed and tested. The system, called TELLER GUIDE, is intended for use with secondary-school first-year students. In this paper, we discuss the approach followed in order to render the system suitable for helping students to construct concepts through applications.

INTRODUCTION

Knowledge based educational systems seem to offer a new insight into the teaching and learning of mathematics, being computational tools which analyse the cognitive structure of a domain and the difficulties which underlie the learning of its structure Nivana (1993) and Wenger (1987). Thus it is worthwhile to explore their potential to give students the capability of organising knowledge to carry out formal reasoning. In our opinion, such capability is the mathematical basis of the modelling activity. Accordingly we designed TELLER GUIDE, a knowledge-based learning environment suitable for secondary-school students being introduced to the concept of formal reasoning. Three main problems have been addressed via the system.

The first regards the need to devise tools which attract students' attention and get them to work on relevant topics. Consequently the system presents problems expressed as games.

The second problem addresses the need to reduce the gap between the constructive view students are used to in the learning of arithmetic and the formal approach they have to adopt in algebraic problems. In our opinion, this gap is one of the main causes of students' difficulties in passing from arithmetic to algebraic modelling. Thus the system makes students deal with a class of symbolic arithmetic problems which also have an algebraic formulation.

The third problem examines the need to find out efficient methods which make young students reorganise experience in order to understand the general formal concept which models various situations. Hence the system embodies a method which introduces students to formal reasoning by making them deduce mathematical concepts from applied examples. The method, which is based on a knowledge-based model of reasoning, will be discussed in the following sections. An overview of the overall organisation of the system, which is detailed in Forcheri and Molfino, (to appear) will precede this discussion. The analysis of the experiment we constructed around our system will conclude the paper.

OVERVIEW OF THE SYSTEM

TELLER GUIDE supervises the activity of a student who solves mathematics problems expressed in the form of one-person games. The game consists of associating a meaning with all symbols which appear in a symbolic sum so that it can be interpreted in the arithmetic domain. The symbolic sum (hereafter called relation) is presented in a column. The meaning of each symbol is a value belonging to the set $\{0,1,2,3,4,5,6,7,8,9\}$. Different symbols must be assigned different meanings.

To play the game, two kinds of moves can be performed: application of a symbolic arithmetic concept to a column of the given relation (deduction or blind substitution of a symbol with a value (attempt). The system prevents the student from making attempts when an arithmetic concept can be applied. The effect of a move is to narrow down the set of values which can be associated to the symbols. The game ends when a set of no more than one element is associated with each symbol.

Fig. 1 shows a portion of the dialogue referring to the solution of the relation $AA+AB=DCA$. The move is signified by pressing a corresponding button on the screen and by selecting its parameters with the mouse. The student signifies their intention to perform an attempt by pressing the button *Chance*, and by choosing a symbol (or a carrier) and a value for it. The intention to perform a deduction is indicated by pressing *Rules* and by choosing the column on which to operate. If the move is a deduction, TELLER GUIDE displays a series of one-column relations which constitute examples (correct or incorrect) of the concept underlying the deduction. The student is then required to select some symbol/example pairs and to confirm their choice by pressing the button *OK*. The examples displayed in Fig 1 refer to the concept of even numbers when applied to the tens column. The choices made by the student are surrounded by four points.

If all choices represent correct examples, as in this case, the system narrows down the set of values for symbols according to the deduction. Otherwise errors are pointed out and the system gives a brief explanation depending on the student 's action.

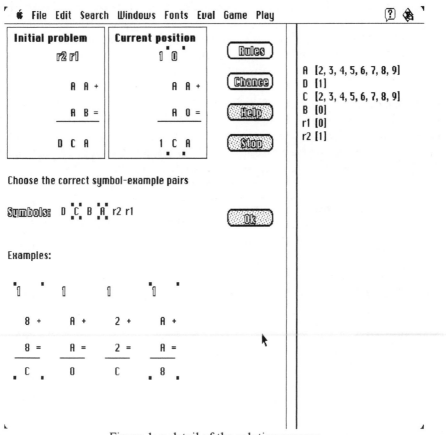

Figure 1: a detail of the solution process

THE EXAMPLE-GENERATION PROCESS

The examples displayed are dynamically generated by the system. The process of generation relies on knowledge-based reasoning, which models the students' approach to the problem, and on a classification of errors which derives from the observation of the behaviour of a sample of students.

The knowledge-based model of reasoning

The way students perform arithmetic moves can be viewed as a knowledge-based process for solving a system of integer linear equations, whose variables satisfy some constraints, represented by a family of finite numerical sets. Every set expresses the constraints which must be satisfied by a variable, for every variable which appears in the system. The problem is solved when, for each variable, the set of values is reduced to a set of only one element. The problem is inconsistent if there exists at least one variable such that the corresponding set of values is reduced to the empty set.

Within this model every column of the relation corresponds to an equation of the system, symbols in the relation and carriers of columns correspond to variables and, for every symbol in the relation and carriers of the columns, the constraints on the numerical values which can be admitted for the symbol (or carrier) are represented by a set of values. Figure 2 shows the representation of the relation of Fig 1.

Variables involved in the problem
          The symbols A, B, C, D and the carriers of the units and tens column indicated, respectively, with r_1 and r_2.

Equations corresponding to columns
          $A+B=A+10* r_1$
          $r_1+A+A=C+ r_2 *10$
          $r_2 =D$

Sets of values
          A, B, C, D: $\{0,1,2,3,4,5,6,7,8,9\}$
          $r_i : \{0,1\}$ $i=1,2$

Figure 2: representation of the relation AA+AB=DCA

The solution of a system of the above kind can be reached in several ways and at different levels of efficiency. For example, it can be handled by modelling the problem as the search for a path in a tree and using backtracking techniques, Horowity and Sahmi (1978), or the problem can be solved by using some suitable Constrained Logic Programming Scheme implementation such as Fruhwirth *et al.* (1992), or by explicitly encoding the arithmetic knowledge needed to reach the solution.

Our interest is not in the efficiency of the solution method, but rather on its expressive power with respect to the task which should be handled by students. Consequently we regard an 'arithmetic' move as a step of reasoning based on facts and rules. The facts of the problem are equations and sets of values. Rules represent arithmetic concepts and are aimed at narrowing down the current sets of values and/or modifying equations. For example, referring to Fig. 1, the rule which implements the concept of even number reduces the sets of values associated to A and C to the sets {6,7,8,9} and {2,4,6,8} respectively.

The solution process is seen as a sequence of transformations, generated by application of a rule to the current situation. The completeness of the process is guaranteed by allowing the blind substitution of a symbol with one of its admissible values when no rule can be applied in the situation at hand. Inconsistency, possibly introduced by a blind substitution, is controlled by backtracking.

Automatic generation of examples

To generate examples automatically the problem is split into sub-problems, each referring to only one column and taking into account the 'position' of the column in the relation. Examples are dynamically generated.

The process takes the given column, associates the variables of the column with the corresponding initial sets of values, and applies the rule to the state s_p corresponding to this sub-problem.

For each variable affected by the rule in the initial state we generate the two following classes of relations:

C1. relations which represent correct applications of the rule are obtained by substituting all occurrences of the variable in the column with a value, one for each admissible value resulting from the application of the rule to s_p.

C2. relations which represent erroneous applications of the rule are generated by substituting all occurrences of the variable in the column with a value, from one of the inadmissible values resulting from the application of the rule to s_p.

Depending on the current situation, relations pertain to different subclasses which give information on the level of understanding of the problem. The subclasses are as follows.

C1.1 relations which correspond to correct applications of the rule to the current situation (*analogies*),

C1.2 relations which correspond to the lack of application of the rule to the current situation (*mis-inferences*),

C1.3 relations which correspond to correct applications of the rule and do not take into account the current situation (*over-generalisations*),

C2.1 relations which correspond to erroneous applications of the rule with respect to the current situation (*mis-interpretations*),

C2.2   relations which correspond to erroneous applications of the rule and do not take into account the current situation (*mis-interpretations and over-generalisations*).

TELLER GUIDE, taking into account the relations generated and the different subclasses to which they belong, automatically chooses a series of examples and displays them during the dialogue with the student.

For example, by referring to the problem in Fig. 1, the relation *2+2=1C* is a *mis-interpretation*, the relation *A+A=10* is an *over-generalisation*, while both the relations *8+8=1C* and *A+A=18* are *analogies*, with respect to the tens column and the concept of even numbers.

THE STUDENTS' REACTION TO THE SYSTEM

TELLER GUIDE has been tested with a class of first-year students in secondary school. The experiment was preceded by a preparatory phase consisting mostly a demonstration and an explanation of the system. We presented some relations and showed how to solve them through interaction with the system. We gave a general explanation including the organisation of the game and of the interface, the meaning of the various pieces of information displayed, their distribution on the screen, the inputs to be given, and the corresponding output. We then asked students to try, on their own, the example relations already seen, and we intervened only when they had a specific question. The majority of students did not and so technical difficulties in using the system, find we decided to start the real experiment in order to obtain feedback on the pedagogical choices we had made in the design.

Organisation of the experiment

We agreed, with the mathematics teacher, a set of relations to be proposed to students. They were divided into groups and were asked to solve them via interaction with TELLER GUIDE. We supervised the work together with the teacher, noted the discussion among the students and intervened with suggestions and explanations when necessary.

The domain we investigated was the sum of two numbers less than 100. Examples of relations are: AC+B=AC, which models the sum of a number with the null element with respect to the sum; A+B=9, which models the addition of an even and an odd number, with the additional constraint that at least one of the numbers cannot be greater than 4; AC+AC=BA, which models the sum of two equal numbers less than 50, and so on. Impossible problems, such as A+A=7, were also considered.

To choose the relations for the test we subdivided them according to the arithmetic concepts needed for the solution and to their complexity. As to the arithmetic concepts, we considered the following: the arithmetic properties (for instance 0 is the null element with respect to the sum, the sum of two even numbers is even, and so on) which must be applied to handle the relation; the number of such properties; the number of solutions and the number of computational steps. As to the complexity, we took into account the quantity of symbols, their position and the number of occurrences each symbol.

The detailed organisation of the relations aims to analyse several kinds of capability (and, consequently, the corresponding difficulties) and the ability to apply them in contexts of increasing complexity. Moreover, this organisation allows us to tailor the test quickly to different situations.

As regards the arithmetic concepts, it must be noted that relations involving a single concept help develop the ability of passing from a concrete example of a rule to its formalisation. Relations which require more than an arithmetic concept allow us to verify if students are able to organise them using deductive reasoning. Unsolvable relations aim at testing the capability of dealing with contradictions while relations verified by a set of additions were used to assess the skill of generalisation. Finally, multicolumn relations aim to verify the ability to recognise sub-problems and their interrelation.

The 'complexity' parameter analyses the capability to handle the concept of variable. In this respect we observe that the complexity of a relation allows us to distinguish between relations representing explicit or implicit additions. In explicit additions the elements of the sum are numbers and the result is represented by one or more symbols. In implicit additions, at least one element is a symbol. In the case of sums less than 10, explicit additions represent, in a way, equations in which the variable is already isolated, thus their solution requires a numerical sum and a substitution. As a consequence, explicit additions of this kind can be used to analyse the feasibility of introducing the concepts of variable and substitution to students within the age bracket in question. The subdivision of implicit additions depending on the number of symbols and their position lets us analyse to what extent the concept of variable is handled by students. Implicit additions allow going deeper, at least at an informal level, into the concept of equation. Moreover, as noted in the previous section, relations, explicitly or implicitly representing sums greater than 9, may be viewed as a system of equations: each equation corresponds to a column, taking into account the carriers . These relations can thus be employed to introduce new topics and basic ideas about mathematical modelling, in particular the solution of a system of linear equations, equivalence of representation, choice of a model, solution of a problem with respect to a representation, and so on.

Discussion

The observation of students' behaviour during the interaction with the system confirmed the idea that the concept formation process, at the age in question, is mainly achieved through working with examples applied. Moreover, the relations generated by the example-generation process seem quite appropriate to help students single out the different concepts involved in a deduction. Qualitative aspects of the students' performance are now reported.

Students mainly used arithmetic knowledge in a quite informal way. For example, the majority of them were able to deduce, from $A + B = 7$, that $A \leq 7$ and $B \leq 7$; notwithstanding this fact, they found difficulty in deriving $A \leq C$ and $B \leq C$ from the relation $A + B = C$. Moreover, they were able to use some 'empirical' rules for handling simple equations that contained small numbers. For example, students were generally able to solve relations of the form $17 + XY = 29$.

Students who found difficulty in carrying out symbolic reasoning benefited from working mainly on examples - after the same concept had appeared a number of times in different situations they were able to express, almost formally, the rule applied. For example when presented with the relation $AC+B=AC$, several students were perplexed. The same students, when given relations of the form $A3+B=A3$, were able to deduce that B is 0; after a number of exercises of this kind they were able to solve the problem $AC+B=AC$ without difficulty, as they fully understood the underlying rule.

As to the difficulties faced by students we noted that, in several cases, wrong choices were not due to arithmetic misconceptions or to the lack of mathematical knowledge. In particular a common cause of error was the fact that students did not take into account the current situation, *i.e.* the context in which the rule applies. For example, given the relation $A+A=B$ and knowing that A is different from 4, they were usually able to recognise that B is even, but they had difficulty in observing that B is different from 8 as 4 is not permitted for A. Moreover several students were unable to recognise that in a single step of deduction more than a conclusion can be derived. For example the majority of students were able to conclude that B is even from $A+A=B$. However, they only concluded that $A < 5$ after intervention by the teacher. Another problem we observed was the confusion of the concept of validation with that of deduction. For example students, when given a relation such as $A+A=B$, tended to say that B is even, even when this fact was already known.

Other causes of difficulties were mainly: handling the carriers; dealing with unsolvable relations; the possibility of multiple solutions; taking into account the context; the presence of a high number of variables. As to the carriers, the chief problem appeared to be their formalisation and the role they play in determining the value of the symbols at hand. As to problems with more than one solution and contradictions, students were perplexed, probably because they were completely unfamiliar with this type of problem. As to relations containing more than one symbol, the difficulty increased with the number of symbols.

CONCLUSIONS

Two main pedagogical considerations underlie the example-based reasoning process, namely the students' attitude towards symbolic problem solving and the choice of the examples.

As to the first aspect, some authors point out that, at the age in question, abstract abilities are not developed (Ausubel, 1968), and the capability to use concepts usually depends on the particular problem. Experiments carried out with these students led us to formulate the hypothesis that the concept formation process, at the age in question, is mainly based on examples and counter-examples of application. These observations explain why the system has been discussed via examples which illustrate how students determine the meanings of symbols, both correctly and incorrectly.

As to the second aspect it must be noted that an accurate classification of examples can help students attain the capability to manipulate formally the concept at hand, at least in prescribed situations. The selection of examples made by the user, moreover, gives an indication to the tutor about the level of understanding of the concept itself. This has been taken into account when developing the example-generation process. In fact it can be noted that different errors are related to different parameters embedded in a deduction, namely the arithmetic concepts involved in the rule, the number of symbols affected by the rule and the current set of admissible values for the symbols to which the rule applies. Accordingly the example-generation process has been refined by varying these parameters one at a time, whenever possible. Moreover, for each relation, all examples are shown together but students are asked to indicate variables (and corresponding examples) one at a time. For each variable we show an example and a counter-example; when appropriate, an example is shown which corresponds to the lack of application of the rule to the variable at hand.

As indicated earlier, the results of the experiment confirmed the validity of these choices. However the difficulties faced by some students during the work showed the limitations of a system which obliges students to follow a single learning path. In particular, in some cases, reasoning by examples was not sufficient to lead to the abstraction of general rules underlying the deduction. The teacher who assisted the experiment had then to suspend the exercise and to explain in depth the concepts involved. This observation suggested an improvement to the educational performance of the system by adding an explanation utility, this being organised on the basis of students' reactions and taking into account the kind of help they needed. The explanation utility allows attention to focus on the concept of current interest by setting aside a relation and asking for the analysis of a single column. The explanation consists of a picture containing all relevant information on the deduction to be performed. Such a picture is organised in three parts: a headline marks the starting point of the reasoning; the column selected by the student is displayed in the centre of the picture in the form of an equation; one or more statements regarding the deduction are displayed. These statements, which are situated at several places within the picture, consist of a short textual observation. Arrows point to the symbols of the relation referred to in the statement (see Fig. 3).

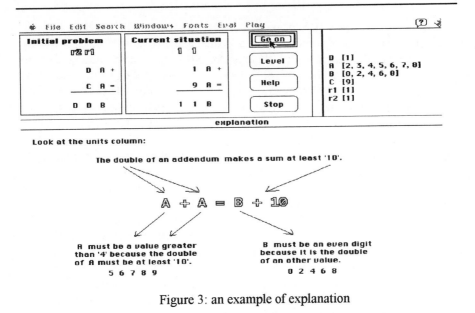

Figure 3: an example of explanation

The explanation is obtained by selecting the column and pressing the button *Help*.

As to the other difficulties we observed, they seem to indicate that the external representation of the problem does not give enough information regarding the different meanings of the objects under examination. We are investigating this limitation and we are studying corrections and improvements to the interface.

REFERENCES

Ausubel D P,1968, **Educational Psychology: A Cognitive View**, Holt, Rinehart and Winston, New York.

Forcheri P and Molfino M T, 1995, *Teaching Concepts Through Examples of Application*, **CNR-IMA Technical Report number 13**.

Fruhwirth T, Herold A, Kuchenhoff V, Le Provost T, Lim P, Monfroy E and Wallance M, 1992, *Constraint Logic Programming - An Informal Introduction*, **Proceedings of AI*IA.Tecnologie Intelligenti e Loro Impatto Nelle Applicazioni Industriali**, pp 1-27.

Horowitz E and Sahmi S, 1978, **Fundamentals of Computer Algorithms**, Computer Science Press, Rockville, Maryland.

Nwana H S, 1993, *An approach to developing intelligent tutors in mathematics*, **Computers and Education**, **20**, 1, pp 27-43.

Wenger E, 1987, **Artificial Intelligence and Tutoring Systems**, Morgan Kaufmann Publishers., Los Altos, California.

13

The Application of Mathematics - an Essential Component of New Vocational Qualifications in the UK

John Gillespie
Shell Centre for Mathematical Education, University of Nottingham, England.
e-mail: john.gillespie@nottingham.ac.uk

ABSTRACT

New college-based or school-based vocational qualifications - called General National Vocational Qualifications (GNVQs) - have been introduced in the United Kingdom and have proved very popular. The application of mathematics - called the Application of Number in GNVQ specifications - is one of three mandatory Core Skills and forms part of all GNVQ qualifications. A student must demonstrate, among other requirements, his or her abilities to apply mathematics in solving appropriate problems in the context of that GNVQ.

An objective of the Core Skills mathematics requirement is the development of widespread, genuine, practical and transferable mathematical competence. This requirement has led to new demands being placed on teachers and students, as new styles of learning have to be developed to suit the application-led approach to mathematics. Both teachers and learners are having to modify traditional views of mathematics learning.

In this paper the author

- *describes briefly some of the distinct features of Application of Number Core Skills and their implementation,*
- *identifies some of the competing (and perhaps mutually exclusive) demands for features of the specifications from different interest groups,*
- *provides examples of the application of mathematics in use in a vocational setting.*

INTRODUCTION

Since 1992 major changes have taken place in the United Kingdom in the framework of vocational courses and qualifications, including those available in colleges and schools. The new General National Vocational Qualifications (GNVQs) are proving very popular, and over one hundred thousand students are now involved with GNVQ studies throughout the UK.

In order to achieve any GNVQs, individuals must demonstrate their ability to apply mathematics to help solve problems in realistic vocational settings, not just 'do mathematics'. This opens up opportunities for the development of genuinely useful mathematical aptitudes in widely diverse fields. Common standards are required for this core skill irrespective of the vocational setting of the main studies. One objective is the development of widespread practical and transferable mathematical competence.

Understandably, many teething problems have arisen in the implementation of this compulsory applications-centred approach to mathematics learning in all GNVQ courses, but at the same time many opportunities for fruitful mathematics learning for all now exist which were not present before.

The author has been closely involved with many of these problems, including playing a major role in the current redrafting of GNVQ Application of Number Core Skill specifications, within which the practical application of mathematics is called Application of Number.

BACKGROUND

Over the last fifty years, secondary education in England and Wales has been in a state of almost continuous change. The Butler reforms in 1944 provided the chance of free selective secondary education up to the age of 18 for any who qualified on the basis of aptitude, as measured by tests at age 11. The comprehensive schools of the 60s and 70s extended sixth-form education to more pupils, so that by the early 90s about 30% of the age group were obtaining 18+ qualifications (A-levels in the main), compared with the 10% who were staying on at school beyond school leaving age in the 50s.

However, there is convincing evidence and concern that these improvements in educational achievement lag behind those of other developed countries with which Britain is competing (Table 1).

Another major concern has been the problems faced by industry when recruiting school-leavers who were lacking in communication, mathematical and other essential skills. For instance the analysis of a representative group of 21 year olds revealed that 20% had only very limited competence in basic maths (ALBSU, 1993).

Country	Percentages
Germany	68
France	48
Japan	80
England	29

Table 1. Young people obtaining a comparable upper secondary school qualification at 18+ in 1990 (Green and Steedman, 1993)

This poor performance is not a recent phenomenon. For instance, James Callaghan's Ruskin College speech in 1976, introducing the so-called Great Debate in education, was chiefly concerned with these issues.

The research leading to the Cockcroft Report in 1982 made it clear that basic skills were not enough on their own:

". . .most important of all is the need to have sufficient confidence to make effective use of whatever mathematical skill and understanding is possessed, whether this be little or much, ... our concern is that those who set out to make their pupils numerate should pay attention to the wider aspects of numeracy and not be content merely to develop the skills of computation." (Crockford, 1982: page 11.)

GENERAL NATIONAL VOCATIONAL QUALIFICATIONS

Decline in employment opportunities and in training opportunities in the workplace, allied with the demand for a parallel route for post-16 education which was equally demanding but more vocationally oriented and less academic than A-levels, led to the development of General National Vocational Qualifications (GNVQs). Introduced from 1993, with specifications in 14 areas such as Health and Social Care, Leisure and Tourism, Business and Art & Design, and with provision for 5 levels, the new qualifications have been a tremendous success in terms of student uptake (see Table 2).

Year	Students	Centres
1993	10 000	108
1994	82 000	1400
1995	144 000	2000

Table 2. Increase in number of GNVQ students 1993-4.

Now, while approximately 50 % of post-16 students are following A or AS courses, a further 20 % are following GNVQ courses, leading either into industry or on to higher education.

CORE SKILLS

A distinctive feature of GNVQ qualifications is the requirement to demonstrate competence in the three Core Skills of:

- Communication,
- Application of Number,
- Information Technology.

Core Skills (as distinct from the core curriculum) are concerned with the application of language, mathematics or IT as tools in solving problems in vocational or other contexts. They are seen as part of a programme of raising both general and transferable skill levels. The skills

- should be transferable between areas of work and employment,
- should be of benefit to many aspects of employment and life,
- should support other aspects of development - not stand alone.

APPLICATION OF NUMBER

Application of Number (or more correctly application of mathematics) thus adopts the Cockcroft, as distinct from the narrower basic skills, view of numeracy. Its assessment has no separate examinations or tests - all assessment is based on evidence from work in vocational and other contexts, thus requiring the demonstration of skills in context. It does not set out to be a maths 'subject' - rather it infuses other areas of study. It is about using mathematics as a tool to help solve problems in vocational contexts - not about demonstrating fragmented mathematics skills in isolation.

There are three elements at each of five levels:

- collect and record data,
- tackle problems,
- interpret and present data,

which broadly correspond to 'mathematising', 'doing calculations' and 'demathematising'.

There are three dimensions of increasing demand as levels rise:

- increasing autonomy of the student,
- increasing technical demand,
- increasing complexity of problem.

Table 3 shows introductions to the elements at levels 1 and 3, with differences at level 3 underlined.

	Introductions to Level 1	Introductions to Level 3
Collect and record data	The student decides which technique to use in a range of simple data-collection and measuring tasks. The student is told which level of accuracy to use and, when collecting data, what sample should be used.	The student decides which technique to use in a range of data-collection and measuring tasks which they have designed themselves, including determining an appropriate sample size. The student decides what level of accuracy and tolerances to use. The student also identifies sources of error and the impact of these errors on both large and small data sets.
Tackle problems	The student decides which techniques to use to solve problems. In each case, solving the problem should involve the use of a single technique and not a series of techniques. The student uses both calculator and non-calculator methods to make calculations correctly and to check the results. The student is told which level of accuracy to use.	The student decides which techniques to use to solve problems, including problems in three dimensions. In each case, solving the problem should involve using a series of techniques with numbers of any size, including negative numbers. The student uses both calculator and non-calculator methods to make calculations correctly and to check the results. The student decides which level of accuracy to use.
Interpret and present data	The student identifies the main features of the data and relates these to the problem tackled. The student decides which techniques to use to display data and which titles and labels to use. The student is told what level of accuracy and which scales and axes to use.	The student identifies the main features of the data and relates these to the problem tackled. The student decides from a wider range of more complex techniques which to use to display data and which titles and labels to use. The student decides what level of accuracy and which scales and axes to use.

Table 3. Introductions to level 1 and level 3 specifications for Application of Number

The specifications for each level are based on

- **Performance Criteria** (PC) - relating to qualities of performance which have to be met
- **Range Statements** - showing the extent of techniques and other dimensions of performance which have to be demonstrated .

Supporting these are **Evidence Indicators** to show some ways in which assessment demands can be met, and **Amplification** which spells out meanings of words and phrases in PCs and Range.

This style of specification is that adopted for all GNVQ specifications (see Table 4). The student is required to demonstrate mastery in all aspects of the specifications - not just to perform well on a sample of tasks. All these skills are assessed through their demonstration in the course of vocational work and assignments. Deliberately there are no stand-alone tests, since it is the application of mathematics skills in vocational contexts which is being assessed, rather than the skills in isolation.

PERFORMANCE CRITERIA

A student must:
1. make decisions about what data should be collected
2. choose and use **techniques** which suit the task
3. perform the **techniques** in a correct order
4. choose and work to an appropriate level of **accuracy**
5. record data in appropriate **units** and in an appropriate format
6. make sure that records are accurate and complete
7. identify sources of error and their effects

RANGE

Techniques:
- **Number:** describing situations (using fractions, using decimal fractions, using percentages, using ratios, using negative numbers); using estimation (to judge quantities, to judge proportions, to check results, to predict outcomes).
- **Shape, space and measures:** choosing and using appropriate measuring instruments and appropriate units of measurement for the task.
- **Handling data:** selecting an appropriate sample and designing and using a data collection procedure; obtaining data from written sources, obtaining data from people; discrete data, continuous data; handling large data sets.

Levels of accuracy: choosing and using an appropriate level of accuracy for each measuring task, choosing and working within appropriate tolerances for each measuring task.

Units: of money, of physical dimensions, of another property, of rates of change.

AMPLIFICATION

Choose and use techniques (PC2)
At this level, the student should be able to design and successfully undertake data collection procedures, deciding on appropriate levels of accuracy to work to…They should also be able to select an appropriate sample in order to obtain adequate, reliable data (*e.g.* decide at what stages of production, and how frequently, to sample items being put together on a production line).

Estimation to judge quantities and proportions (PC2 and PC3 range)
The student should be able to estimate quantities and size when collecting data (*eg* "this corridor is about 8.5m long", "if we stack up six of those boxes they will come to over 2m", "each file is just over 2cm thick"), and estimate fractions (*eg* "that's about a quarter") and percentages (*eg* "that's about 30%").

Estimation to check results (PC2 and PC3 range)
This should be used by the student to check whether results are reasonable, using knowledge of the context and/or the size of numbers which might be expected.

Data collection procedure (PC2 and PC3 range)
The emphasis is on designing valid and effective procedures for exploring problems and questions: the precise procedures can include use of a questionnaire (*eg* of people's reactions to the level of service an organisation provides); a survey using observation (*eg* the routes people take through a building); or an experiment (*eg* how a material responds to changes in temperature).

Errors (PC7)
The student should be aware of how errors can occur when information is collected: for example through inaccuracies in measuring procedures, the cumulative effects of using tolerances when measuring and recording, or through bias in the samples used. This level should give the student the opportunity to explore the effect of errors within large and small sets of data.

Table 4. Extract from the specification for Element 3.1 (level 3 element 1) - Collect and record data

SOME ISSUES

Underlying the drafting of these specifications are some important issues, including the following:

Mathematics first *v* applications first

Much mathematics learning is traditionally of the style 'learn the technique first, then look for applications to practice it'. Even in GCSE mathematics coursework or portfolio work, with its accent on using and applying mathematics, the bottom line is usually the demonstration of mathematics skills and processes in 'mathematics rich' and somewhat contrived contexts.

Such an approach is almost the reverse of the problem-solving approach of Application of Number, where the student becomes more and more responsible for selecting an appropriate mathematical technique to solve a particular problem. A contrived (though seemingly clever) use of mathematics in a vocational context can be demolished by comments such as ". . .but you would never do it that way - it doesn't make sense. . . . It is much easier just to do it this way. . .". For GNVQs, the selected mathematics should be a tool that enables a vocational problem to be tackled more effectively.

Union of skills *v* intersection of skills - the compromises of entitlement and usefulness

Each vocational area has its own set of mathematical skills which readily spring to mind as useful tools. However, as more vocational units are developed, the intersection of these sets becomes smaller and, if core skills are to be genuinely transferable, they should form a sufficiently substantial 'toolkit' in a variety of contexts. Hence there needs to be a compromise between the minimal 'intersection of skills' and the all embracing but less obviously applicable 'union of skills'.

Add to that another common expectation of mathematics learning - that students have an entitlement to certain aspects of mathematics as part of their education, even if these are not of immediate application, for example trigonometry or particular aspects of geometry because of the demands of some possible future courses. This may be in higher education, or because it is felt that students 'ought to know' some facts - and hence there are further arguments for including more and more mathematics content in the toolkit.

This, however, is to confuse a core curriculum with core skills. The more techniques are introduced which cannot be used 'naturally' then the more the Application of Number specification becomes a mathematics course and the less it remains a readily applicable core skills 'toolkit'.

The pressure for equivalence with GCSE and A-level

There is pressure to lay down equivalences, in order to give status to the new GNVQ qualifications, and to enable students to progress up the qualifications ladder - but equivalence in level of demand does not mean that the demands should be the same.

The specifications for Application of Number can appear mathematically undemanding compared with those for GCSE. However, this fails to take account of the enormous difference between demonstrating mathematics in the mathematics classroom or examination and using mathematics fluently as a tool in, for example, Art and Design. It also ignores the fact that the use of a mathematical technique is only part of Application of Number; the preparation of data and selection of techniques, the constant checking, the interpretation and presentation of results in the practical context - all these are just as important as the use of the techniques themselves.

Integration of Application of Number with vocational units - the human dimensions

The integration of core skills with vocational learning is not easy - there are many human tensions. For example, the art and design teacher - whose self image may be antipathetic to the perceived world of mathematics - and the mathematics teacher may find themselves close colleagues. Each has contributions to make to Application of Number, but each has to learn from the other so that student assignments can genuinely integrate vocational and mathematics learning. The mathematics teacher needs to see their subject as a tool for vocational work, which requires understanding of the objectives of the vocational tasks themselves, while the vocational teacher needs to acquire a broader view of what mathematical skills can include. Such readjustments take time and effort, and are demanding.

STYLES OF DELIVERY FOR APPLICATION OF NUMBER

There is a variety of teaching and learning styles, each with their strengths and weaknesses. The following are examples.

(1)   Students attend separately planned vocational and mathematics courses. In such arrangements the basic mathematical skills can be taught, but the crucial linking between the two courses is often weak. This is very much a 'mathematics first' style.

(2)   Students attend 'drop-in' mathematics workshops and have individual help from tutors when they encounter mathematical problems in their vocational courses. The help is more relevant than in (1), but there can be high demands on mathematics staff for time and knowledge of the vocational problems being tackled.

(3)   Vocational and mathematics staff jointly plan a unified course, where vocational assignments are designed together so that opportunities for the application of mathematics can arise naturally in them. This has the advantage that mathematics and vocational staff come to understand the underlying framework and approach of each other's learning programmes and work co-operatively. Each can   incorporate learning opportunities at the planning stage and students thus experience a unified course.

Although more difficult than the others to implement, because of the need for a commitment to working cooperatively, style (3) can be very successful.

AN EXAMPLE IN PRACTICE

The example in this section illustrates how hard it may be for students (and staff) to see mathematics as a genuine help in a vocational task. In one Art and Design course several assignments were concerned with marketing a proposed Nature Park. As one assignment, students were asked to design and make a small ceramic item (such as a coffee mug) together with its packaging, for sale at the Park. The item had to be attractive to buy, and be costed to generate income for the Park.

Students were asked to include as much of the required Application of Number as possible, with tutorial help from an experienced mathematics teacher; the learning style was an amalgamation of (1) and (2) above.

What happened in practice was that the students produced examples to illustrate particular mathematical techniques in use in the general context of the vocational topic, but without fully integrating the mathematics into their vocational assignments. Hence some of the examples appeared as 'add-ons', where the conclusions had no immediate consequences for the main vocational work.

The work of one student illustrates these points.

For instance, although there was a clear appreciation of the practical problems of packaging:

- calculations of wastage of card for packing were made but this did not appear to lead to any judgements about how, or even if, the packaging should be redesigned,
- area calculations demonstrated the use of a formula but did not appear to lead anywhere,
- calculations of volume of the ceramic item were carried out but it was not clear where those calculations were leading,
- misuse of formulae was not noticed, since the results had no consequences.

With this experience behind them, the staff concerned are now redrafting the vocational assignments so that mathematics can be used more effectively as a tool and integrated into the vocational assignments. Such experiences are to be expected as staff and students come to terms with a very different image of the role of mathematics from the customary school one.

CONCLUSIONS

The development of GNVQs, and the core skills requirements in particular, have been very rapid. It is easy to point out shortcomings in present practice, and use these as a justification for reverting to more tried and tested approaches. However, this would be to forget that the goal is to enable students to develop their confidence and ability to apply mathematics as a useful tool in a variety of contexts, rather than simply to develop the tools.

REFERENCES

ALBSU, 1993, in Dearing R, 1993 **The National Curriculum and its Assessment: An Interim Report**, Schools Curriculum and Assessment Authority, HMSO. London

Cockcroft WH, 1982, **Mathematics Counts**, HMSO, London.

Green A and Steedman H, 1993, **Education Provision, Education Attainment and the Needs of Industry: A review of research**. National Institute of Economic and Social Research. London

National Council for Vocational Qualifications, 1995. **Application of Number specifications**, NCVQ, London.

14

Mathematics as Orientation in a Complex World

Hans-Wolfgang Henn
Staatliches Seminar für Schulpädagogik, Karlsruhe, Germany.
e-mail: za164@lehrer1.rz.uni-karlsruhe.de

ABSTRACT

Besides the teaching of basic numeracy, the most important goal of mathematics education is to develop the ability to see the possibility of using mathematical methods to solve the problems which arise in all aspects of life. After school, everyone should be able to use mathematics as a tool to understand the world better, and as an aid to more intelligent action both privately and professionally. A survey of the present situation in Germany relating to modelling and applications at secondary level, especially in Baden-Württemberg, is given. Three examples illustrate how mathematics can improve the orientation in our world.

INTRODUCTION

About 20 years ago, Heinrich Winter's seminal paper *Allgemeine Lernziele für den Mathematikunterricht* was published (Winter, 1975). This formulated four primary objectives for a mathematics lesson. The lesson should give the pupil opportunities

- to be active in a creative way,
- to present rational arguments,
- to learn the practical applications of mathematics,
- to acquire the ability to formalise mathematics.

These considerations were given a great deal of attention in German speaking countries. Erich Wittmann proposed in connection with this that pupils should learn how to 'mathematise' (Wittmann, 1981). The Kassel team, based around Werner Blum, has influenced the didactic discussion in Germany for many years, particularly regarding an applications orientation for mathematics lessons. The ICTMA series of conferences reflects the international state of the debate (Blum, 1995). Applications and modelling are gradually entering into everyday lessons but, as yet, they do not have the necessary impact.

Mathematics lives and develops through its connections with reality. Mathematical concepts and objects generally arise due to intra- or extra-mathematical questions, often emerging from a desire to understand nature better. The strength of mathematics lies in its twin pillars of formal structures and concrete applications. The Swiss mathematician Armand Borel drew a very beautiful picture during the International Congress of Mathematicians in Zurich in 1994 (Borel, 1994):

> Mathematics resembles an iceberg: beneath the surface is the realm of pure mathematics, hidden from the public view... Above the water is the tip, the visible part we call mathematics.

The declaration of the UNESCO conference World Mathematics Year 2000 held in Rio de Janeiro, stated that "Pure and Applied Mathematics is one of the main keys to understanding the world and its development".

Reality and mathematics are two distinct worlds. The process of modelling always results in the representation of a part of the world of experience as a mathematical model which exists in the world of the intellect. In modelling, not all aspects of the reality can be modelled - considerable facets of the reality are lost. Every model must be checked and validated critically, modified if necessary and evaluated. An essential characteristic of modelling is that the phenomena under study become more and more complex and interconnected. However, hardly anyone is willing and able to apply mathematical methods, even in the simplest cases.

A typical example is making comparisons and choosing between different ways of financing a loan. A transfer problem becomes obvious here: even though students have studied a particular mathematical method or algorithm, they are not necessarily able to apply it, *ie* to transfer the acquired theoretical knowledge to the practical situation. Mathematical lessons must allow diverse experiences of how mathematics can be called upon in modelling to achieve better understanding of phenomena which are not primarily mathematical. Only then can students develop the ability to transfer between reality and mathematics; only then can they make their own judgments about relevant questions and contribute as future discussion makers, knowing that the process of 'mathematisation' is compatible with environmental and social structures.

In schools in Germany, and in most other countries as well, the situation is considerably different from this ideal. "A ridiculous thing anyway, this mathematics at school - of no value to most people, a remote thought game, a pain, missing an attitude, missing a goal, missing a relationship with life", remembers the writer Alfred Döeblin (Endres, 1990). The mathematics lesson is often an experience separated from everyday life. Mathematical reasoning is established there which has nothing to do with everyday reasoning. Often pupils experience mathematics as meaningless calculations which one forgets as quickly as possible after the examination. Syntactic transformation without semantic relationship is, however, senseless. Presumably owing to such experiences, Schopenhauer understood arithmetic as the lowest of all mental jobs.

Calculus lessons are a typical example of this. The aim seems to be to make pupils into extremely slow and faulty computer-algebra systems who handle examination questions that like this are typically: a given function is differentiated, the derivatives are examined for zeros, and from this the extrema are obtained. A question asking, for example, for the derivative of $y = x^3 - 3x^2 + 1$ is answered immediately with the correct answer $3x^2 - 6x$, but one dare not ask what the derivative really means! Only empty formalisms are remembered, but these are adequate for survival at school. This is similar to learning to read the Greek language without understanding the meaning of the words. Newton did not invent calculus so that he could discuss curves but to be able to deal with his mathematical models of nature, namely differential equations.

Freudenthal rightly demands that pupils should not learn applied mathematics, but should learn how to apply mathematics. Only their own active engagement can, in the end, make a difference for pupils. This requires a move away from teacher-orientated lessons to pupil-orientated lessons, as the well-known quotation from Confucius (551 - 479 BC) points out:

> Tell it to me and I forget it;
> Show it to me and I recall it;
> Let me do it and I remember it.

We should motivate pupils to ask questions, and to experience for themselves the more inductive aspects of mathematics like investigation and experimentation, generalisation and specialisation. Therefore not the variety, but the quality of the examples and their analysis, is crucial.

MODELLING AND APPLICATIONS AT GERMAN HIGH SCHOOLS, ESPECIALLY IN BADEN-WÜRTTEMBERG.

Since the 1970s, suggestions and proposals have been made (especially by British educationalists) to teach mathematics in such a way as to ensure that pupils learn to make connections between mathematics and reality. Even so, the idea of a mathematics lesson involving real-world problems enters the teaching curriculum only very slowly. The Federal Republic of Germany comprises 16 federal states which are independent in matters of education and culture. This means that in Germany there are 16 different school systems, teaching curricula and degree examinations. Only very general conditions are agreed by the state ministers of education. This paper reports especially about the conditions regarding applications-orientation in Baden-Württemberg. More precisely, the paper refers to the 'Gymnasium', which pupils attend for nine school years after the four-year elementary school. The 'Abitur' is the final examination of the Gymnasium.

In Baden-Württemberg we are in the fortunate situation that our new curriculum, which has recently been introduced, sharply emphasises the importance of applications-oriented mathematics lessons. Four or five curriculum units (for example, 'fractions' and 'differentiation') are prescribed for every school year. One of these is called *Mathematik in der Praxis* (the application of mathematics in everyday life). In this unit, in a manner appropriate to the level of the pupils, one has to deal with the mathematisation of real problems and the interpretation of mathematical statements in concrete situations.

However, the teacher cannot rely on text books in order to fill the curriculum with real-life situations. While it is true that, in most German text books, one can find 'applications' after 'theory' and 'mathematical problems', usually these are highly structured pseudo applications. The process of modelling has taken place already, the methods of solution are explicitly or implicitly given, and any discussion that takes place is almost all about the mathematics. Therefore extensive continuing professional development courses for teachers are required. I am responsible for a work group, Applications Oriented Mathematics Lessons, whose members contribute case studies from their own experiences. This work group is part of the German ISTRON section (ISTRON, 1994). We seek to find out how average pupils can apply mathematics to real-world problems in the situations presented in everyday classes. Teachers must be willing to ask their pupils 'more open questions', and it must not be certain from the beginning which method they shall use. Different answers of different quality must be possible, and we consider how to incorporate such questions into normal written class tests. This requires different assessment competences on the part of the teachers, but no more than is required of the teacher of the mother language. There are some problems with the transition to this way of doing things. Due to federal budget cuts only a few new teachers were employed within the last 15 years, and this has resulted in bigger classes with an ageing teacher population. Many teachers feel over taxed and unable to handle problems outside their traditional domain.

A particularly relevant obstacle in Baden-Württemberg, and four others states, is the fact that central Abitur examinations are held. This means that pupils throughout the country are taking the same examination questions on the same day. Central examinations lead to cramming and to question spotting, with many teachers only teaching material that will be assessed. The Abitur questions of the last ten years can so easily become the hidden teaching curriculum. Preparing pupils in such a way leads to good grades in the Abitur, and so there is no trouble from parents or directors. However pupils quickly forget techniques taught in this way, no matter how well they were justified they have not learned any skills which could be useful to them in their working lives. In addition we have, in the past, educated ministers, presidents and tutors of universities who, when they open conferences and congress, are not ashamed to say that they were bad at mathematics. At the moment our working group is writing new style Abitur questions which are more open and closer to real life. Such problems will be set increasingly in years to come.

A LOOK INTO THE CLASSROOM

In the following I will show how mathematics gives orientation or direction in a complex world. You can find many excellent examples in newspapers. On the one hand, the disregard of elementary techniques leads to grotesque misinterpretation; on the other hand, simple mathematical knowledge suffices to check statements, for example in advertisements, for credibility.

Three examples of modelling and world problems will follow, which we currently use in calculus lessons at upper secondary level (age 17 to 19). The examples belong to the students' world. Mathematical models are able to make the connections more obvious, and abstract mathematical concepts are illustrated and made more concrete. All three examples illustrate typical modelling problems. The present German Income Tax law shows how a small modification on the mathematical side can cause a large violation of the model constraints. In a model of fuel consumption one has to go round the modelling cycle several times. Finally in a model of motor car crash tests, the simplifying assumptions made in modelling the head injury criterion are such that the results are unproductive, despite considerable mathematical effort.

Mathematics is everywhere

The German journal *Wochenpost* reported in 1995, "Every third marriage in Germany ends in divorce; indeed in big cities it is very fourth marriage".

The former German minister for the economy, Martin Bangemann (now an EU commissioner) explained to an astonished audience in 1987, "The peak level of the dollar is now 3.47 Marks but two years ago, before the events on the stock exchange, we had an exchange rate of 1 dollar to about 1.80 Marks. This means the dollar has lost nearly 100% of its value."

These examples show how the use of everyday reasoning stands in opposition to mathematical reasoning.

The handling of significant figures is another thing that is done poorly. In a press release dated 28th April 1985, the USA Post said that the size of its eagle head stamp (Fig. 1) is 48.768 × 43.434 mm. Apart from the incorrect unit of measurement for area, it is senseless to describe the dimensions of a piece of paper to the thousandth part of a millimetre - even the variation in length caused by moisture in the air is greater than this. The reason is the thoughtless conversion of the original size of 1.92 × 1.71 square inches into metric units by multiplying each dimension by 25.4. Similarly one could give many more examples of this sort of abuse, including instances in chaos theory.

Figure 1: USA postage stamp of 1985.

Manipulated graphs which appear to illustrate objectively the desired statement are often published. This advertisement of the Peugeot company (Fig. 2) misleads with a clever scale on the y-axis which suggests that Peugeot cars need significantly less fuel than their competitors' cars.

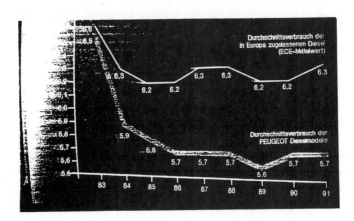

Figure 2: fuel consumption illustrations.

The graph in Fig. 3, showing the origin of the y-axis, presents a somewhat different picture.

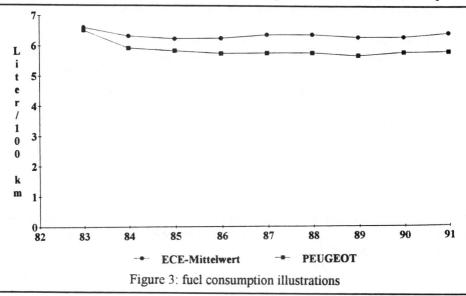

Figure 3: fuel consumption illustrations

The German Income Tax function

There are only a small number of principles governing a fair tax system. Two of these are as follows.

- Taxation should be progressive - the amount of tax should grow with increasing income both in absolute and relative terms.
- Taxation should take family status into account.

The German Income Tax tariff has been defined in terms of polynomial functions, making it unique in the world, I believe. Calculus can be used to check if the above-mentioned principles are embodied in the mathematisation of the tax problem (Henn, 1988, 1995a). The tax to be paid, in DM, is a function $t(x)$ of x, the yearly income in DM.

$$t(x) = \begin{cases} 0 \\ 0.19x - 1{,}067 \\ 151.94\left(\dfrac{x-8{,}100}{10{,}000}\right)^2 + 1{,}900\left(\dfrac{x-8{,}100}{10{,}000}\right) + 472 \\ 0.53x - 22{,}842 \end{cases} \text{for} \begin{cases} 0 \leq x < 5{,}617 \\ 5{,}617 \leq x < 8{,}154 \\ 8{,}154 \leq x < 120{,}042 \\ x \geq 120{,}042 \end{cases}$$

The graph in Fig. 4 is a monotone increasing curve that does not tell us very much. Neither does the diagram show the small gaps at the borders of the intervals.

If, for example, someone earns 100,000 DM, they have to pay 30,765 DM in tax, which corresponds to a mean rate of 30.8% . This is the gradient of the secant through the origin in Fig. 4. This secant gradient, $s(x) = t(x) / x$, is therefore the mean tax rate. If the same tax payer earns 10,000 DM more, then they have to pay 35,610 DM in tax; the mean rate only rises to 32.4%. However, if we consider the increase $\Delta x = 10,000$ DM in income, the increase Δt in taxes is

$$\Delta t = t(110,000) - t(100,000)$$
$$= 4845 \text{ DM}$$

and this corresponds to a tax rate of 48.5% on the increase. This is the gradient of the second secant in Fig. 4.

Taking the limit $\Delta x \to 0$ leads in an obvious way to the definition of the marginal tax rate $s^\wedge$ with $s^\wedge(x) = t'(x)$. Thus $s^\wedge(x)$ shows, in an intuitive way, the extra amount taken in tax for each extra Mark earned.

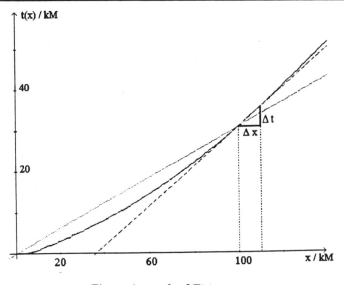

Figure 4: graph of $T(x)$ versus x

The graphs of s and $s^\wedge$ are drawn in Fig. 5, and we see that the first principle of a progressive tariff is met.

In order to satisfy the second principal, regarding family status, German tax law prescribes the splitting system which came originally from the United States. In the splitting system the incomes x_1 and x_2 of a married couple are added together and then tax is calculated by using the splitting formula.

$$2t\left(\frac{x_1 + x_2}{2}\right)$$

To ensure that married people are not disadvantaged, the splitting tax must be less than or equal to the sum of the individual taxes $t(x_1) + t(x_2)$.

Thus $\qquad\qquad\qquad t\left(\frac{x_1 + x_2}{2}\right) \le \frac{t(x_1) + t(x_2)}{2}$

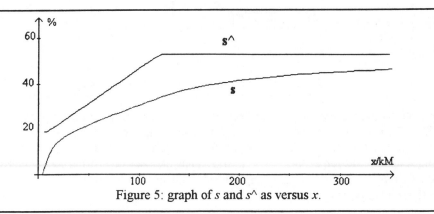

Figure 5: graph of s and $s^\wedge$ as versus x.

The inequality implies that the graph of $t(x)$ must be concave upwards and the diagram confirms this.

So far so good!

Early in 1994 the German Income Tax laws were changed a little. Under the previous tariff, an unmarried citizen had to start paying taxes with an annual income of 5617 DM. However this tax threshold was less than social welfare subsidies and, in order to rectify this inequity, the highest German court, the *Bundesverfassungsgericht*, required a modification of the tax law. The Minister of Finance had to comply with this demand, but he tried to do this in a way which lowered the tax yield as little as possible. Thus the new tax tariff includes some inconsistencies for small incomes which are unique in German taxation history. The tax function has been changed only for $0 \le x < 15{,}012$:

$$t(x) = \begin{cases} 0 \\ 0.612(x-12,041) \end{cases} \quad \text{for} \quad \begin{cases} 0 \le x < 12,042 \\ 12,042 \le x < 15,012 \end{cases}$$

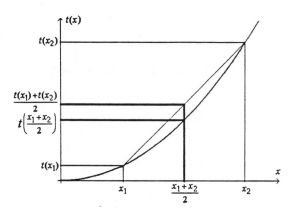

Figure 6: graph illustrating the splitting system.

We see that the graphs of t and s become concave downwards in the lowest income band (Figs. 7 and 8). A consequence is that a married couple with low earnings employing the splitting function have to pay more tax than an unmarried couple. The graph of $s^\wedge$ even jumps up to 0.612 for $12,042 < x < 15,012$. Thus the poor low earner has a relative tax rate of 0.61 (Fig. 8). The seemingly insignificant mathematical change has lead to a violation of both essential principles.

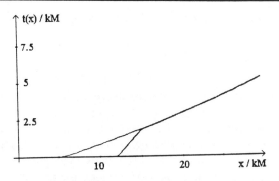

Figure 7. graph of new situation - $t(x)$ versus x

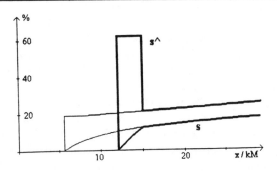

Figure 8: graph of new situation - s and $s^\wedge$ versus x.

Consumption of Fuel

The analysis of the fuel consumption of a car provides a good example of a situation where the pupil can test their own 'theory' and thus go through the complete process of 'mathematising' (Henn, 1995b). Students are prompted to consider both fast driving, which damages the environment, and also their finances.

We start with a fuel consumption diagram (Fig. 9). Diagrams like this are given in the operator's manuals of some cars. This diagram shows the consumption curves for three different models of the same car. It indicates the fuel consumption (in litres per 100 kilometres) when the car is driven at constant speed v km/h in the highest gear. The curve only begins at 40 km/h since average driving is only possible in this gear from this speed on.

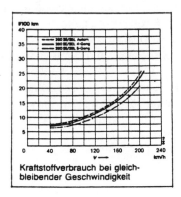

Figure 9: graph of fuel consumption versus speed.

Model I: air resistance

It is rather senseless to postulate the equation of the curve without a 'consumption theory'. This theory should be as simple as possible, and should require a knowledge of only elementary physics. Fuel has 'stored energy' which the engine transfers into 'usable energy'. Energy stored in the fuel is used to do work W on the car, where $W = force\ F\ times\ distance\ s$. The force F is divided into two parts F_1 and F_2.

F_1 is the basic part, which is needed to compensate for the friction of the moveable parts of the engine, the transmission, the axles, the wheels, and so on, and does not depend on speed. F_2 is the part which is needed to compensate the air resistance to the car during the drive. Unlike F_1, which is independent of speed, F_2 increases (approximately) with the square of the speed. The corresponding approximation formula is in secondary-level physics books. Pupils who do not know this formula can collect data regarding maximum engine power and maximum speed for as many cars as possible. Using these data, they can check that an increase in engine power by a factor k only causes an increase in the maximum speed of about $\sqrt{k}$. Thus we are able to start with an empirical model. The fuel consumption $B(v)$ at a constant speed v is given as $B(v) = a + b\ v^2$ where v is measured in km/h and B is measured in litres/100km. The constant a corresponds to the constant part $W_1 = F_1\ s$ while the second term corresponds to $W_2 = F_2\ s$, the work done against air resistance.

Next we need to examine how well this theoretical model describes the actual, empirically-measured consumption. The constants a and b have to be found so that the corresponding parabola is a good fit to the consumption curve. With the possibilities available in school, there is no meaningful method which allows us to check this fit of the curve - with one exception, that is when the curve is a straight line. A ruler is all that is now needed to compare the data with the curve. Here we start with the points on the consumption curve. We cannot judge the possible position of these points on a parabola, so we have to produce points by linearisation, which is an important method of data analysis. To do this we read from the diagram the values of B corresponding to $v_1 = 40$, $v_2 = 60$, and so on. The points (v_i, B_i) are on a quadratic parabola with an equation corresponding to our theory if and only if the linearised points (x_i, y_i), with $x_i = v_i^2$ and $y_i = B_i$, are on a straight line.

Fig. 10 shows the graphical representation. The linear model for the points (x_i, y_i) is justified in a first approximation so that we can draw a best-fit line from which we can deduce a (the y-intercept) and b (the gradient). With this our simple fuel consumption theory has passed its first validation test.

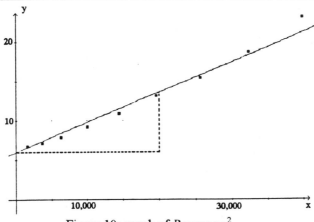

Figure 10: graph of B versus v^2.

Model II: engine efficiency

A more exact analysis of the above illustration shows that the measured points differ in a systematic way from the best-fit line. This indicates that our theory is still incomplete. We modify our model by considering engine efficiency: car engines are so structured and optimised that they are more efficient at high revolutions. Lower speeds, and thus lower revolutions, result in poorer engine efficiency. We can model this, to a first approximation, by inducing a negative term proportional to the speed:

$$B(v) = a - b\,v + c\,v^2$$

The constant corresponds to the constant frictional forces, and the v^2 term to the air resistance, as included in the model above. The linear term attempts to take into account the diminishing influence of increasing speed on the fuel consumption by the improving efficiency of the engine.

The estimation of values for a, b and c is, however, more difficult. One possibility is again to consider a parabola as a fit to the empirical curve, but to take the vertex at the first point $(\frac{40}{6}, 8)$ as suggested by Fig. 9. This selects one of the parameters and we now have

$$B(v) = d + e\,(v - 40)^2$$

This is linearised as above, by drawing the points (x_i, y_i) with $x_i = (v_i - 40)^2$ and $y_i = B_i$ (Fig. 11).

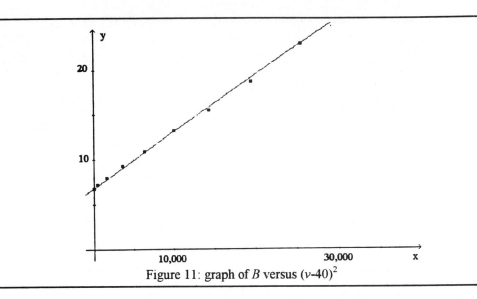

Figure 11: graph of B versus $(v-40)^2$

Fig. 12, which contains the parabolas for model I and model II, shows the improvement.

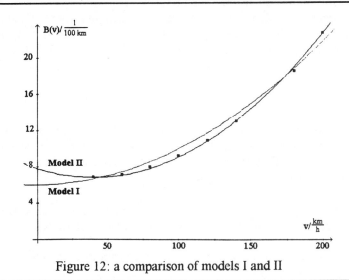

Figure 12: a comparison of models I and II

Crash tests and head injury criterion

Fortunately many consumers pay more attention to the safety of cars then to spoilers and fancy trim. Crash tests can give valuable hints on the advantages and disadvantages of the way a car is constructed. An index number, the HIC (Head Injury Criterion), is often quoted in tests and reports (Table 1).

Dummy-Results	VW Golf III	Opel Astra	Renault 19	Seat Toledo	Citroen ZX	Ford Escort	Fiat Tipo
Head injuries, driver (HIC)	455	901	914	709	1368	516	818
Head injuries, passenger (HIC)	287	429	265	694	893	710	471

Table 1. HIC for various cars.

Not only can this number serve as a valuable eye-opener for consumers, but it can also be used to motivate a Riemannian integral (Henn, 1994).

The HIC is used to assign a numerical value to the risk of head injury. The force on the head in an accident results from a high (negative) acceleration during the crash. The dummies in crash tests have several sensors in the head area which record the deceleration as it varies with time. It seems plausible that the force on the head is greater for higher values of deceleration and if the deceleration lasts for a longer time. Some empirical values are known from tests with animals, cadavers (with some ethical questions to be answered!) and from the evaluation of head injuries to boxers.

Fig. 13 is an acceleration-time diagram and, when shown to students, they usually propose, (as a first model) to measure the area below the curve. They often do this before integration has been taught to them.

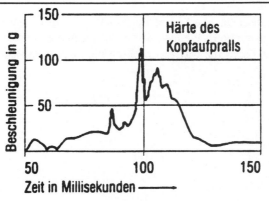

Figure 13: acceleration versus time.

However, looking at the two graphs in Fig. 14, which contain the same area, the spiky graph looks 'more dangerous', and thus they will recognise the problem of the weighting of the acceleration.

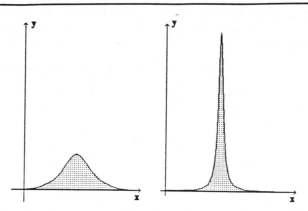

Figure 14: distorting graphical output.

The first model used in practice was the Severity Index (SI) model. The value of the acceleration was raised to a power n, and the value of n depended on the part of the body. For the head, $n = 2.5$. The SI was defined in mathematical terms as

$$SI = \int_0^T \{a(t)\}^n \, dt$$

where a(t) is the acceleration at time t and T equals the length of time that the deceleration has an effect on the head.

However the validation of this model was not satisfactory with regard to different car types and different accident situations. We look first at the mean acceleration between times t_1 and t_2 (Fig. 15). This also paves the way for teaching the relationship between integration and mean value.

$$\bar{a} = \frac{1}{t_2 - t_1} \int_{t_1}^{t_2} a(t) \, dt$$

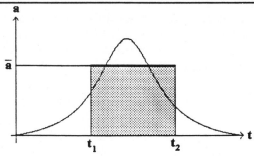

Figure 15: illustration of a and $\bar{a}$

The factor $(t_2 - t_1)(\bar{a})^{-2.5}$ takes into account both duration and weighted value of the acceleration for the time interval from t_1 to t_2. The maximum of all such numerical values is the Head Injury Criterion:

$$\text{HIC} = \max_{t_1, t_2} \left[(t_2 - t_1) \left\{ \frac{1}{(t_2 - t_1)} \int_{t_1}^{t_2} a(t) \, dt \right\}^{2.5} \right]$$

Besides the condition $0 \leq t_1 < t_2 \leq T$, the beginning t_1 and the interval $(t_2 - t_1)$ are arbitrary. To be able to evaluate the curve, one must make simplifying assumptions as follows.

- We require $t_2 - t_1 \leq 36$ ms; according to experience, longer acceleration times do not increase the risk of injury.
- Top acceleration values must be held for at least 3 ms.
- It is believed that accelerations of shorter duration do not have an effect on the brain and, besides, it is impossible to measure them.

The computation of the HIC from an empirical curve is, however, still very laborious.
The curves in Fig. 16 show the results of crash tests for two E-class Mercedes Benz, one without an airbag and one with an airbag. In each case the rectangle corresponds to the interval used to determine the HIC.

The airbag reduces the HIC from 682 to 308. Also the maximal acceleration value which lasts for more than 3 ms (the A-3ms value in Fig. 16) decreases from about 90g to 43g (where g is the acceleration due to gravity).
Despite the complicated formula and the huge computer used for the calculation, the very precise numerical results imply a meaning which crash-modelling using the HIC does not have. The real situation was simplified considerably in the modelling. In fact only the

acceleration curve, obtained from the use of dummies, has influenced the model. Experts agree, however, that HIC values above 1000 are definitely life threatening . Furthermore the range of variation of the HIC is very large. HIC values are suitable only for a rough comparison of two cars, and they do not predict the nature and severity of injuries. In short the HIC only gives a rough estimation of the general injury risk.

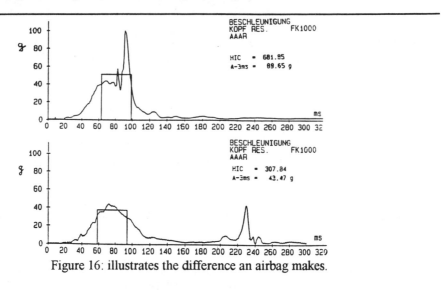

Figure 16: illustrates the difference an airbag makes.

CONCLUSIONS

To make orientation in a complex world possible, mathematics curricula must include applications. One should always consider, however, whether or not the real-world problems set are accessible to students, given their present state of mathematical knowledge and understanding of the world. For example, the models on fuel consumption could be presented as applications of quadratic equations to pupils in class 9, but our experience indicates that pupils of this age know little about driving and so are not enthusiastic about this problem. The pupils of class 11 (age 17 to 18) usually have a driving licence and are interested in questions which relate to cars and traffic. The same applies to the Income Tax problem. Although mean tax rates and marginal tax rates are an ideal example of the transition from mean to local rate of change, our experiences from teaching this suggest that it should not be introduced into the calculus course. The pupils of class 11 do not pay taxes, the reality of employment is still far away, and so they are not interested in the topic. On the other hand, this topic is very suitable for class 13 (age 19 to 20) as, for example, an inter-disciplinary project. These pupils are beginning to choose a career, and are also becoming aware of politics and economics. These well-known mathematical methods open their eyes to aspects of economics.

In our experience the modelling of crash test data using the HIC with pupils in class 12 proved to be highly motivating, and forged a useful link with the integral calculus.

In all lessons, pupil participation must be paramount. Teamwork is often possible and this is an important preparation for professional life. Extensive examples, like the three presented in this paper, must not be isolated. A wide field of analogous exercises, with similar or more complicated questions, must be at the teacher's disposal. For example, with the fuel consumption model, this can be an analysis for other types of car, and with the tax example it could be the analysis of other tax functions (which are readily available due to frequent changes in the law.)

Again and again, it is pleasing to see how pupils, who are normally rather uninterested in mathematics, participate critically and actively in the lessons. Of course, it is necessary that the teacher takes all aspects seriously, both the application and the associated mathematics. The discussion that takes place must be about the modelling as well as about the mathematics. These pupils will then experience mathematics with all its consequences and manifold applications.

REFERENCES

Blum W, 1995, *Applications and Modelling in Mathematics Teaching and Mathematics Eduction - Some Important Aspects of Practice and of Research,* in Sloyer C, Blum W and Huntley I (eds.)**Advances and Perspectives in the Teaching of Mathematical Modelling and Applications**, Water Street Mathematics, Yorklyn, Delaware.

Endres EA, 1990, **Mathematik mit Methode**, Beltz Weinheim.

Henn H-W, 1988, *Einkommensbesteuerung aus mathematischer Sicht*, **Zentralblatt für Didaktik der Mathematik**, 20, pp 148 - 163.

Henn H-W, 1994, *Auto und Verkehr. Beispiele zum realitätsnahen Mathematikunterricht*, **Berichte über Mathematik und Unterricht**, ETH Zürich, Bericht No. 94-04.

Henn H-W, 1995a, *Einkommensbesteuerung als mathematisches Problem*, **Mathematik in der Schule, 33**, pp 167 - 175

Henn H-W, 1995b, *Benzinverbrauch eines Pkw*, **Mathematik Lehren, 69**, pp 54 - 57

ISTRON: **Publications of the German ISTRON-section:**
        Anwendungen und Modellbildung im Mathematikunterricht:
                Blum W. (Hrsg.): 1994;
        Materialien für einen realitätsnahen Mathematikunterricht:
                Blum, W. u.a. (Hrsg.): Band 1, 1994;
                Graumann, G. u.a. (Hrsg.): Band 2, 1995;
                Bardy, P. u.a. (Hrsg.): Band 3, 1996,
Franzbecker-Verlag, Hildesheim.

Winter H, 1975, *Allgemeine Lernziele für den Mathematikunterricht*, **Zentralblatt für Didaktik der Mathematik**, 7, pp 106 - 116.

Wittmann E, 1981, **Grundfragen des Mathematikunterrichts** (6. Auflage), Vieweg-Verlag, Wiesbaden.

15

The 'Mathematical Modelling' Course for Russia's Schools: its Aims, Methods and Content

EK Henner and AP Shestakov
Pedagogical University of Perm, Russia
e-mail: henner@pspi.perm.su

ABSTRACT

The course in Mathematical Modelling is designed for the third level of school(forms 10-11, high school) and for the training and re-training of teachers of Maths and Computer Science (Informatics) in higher pedagogical establishments. The textbook and corresponding software have been developed for the IBM PC and for the small Russian-made educational computers UKNC and KORVET. The school course is most effective at those schools (Lyceums, Gymnasiums) which are oriented towards in-depth mathematical education. Such schools offer applied courses which integrate various subjects and thus help to develop students with a broad range of knowledge. The main idea of the course is the deepening of students' knowledge in mathematics, physics, ecology, economics and in some other fields, as well as developing their general IT skills. All models are implemented by using the methods of applied mathematics.

INTRODUCTION

During 1992-94 we developed two courses in Maths Modelling - one for the third level of school (forms 10-11, high school, 16-17 year-olds) and the other for training and re-training teachers of mathematics and computer science (Informatics) in higher pedagogical establishments. These establishments are named 'Pedagogical Institutes', and there are about 70 of them in Russia. The best, about 15, are named 'Pedagogical Universities'. The duration of study is 5 years, and most of the students are 17-24 years old. We also did experimental testing of the courses and developed a textbook and corresponding software for the IBM PC and for the small (Russian-made) educational computers UKNC and KORVET. This work was supported by International Soros Foundation.

The development of this former course was undertaken for the following reasons. The course in Informatics (Computer Science) is a compulsory one at most Russian schools of the second level (forms 7-9). It is necessary to continue this course in the schools of the third level, and Mathematical Modelling is one of several possible ways of doing this.

The aims of our school course are:

- to promote the development of students' mental abilities,
- to gain hands-on experience with computers including the creation of simple software,
- to deepen students' mathematical knowledge,
- to demonstrate the existence of universal methods of cognition in different areas of human activity.

The foundations of the school course in Mathematical Modelling

A.   Mathematics

Traditionally, the level of mathematical education in Russia is high. There are two compulsory mathematics courses in the 10th and 11th forms. The first of these is Algebra and Mathematical Analysis, the second is Geometry. From the 10th form we have two courses, recommended by the Federal Ministry of Education, at two different levels. Course A is for ordinary mathematical education, course B for in-depth study.

B.   Informatics (Computer Science)

The first experiments in using computers in education and in school management began back in the sixties. In specialised schools, a new subject 'Programming' was taught. Universities and other higher educational institutions invented a computer management system. However, such activities were not common, especially in schools.

In 1985, with the beginning of perestroika, the first Federal Programme of Computerisation of School Education was adopted. Its aim was to improve IT provision within high schools and hence introduce a new compulsory subject called 'Principles of Informatics' into the curriculum. The programme also dealt with the development of methodology and software, both for that subject and for the introduction of computers into the study of other disciplines.

In the short period of time from 1985 until 1994 the subject 'Principles of Informatics' developed greatly. A third generation of textbooks and other educational and methodological materials is now appearing, which reflects modern trends in the organisation of the course.

Beginning from the academic year 1994/95 this compulsory Informatics course was shifted from the high school to the secondary school (that is, for 7th to 9th formers). The federal standard requires a short compulsory course of approximately 70 academic hours. The question of further computer education and its development is left to the discretion of the local schools and educational authorities. This has led to a heated discussion about the purpose and the content of the course.

This compulsory course must represents the main directions of modern Informatics:

- information,
- computer technique,
- algorithms
- programming,
- modelling,
- cybernetics,
- social,
- natural science,
- mathematics.

We are sure that such a course will aid the development of the basic practical skills and intellectual qualities of a person. One variant of this course ('The Perm version') was developed by a group of authors which included EK Henner. This work was also supported by the International Soros Foundation and was published as a teacher's book in 1995.

This basic course makes it possible to study one of several optional courses at the second level in high school (the 10th and 11th forms) namely Programming, Computer Graphics, Databases and some others. One option is the course in Mathematical Modelling which is discussed in this report.

THE SCHOOL COURSE IN MATHEMATICAL MODELLING

Basic principles

The courses in mathematics and computer science mentioned above are the basis for our high-school course 'Mathematical Modelling' We consider it to be one of the many possible courses which serve as the continuation of the second level Informatics. It is intended for the high school and can be successfully undertaken at the schools (Lyceums, Gymnasiums) which are oriented towards in-depth mathematical education. By its nature it is an applied course, which may integrate different subjects and thus help to develop students with a broad range of knowledge.

The main idea of the course is the deepening of students' knowledge in mathematics, physics, ecology, economics and in some other fields, as well as developing their general IT skills. All models are implemented by using the methods of applied mathematics.

The integrity of the course is assured by the following.

1.   Methodological integrity. During the course we try to maintain the traditions of mathematical modelling at every stage of the procedure.

2.   Computer realisation of models. All of us are aware that a computer is not always needed for the realisation of a mathematical model. Professionals prefer analytical solutions of mathematical models, but this is usually associated with a very high level of mathematical training. Such a level is not applicable at the high school so we use the simplest numerical procedures and implement them on a computer or use pre-written software.

We have provided two variants of this course. The first includes the original development of software by every student (including programming of mathematical methods). This is possible only for schools and/or individual students that are oriented towards in-depth maths and computer science education. The second variant allows the use of pre-written programs. Usually, however, it is necessary to combine these approaches.

Content of the course

We now provide some details of our school course.

Definitions: what is a model and what is a mathematical model?

Aims of mathematical computer modelling:

- understanding the organisation of a concrete object or process, its structure, properties, and its interaction with its environment,
- learning the methods of control for an object or process, and the optimisation of control,
- prediction of the change in condition of an object or process in the course of its evolution.

Different classifications of mathematical models:

- branches of sciences,
- mathematical methods,
- aims of modelling (descriptive models, optimisation models, multi-criterion models, games models, imitation models).

Stages of mathematical modelling (using a computer):

- initial object or process,
- ascertaining the aims of the modelling,

- gradation of parameters,
- searching for a mathematical description which leads to the creation of a mathematical model,
- choosing the method of investigation,
- choice or construction of software,
- debugging and testing the program,
- calculations on a computer,
- analysing the results,
- more precise definition of the model.

We include in our course some additional mathematical information. The standard course in school mathematics does not contain some of the topics which are necessary for our course. The method of presenting teaching material is not the same as in the usual mathematical courses. It is not an insipid mathematical text but rather contains many remarks, pictures and examples. We recommend that school teachers do not study this material independently but use it only when it is necessary for the basic part of the course. The corresponding topics are: differential equations and the simplest numerical methods of solving them, numerical methods of solving linear algebraic equations, approximation of functions (interpolation, least-squares method), introduction to probability and mathematical statistics.

For schools with an in-depth mathematical education programme our course has a duration of 2 years, with approximately 70 hours per year.

The first year is devoted to mathematical modelling of physical processes and to mastering the methodology of computer modelling as a whole. It is well known that in physics computer modelling is now an important method. All problems are considered on the basis of the usual school course in physics but without some of the limitations which are typical in such a course. The main mathematical tools are differential equations and numerical methods for their solution. Approximately two thirds of the time allotted for this course is occupied by computer-based experiments.

The first problem which we consider is the descent of bodies in a medium. Students try to assimilate the methodology of mathematical computer modelling. One of the main difficulties which it is necessary to overcome is the classification of factors according to their significance.

After that, it is necessary to make up a mathematical description of the movement. The students understand that the basis of this is Newton's second law but now they need it in differential form. This situation is unusual and it is necessary to understand the reasons and consequences of it.

As was mentioned before, we use only the simplest numerical methods for solving differential equations and systems of equations. We practise programming suitable algorithms, and the simplest methods of debugging and testing the programs. For the task which was discussed above, the testing is not difficult because it is possible to use the simplest analytical form of the mathematical law of falling bodies without friction.

We have included the following topics in this part of the course:

- dynamics, the movement of bodies with friction,
- the movement of bodies with a changeable mass,
- the movement of heavenly bodies and Kepler's laws,
- different types of pendulum oscillations,
- electrical fields and the movement of charged particles,
- one-dimensional heat conductivity.

Every student may choose definite tasks and produce mathematical computer models at their own rate.

A very important part of the work is the construction of a convenient user interface . Of course, the use of pre-written software makes such work unnecessary but we think it useful to develop in students initial skills in programming tables, trajectories, bodies in movement and suchlike. This work is interesting for most students. Hence this part of the course includes additional elements of programming, and is designed for students who will such develop their software independently. They know the foundations of programming, so we include only two topics in our course - interactive human-machine interface and scientific computer graphics.

A critical stage of the work is the analysis of physical results which were obtained during the modelling - are they realistic or not? how do they correspond to our intuitive ideas? It is necessary to find answers to these questions because otherwise the modelling would be only an exercise in programming. The skills necessary for such analysis have been refined during the mathematical computer modelling undertaken at the first stage of the course. Afterwards, students may help the teacher to solve other problems or do it independently.

The second year is devoted to constructing models in different branches of sciences. Our course helps meet two main challenges that Russia's educational system is facing today: ecological education and economics education.

Firstly, economics. The simplest method in the mathematical theory of optimal planning is the linear programming. We consider the geometrical method, applied in the case with two variables, and the simplex method. The mathematical method of discrete transformations is used. It is new to the students and gives them a good preparation for further mathematical education.

Very interesting problems are connected with the study of mathematical models in ecology. We consider the fundamental ideas of classical ecology - competition and co-existence of species. We consider the important role that mathematics and computer modelling play in the study of nature. The following popular models are considered: interspecies competition with discrete reproduction; interspecies competition with continuous reproduction; the Lotka-Volterra model of interspecies competition; the Predator-prey system. In this section our students acquire such new mathematical knowledge as phase diagrams of different kinds of evolution, limit cycles, dynamical chaos, and of a process with after-effect.

The next division of our course is modelling queuing systems. We consider Erlang's task and some of the simplest problems of queues. This field of activity contains many interesting tasks for imitation modelling. Pupils obtain new mathematical knowledge about random processes, methods of treatment of the results of numerical experiment with random variables and new skills in programming.

Methods of teaching

The usual procedure with students in schools is as follows. First of all they attend some lectures on relevant problems - the share of lectures during the entire period of studying this subject is approximately 30%. Only the first task is used for collective work, thereafter individual work or work in small groups begins. Every student is given a task or may choose it from the list of optional tasks, which they may do at home or in class. During this time, the teacher works as a consultant.

A student must write a special report for every task. Such a report typically consists of the following chapters:

- content of the problem,
- description of the aims of modelling and a list of the more important parameters,
- mathematical form of the problem - mathematical model,
- description of mathematical methods which are used,
- flowchart of algorithm,
- computer program and a short description,
- results of computer experiment (numbers, tables, diagrams, pictures, and suchlike) and its analysis.

Of course, in every group there are students with different levels of preparation and skills. We adjust the individual tasks to suit them. Examples of simple tasks are:

- the movement of a body which is projected at an angle to the horizon,
- the fall of a parachutist.

Examples of difficult tasks which may be used as examination work for a semester are:

- investigation of the boundaries of instability zones of pendulum oscillations in parametrical excitation,
- investigation of actions of small perturbations on the movement of a heavenly body,

- investigation of different evolutionary patterns of two interacting species of animal.

CONCLUSIONS

The experience which we have now makes it possible to assert the following. The number of very active students during this course is about 25% - they read additional books and try to carry out most of the work independently. Approximately 40% of the students work hard, but without initiative; the others are passive.

The course described above is not feasible in its entirety for every high school. We have implemented it in some schools which are oriented towards in-depth mathematics and/or computer science education. By the end of the two-year course a change in our students was very noticeable. They had acquired new knowledge in mathematics and other sciences and a wide scope of knowledge and professional skills in programming. Most of them are now continuing their education in corresponding fields in higher educational establishments.

We have developed an analogous, but much broader, course for students of the Mathematics Department of our Pedagogical University. Its duration is the 2 semesters of the final year. It includes 24 hours of lectures and 46 hours of practical work, in which every student must carry out 4 or 5 projects.

ACKNOWLEDGEMENTS

The authors are grateful to the Culture Initiative Foundation (International Soros Foundation) for support of their work and to the British Council for partial financial support for one of the authors which enabled him to participate at ICTMA-7.

16

On the Use of Open-ended, Real-world Problems

Ted Hodgson
Montana State University, Bozeman, MT 59717, USA
e-mail: hodgson@math.montana.edu

ABSTRACT

With increasing frequency, mathematics teachers in the United States are using real-world situations as contexts for developing students' modelling skills. The mere use of real-world settings, however, does not guarantee students' development as modellers. If, as is often the case, students are excluded from essential stages of the problem-solving process - such as defining the problem or investigating the viability of their assumption - they will never develop truly usable modelling skills. On the other hand, open-ended, real-world problems may help facilitate the development of these skills. This paper describes an initial effort to use open-ended problems in the mathematics classroom. The paper offers observations regarding students' behaviour in real-world, open-ended situations, gives examples of students' modelling efforts and practising teachers' reflections on the use of open-ended problems.

INTRODUCTION

The Curriculum and Evaluation Standards for School Mathematics (National Council of the Teachers of Mathematics (NCTM), 1989) maintains that mathematics instruction should prepare students to use mathematics to explore real-world situations. Specifically, the Standards recommend that students have "numerous and varied experiences related to the cultural, historical, and scientific evolution of mathematics so that they can appreciate the role of mathematics in the development of our contemporary society and explore relationships among mathematics and the disciplines it serves" (NCTM, 1989: page 5). To accomplish these objectives, students are building and using mathematical models.

Like problem solving, modelling is a difficult skill to teach. For one, modelling is complex and multi-faceted (Galbraith and Clatworthy, 1990; Hirstein, 1995; NCTM, 1989). In authentic modelling situations, modellers typically:

- extract the problem from the underlying real-world situation,
- construct a simplified version of the initial problem,
- construct a mathematical model of the simplified problem,
- identify solutions within the framework of the mathematical model,
- interpret these solutions in terms of the simplified problem and
- verify that the solutions generated for the idealised problem are solutions to the initial problem (NCTM, 1989: page 138).

Ideally, students develop these skills through their work with real-world problems. The use of real-world problems, however, provides no guarantee that students will develop modelling skills. In particular, if students experience only *some* of the modelling process, then questions can be raised about the impact of classroom modelling. Do these experiences facilitate students' development as modellers? Does classroom modelling represent 'authentic' modelling?

Unfortunately, the modelling that many students experience cannot be regarded as authentic. Specifically, Hodgson (1995) notes that although many teachers use real-world situations as instructional contexts, the underlying style of instruction remains largely traditional. That is, rather than allowing students to interact directly with real-world problems, many teachers offer simplified versions of real-world problems. They present the real-world situation, pose questions about the situation, identify 'real' models of the situation (as in Maki and Thompson, 1973), and even present data generated by the real model. Subsequently, students identify mathematical models of the situation, solve the resulting mathematical problem, and interpret the solution in terms of the original problem. Although the guided approach is efficient and minimises the chance that students will deviate from intended curricular outcomes, the approach also robs students of valuable problem-solving experiences.

To provide students with authentic modelling experiences and facilitate the development of modelling skills, a growing number of mathematics educators recommend greater student autonomy and involvement in the modelling process (Gurtner *et al.*, 1993; Mason, 1984; Naylor, 1995). This chapter outlines one effort to accomplish these objectives by describing an ongoing experiment in which participant teachers use open-ended problem situations in the mathematics classroom. As opposed to classroom exercises, real-world problems rarely come neatly packaged - on the contrary, real-world problems are often ill-defined and open-ended, and it is the modeller's duty to package them in manageable ways. In the classroom, the use of open-ended problems promises to provide students with authentic modelling experiences and engage them in all phases of the modelling process.

AN EXPERIMENTAL USE OF OPEN-ENDED PROBLEMS

The experiment was initiated during the summer of 1994 and consists of two phases. In the first phase the author administered open-ended, real-world problems to students enrolled on the course Mathematical Modelling for Teachers at Montana State University. Initial observations of students' behaviour in open-ended situations were collected and procedures were developed for using open-ended problems in the mathematics classroom (refer to Table 1). In the second phase, which is ongoing, selected secondary and middle school teachers administer open-ended problems to their students.

Several forms of data were collected. Firstly detailed observations were collected of students' interaction with open-ended, real-world problems. These observations included field notes of students' behaviour and discussion, as well as examples of their modelling efforts. Secondly, participating teachers recorded their observations and reflections. In particular, the teachers reflected on the strengths and weaknesses of each investigation and on the use of open-ended problems.

---

(1)   Students submit topics or areas of interest, from which they (with the teacher's guidance) select an appropriate focus for the investigation.

(2)   Once the topic is defined, whole-class brainstorming sessions are conducted and preliminary questions for investigation are formed.

(3)   Preliminary questions are reviewed and refined.

(4)   Divide students into groups, and have each group select one question to investigate.

(5)   Student groups complete their assigned projects and report the results of their investigations.

(6)   Combine the findings of the various groups and form classroom summaries of the topic of interest.

(7)   Look back over students' investigations to highlight important mathematical content and processes.

Table 1. Procedures for creating and using open-ended problem situations

---

CLASSROOM EXAMPLES

The selection of each investigation was guided by the interests of the students. In one Advanced Algebra class, students were instructed to collect articles that were mathematical in nature or raised questions that could be investigated through the use of mathematics. Several students contributed an article on the bungee jumping efforts of professional basketball players Shawn Bradley and Sharone Wright (Appendix 1). Following some negotiation, bungee jumping became the overall theme of classroom investigation.

To initiate students' investigations, the teacher asked them to brainstorm. What would they like to know about bungee jumping? In a matter of minutes, the board was covered with questions. How does the thickness of the cord affect the jump? How does the weight of the jumper affect the jump? What are the effects of jumping on the jumper? How long does an 'average' jump last? What is the history of bungee jumping? What are the risks? Subsequently, the students (with, it should be noted, some guidance by the teacher) identified appropriate questions to investigate.

At this point, groups of three to four students were formed and each group investigated one of the questions listed. For instance one group examined the relationship between the jumper's weight and the length of the drop. To do so, the group used rubber surgical cord and a variety of weights to simulate bungee jumping. Subsequently the students entered the data into a spreadsheet, graphed the data, and identified best-fitting models using the statistics capabilities of the spreadsheet. The group concluded that within a 'reasonable' range of weights, linear functions accurately model the relationship between weight and the length of the drop. Group members observed that the exact nature of the relationship, however, is dependent on the materials used and the 'scale' of the experiment.

In another class, archery became the theme of investigation. For instance, one group of students examined the effect of bow tension (as measured by the length of string used to string the bow) on flight distance. By holding the angle of elevation and 'draw' on the bow constant, the students discovered the existence of an optimal tension for their particular bow.

OBSERVATIONS AND TEACHER'S REFLECTIONS

In general the teachers followed the format prescribed in Table 1. They allowed students to select real-world situations of interest, formulate questions about the situations, and develop experiments to investigate these questions. From a problem-solving perspective, the teachers' hands-off policy allowed students to participate in critical phases of the modelling process, such as defining and simplifying the initial problem. As a result, the students seemed to take ownership of their investigations and learning ie they maintained high levels of interest, posed questions about the situations of interest, and pursued the questions they posed. This observation corroborates the findings of Cobb and his colleagues (1992), Megnin (1995), and Steele (1993), and is in contrast with students' behaviour in traditional classrooms settings (Steele, 1993).

As a generalisation, students' modelling consisted of the identification of factors (typically two) of interest, the design of an experiment to isolate and uncover the relationship between the factors, data collection and curve fitting. For the most part students were satisfied with the identification of some relationship, be it linear, quadratic, or other. Rarely did students test or revise their models. Although students' conceptions of models and modelling were not the focus of this study, these observations suggest that some secondary students see models as graphs that fit data and modelling as curve fitting.

Some students were hesitant to embark on their own investigations. To these students, the objective of the activity was to investigate an appropriate question or construct an appropriate model. In other words they searched for questions and models that would meet with social (that is the teacher's) acceptance. Although this should not be viewed as a drawback of open-ended problems (eliciting students' beliefs about the existence of correct answers and problem-solving strategies was an unexpected benefit of the teaching experiment), the phenomenon did affect the outcomes of students' investigations.

The teachers participating in the study were asked to reflect on the investigations that were conducted in their classes, and on the use of open-ended problems. Answers to the former question are only applicable to the particular investigation, and are of greatest use to the teachers themselves. With regard to the use of open-ended investigations, however, teachers were cautiously optimistic. For example, all teachers cited the high levels of student participation and interest that open-ended problems seem to elicit:

> This type of activity offers students something to get excited about. They take a vested interest in the process and results.

> Open-ended questions are a valuable strategy for learning. I think they are also more fun for the students than most strategies used in the classroom. These are the experiences that students tend to remember and value.

On the other hand, the teachers also identified factors that inhibit the effectiveness of open-ended problems - the need to follow a prescribed syllabus, the time required to conduct open-ended investigations, and their own subject-matter deficiencies,. Similar obstacles have been noted by Blum (1995), Blum and Niss (1991), and Hodgson (1995). Students' beliefs about proper behaviour in mathematics classrooms was also cited as an obstacle. As an illustration, one teacher noted that the use of open-ended problems tends to contradict students expectations of mathematics and the mathematics classroom:

I think it is difficult to make truly open-ended assignments and still give students a clear idea of what is expected from them. If I were to do this again, I would tend to compromise and give assignments that have one part clearly defined and another part more loosely defined.

CONCLUSIONS AND FUTURE DIRECTIONS

This project represents an initial investigation into the use of open-ended problems. In general it was found that open-ended investigations motivate students and provide them with valuable problem-solving experiences. As a pilot investigation, however, the study raised more questions than it answered. For instance, although students appeared to engage in all phases of the modelling process (with the possible exception of looking back and revising their initial efforts) and take ownership of their investigations, these findings were drawn from data that were largely observational. To allow for detailed analysis of students' behaviour in open-ended problem situations, more video- and audio-tape recordings are needed. Also students could be asked to reflect on their investigations and on the nature of their learning.

The overall goals of the project are to promote the development of 'essential' modelling skills. Even with the open-ended nature of in-class problems, however, one can ask whether or not these skills transfer to genuine real-world situations. From an affective standpoint, one can examine how the use of open-ended problems affects students' beliefs regarding mathematics and the mathematics classroom. Also, is the apparent increase in students' interest and motivation a short-term or a lasting phenomenon? Lastly, the use of open-ended problems raises practical, classroom issues. How should students' work be assessed? How much (or how little) should open-ended problems be used? How can existing classroom constraints be overcome? These and other questions will be the focus of future study within this project.

REFERENCES

Blum W, 1995, *Applications and Modelling in Mathematics Teaching and Mathematics Education - Some Important Aspects of Practice and Research*, in Sloyer C, Blum W and Huntley I, (eds.), **Advances and Perspectives in the Teaching of Mathematical Modelling and Applications**, Water Street Mathematics, Yorklyn, Delaware.

Blum W and Niss M, 1991, *Applied Mathematical Problem Solving, Modelling, Applications, and Links to Other Subjects - State, Trends and Issues in Mathematics Instruction*, **Educational Studies in Mathematics**, **22**, pp 37-68.

Cobb P, Wood T, Yackel E and Perlwitz M, 1992, *A Follow-Up Assessment of a Second-Grade Problem-Centred Mathematics Project*, **Educational Studies in Mathematics**, **23**, pp 483-504.

Galbraith PL and Clatworthy NJ, 1990, *Beyond Standard Models - Meeting the Challenge of Modelling*, **Educational Studies in Mathematics**, **21**, pp 137-163.

Gurtner J-L, León C, Nuñez R and Vitale B, 1993, *The Representation, Understanding, and Mastery of Experience: Modelling and Programming in a School Context*, in de Lange J, Keitel C, Huntley I, and Niss M, (eds.), **Innovation in Maths Education by Modelling and Applications**, Ellis Horwood, London.

Hirstein J, 1995, *Assessment and Mathematical Modelling*, in Sloyer C, Blum W and Huntley I, (eds.), **Advances and Perspectives in the Teaching of Mathematical Modelling and Applications**, Water Street Mathematics, Yorklyn, Delaware.

Hodgson T, 1995, *Modelling in the Mathematics Classroom: Issues and Challenges*, **School Science and Mathematics**, **95**, pp 351-358.

Maki D and Thompson M, 1973, **Mathematical Models and Applications**, Prentice Hall, Englewood Cliffs, New Jersey.

Mason J, 1984, *Modelling: What Do We Really Want Students to Learn?*, in Berry J, Burghes D, Huntley I, James D, and Moscardini A, (eds.), **Teaching and Applying Mathematical Modelling**, Ellis Horwood, Chichester.

McCallum J and O'Brien R, (eds.), 1994, *Scorecard (Free Time, Free Fall)*, **Sports Illustrated**, **81**, 12, pp 13-14.

Megnin JK, 1995, *Combining Memory and Creativity in Teaching Math*, **Teaching PreK-8**, **25**, 6, pp 48-49.

NCTM, 1989, **Curriculum and Evaluation Standards for School Mathematics**, NCTM, Reston, Virginia.

Naylor T, 1995, *The Importance of Student Autonomy in Developing Mathematical Modelling Ability*, in Sloyer C, Blum W and Huntley I, (eds.), **Advances and Perspectives in the Teaching of Mathematical Modelling and Applications**, Water Street Mathematics, Yorklyn, Delaware.

Steele DF, 1993, *What Mathematics Students Can Teach Us about Educational Engagement: Lessons From the Middle School*, **Paper presented at the Annual Meeting of the American Educational Research Association** (ERIC Document Reproduction Service No. ED 370 768).

APPENDIX 1.

The *Sports Illustrated* (19 September 1994) article prompting the in-class investigation of bungee jumping (McCallum and O'Brien, 1994P.

Free Time, Free Fall

With the new rights and freedoms pro athletes are demanding these days, it is surprising that one restructive aspect of their contracts has gone unchallenged. Included in every football, basketball, baseball and hockey player's work papers is a section limiting his free-time activities. Some of the no-nos mentioned in the NBA contract, for example, are riding a moped, skydiving, hang gliding, professional wrestling and - believe it or not - tossing around a baseball or a football. That list is preceded by the clause "including but not limited to", an obvious attempt at a legal catchall.

Still, Philadelphia 76er center Shawn Bradley and rookie forward Sharone Wright recently found a way to get a thrill at Atlantic City's Steel Pier, where they plummeted 250 feet towards the ground in a bungee-like contraption called a Skycoaster.

Had Bradley (who made out his will on hotel stationery before the ride) or Wright been hurt or even killed, the battle over whether their contracts should be honored would have been interesting. While the daredevil duo could claim that they were not, after all, doing something as perilous as, say, running pass patterns, we assume that the 76ers would deem Skycoastering an "included but not limited to" activity.

What is more fascinating is the list of seven pastimes, expressly permitted in the NBA contract: golf, tennis, handball, swimming, hiking, softball and volleyball. You call these innocuous activities? Tennis brings the likelihood of tennis elbow, handball a guarantee of jammed fingers. People drown when they swim and get eaten by bears when they hike. Softball and volleyball (the beach kind, anyway) are played under the specter of skin cancer. As for golf, well, everyone knows that that activity can lead to mental illness.

17

Slow Learners, Mathematics and Future Profession: the search for mathematics on the shop floor of the future

Pieter van der Zwaart,
SLO, Institute for Curriculum Development, Postbus 2041, 7500 CA Enschede, The Netherlands
e-mail: P.vanderZwaart@slo.nl

ABSTRACT

The SLO project, Mathematics Developed Teaching Materials, for students age 15-16 in lower vocational education, is based on a realistic vision of mathematics education. One premise of the project was that mathematics education at this level should be based on the mathematical problems a professional worker meets in the workplace. Investigations were made into current mathematics education in both the vocational streams and the workplace. This led to the development of teaching and background materials to operationalise the premise at a classroom level. The teaching materials were based on observations in the workplace and interviews with workers in related professions, and were extensively tested in the classroom. Based on the investigations and the experiences in the classroom, conclusions are drawn on the conditions that have to be fulfilled in the teaching materials and on pupils' behaviour in the classroom.

INTRODUCTION

In the past decade 'real world' mathematics curricula have been introduced in the Netherlands. (deLange J, 1994) In the period up to 1992, major changes in the secondary education mathematics curriculum took place. In the spring of 1992 SLO considered a useful continuation of this work. In consultation with the Dutch Association of Mathematics Teachers the development of third and fourth form mathematics in lower vocational education was considered. The key issue was: *How can mathematics education contribute to the preparation for a future profession?*

In the Netherlands, some developers were inspired by professional reality and translated this into a series of lessons and a number of articles. (van de Heuvel, 1992; van de Brink, 1992). The volume Mathematics and Work (van der Linden, 1993) by the study group Women and Mathematics was published a little later. In the national core curriculum (Miow, 1994), in the General Aims state: "*In relevant situations, the pupils can see the connection with the practice of some professions…They obtain some insight into the meaning of each subject in subsequent studies and professions*".

On an international level the demand for mathematics in secondary education that contributes to the preparation for a future profession was also increasing. In Mathematics Counts (Cockcroft, 1982) is one goal for mathematics education mentioned as: "*to prepare for the workplace and for further education*". Some inspiring teaching materials (Austwick *et al.*, 1985) and investigations (Bessot *et al.*, 1992), with links to lower level vocational education were also found.

In general education however, attention to the needs of future professions is not (yet) taken into account world-wide. For instance in the Performance Standards (National Centre on Education and the Economy, 1995), Putting Mathematics to Work is one category of the performance descriptions. The use of mathematics in the workplace is, however, not mentioned within this or other categories.

No comprehensive studies on the contribution of mathematics education to the future professions of slow learners were found. Therefore the state of the art in the Netherlands on this issue was taken as a starting point of the project.

THE LINK BETWEEN MATHEMATICS AND PROFESSION

A comprehensive preliminary investigation regarding contents was carried out at the start of the project. One of the main conclusions of this investigation was that mathematics and profession can be linked in many different ways.

- Professional practice can be a source of inspiration for the teaching of mathematics.
- An inventory can be made of the preliminary knowledge presupposed by the teachers in the lower stream of secondary vocational education and the different kinds of professional training (in this case the apprentice system).
- A study of the future shop floor posts filled by lower stream vocational pupils can illustrate what aspects of mathematics are used.

Each of these approaches will be looked at in greater detail overleaf.

Professional practice as a source of inspiration

In cases requiring perpendicular lines and planes, the introduction of Pythagoras seems obvious. Thus you may find yourself mathematically challenged on a building site. However, closer investigation makes clear that in practice not one single construction worker ever uses Pythagoras' theorem. Construction workers know they can construct a right angle with the help of a triangle with sides 3, 4 and 5. For checking perpendicularity they know the diagonals of a rectangle are equally long. Consequently it is hard for a mathematics teacher to answer the question 'What use is Pythagoras' theorem if I want to be a construction worker?' Clearly, this is not a plea to drop Pythagoras. Being inspired by the building site is a very proper argument to motivate mathematical activities. We should realise, however, that justifying Pythagoras by presenting a construction worker using it, is far from the truth. Unfortunately it is often the case that authors of teaching materials walk straight into this trap.

Mathematics as preliminary knowledge of professions

Within the technical professions the image of mathematics is that of the supplier of techniques. Hence the use of mathematics is strongly rule based - if you do it in such and such a way, you will get the right answer.

One of the major problems occurring in this process can be formulated as follows. Each and every pupil can be taught mathematical techniques, for example the substitution of numbers into a formula and the solving of easy equations. It is a much more complicated matter to teach pupils when to apply a specific mathematical technique and in what (professional) situations.

With respect to non-technical professions such as nursing, so far it has rarely been made explicit what mathematical skills, besides basic arithmetic, they actually require.

The search for mathematics on the shop floor

With the above in mind we looked in detail, at the future shop floor posts of lower vocational pupils to find out how and where mathematics was being used.

Interviews with professionals made it clear that, besides some basic arithmetic, they use little mathematics in their jobs. The negative image these professionals have of their former school mathematics is still so strong, it is preferable to have them talk about their work without telling them that it is mathematics you are looking for. Observing professionals at work produces even better results.

The following situations, all of them related to mathematics, occurred in nearly all of the professions that were investigated.

- Situations requiring three-dimensional thinking
- The use of (complicated) tables, manuals, diagrams and other information systems.
- Situations frequently requiring the use of amounts, sizes, prices and other units for calculating or measuring.

Different sets of rules and procedures determining actions to be taken, are of major importance in all three of the above-mentioned areas. Some of these rules require only recall of well known facts.

Clearly three-dimensional thinking is crucial to the building sector while the use of information systems is central to work in the administrative sector.

Within a specific work situation different mathematical aspects are strongly integrated into one another and into the situation itself - isolated skills hardly occur at all. Sometimes one needs to combine three-dimensional thinking with measuring and arithmetic. In turn, this sometimes needs to be subjected to all sorts of regulations, and data filed into a complex system. In the case of a kitchen seller this complex data system is the 36-page line of products showing the manufacturer's range of front panels, colours, wall cupboards, base units, grips, built-in equipment and so on.

We chose to design teaching materials with this last approach in mind. The main aim was to design a practical learning environment in which the pupils could solve practical problems and experience the role of mathematics in it.

TWO INSTRUCTION BOOKS

The SLO project has developed two teachers' books on mathematics and profession, entitled A New Kitchen' (van der Zwaart , 1996a) and Behind the Scenes of the Hospital' (van der Zwaart , 1996b). The teaching materials in these books are based on the investigations on the future shop floor of the pupils concerned. A New Kitchen' includes practical aspects taken from such fields as building, installation and sales. Behind the Scenes of the Hospital presents activities from mechanical engineering, nursing and office work.

The advice given to a customer wishing to design a new kitchen is one of the professional situations used as an example. The instruction sheets are part of this article.

The first exercise stimulates the pupil to connect the plan with the room the customer will be using as a kitchen. Pupils are encouraged to list what the kitchen is used for. This is essential, as most pupils only think of a sink as a place for doing the dishes. It is typical for a salesman/woman to be able to help the customer imagine a spatial situation that is not yet in existence. It is only at a later stage that they will be able to find out whether the advice has been right. In later sheets the pupils turn into real professionals. They find out that the sink will be used for other purposes as well and, in consequence, other requirements need to be considered. The first requirement shows there is a relation between the sink and the cooker. It will take a while before pupils realise that it is safest to work the cooker with your most skilful hand. The customer's original design is tested, being in mind the four requirements listed in the Appendix. By evaluating any changes needed, the spatial imagination of both customer and salesman *ie* the pupil, is constructively challenged.

These sheets are an example of two of the core objectives from the mathematics part of the national core curriculum for lower secondary education. These core objectives (numbers 17 and 18 from the geometry domain) read as follows.

17. The pupils should be able to read and interpret flat 2-dimensional images of spatial figures and objects, such as photographs, plans, maps, construction drawings and imagine such objects as they occur in reality and present them on paper or on a screen.

18. The pupils should be able to perform concrete operations based on representations of spatial figures and tangible objects. They should be able to make outlays, patterns, *etc* and draw and/or copy surfaces derived from spatial figures to scale.

The above example for classroom instruction involves other aspects, such as the selling of kitchens. Since three-dimensional thinking is frequently too complicated a matter for the average customer, kitchen designers avail themselves of the so-called plan table that they have in their shop. With the help of 1 : 6 scale models customers can create their new kitchen, selecting from the range of products that is carried. This plan table has also been included in A New Kitchen, offering pupils with too little three-dimensional imagination possibilities for learning straight from professional practice.

The latest trend in the kitchen trade is the designing of kitchens with the help of the computer. For example you can select the cupboard of your choice from the menu and click in that spot of your kitchen plan where you want the cupboard. The code-number of the cupboard selected, tells the computer whether it is a wall cupboard or a base unit. The computer can also see which way the door should open. The combination of cupboards as a whole is also part of the software, and so the sizes of the worktop is automatically calculated. This aspect of the selling of kitchens has also been included in A New Kitchen. However, this is limited to the computer printout which consists of the plans, perspective drawings and quotations. Clearly the interpretation of these requires considerable mathematical skills.

This example fits into a larger number of lessons which are connected by the story of Tim and Marita, a young couple. The story describes what they go through and what they have to think of once they have decided to buy a new kitchen that has to be installed in an existing house. Three of the many lesson series that can be developed on the basis of this story have been worked out within the project. They are included in the teacher's book for A new kitchen. The same holds for Behind the Scenes of the Hospital, where in this case five examples have been worked out.

CONCLUSIONS

The design of teaching materials

From our experiments and experience it can be safely concluded that it is possible to have mathematics education contribute to a better preparation for professional life. This can be done by carefully integrating well chosen and motivating situations taken from professional life into the teaching materials. These situations serve as a starting point for various mathematical activities.

We found the following aspects to be relevant in the design of teaching materials

- the mathematical skills actually used on the shop floor,
- the mathematical content is perceptible and achievable by slow learners,
- the presented problem situations are applicable to the students' available knowledge,
- the pupils get a clearly define and acceptable role (see below),
- the pupils get the time and the facilities to explore the overall situation,
- different groups of pupils feel involved in the situations presented to them.

Bringing in these aspects created a learning environment in which the student really felt that they were working with relevant problems in which they could use their mathematical skills.

The role of the pupil

The question emerging is, of course, how to work with pupils in the lower stream of secondary education and how to teach them some adequate mathematical skills. We opted for giving the pupil a clear role, if possible that of a professional potential. In such a situation they appear to be able to solve problems using their mathematical knowledge and skills.

The experience has shown that pupils are very serious in carrying out their professional duties and that they feel very responsible for creating a good product and achieving a good learning result. The question *'What do I need mathematics for if I want to be....?'*, has never been heard in the classroom during these pilots.

REFERENCES

Austwick K, Richards PN and Livingstone KM, 1985, **Maths at work**. Cambridge University Press, Cambridge.

Bessot A, Deprez S, Eberhard M and Gomas B, 1992, **Approche didactique des processus de formation de base a la lecture de systeme de vues, destines a des adultes peu qualifies, dans le cadre des metiers du batiment**. Université Joseph Fourier, Grenoble.

van de Brink J, 1992, *Professional mathematics*, (in Dutch), **Nieuwe Wiskrant**, May 1989, pp 5-9.

Cockcroft WH, 1982, **Mathematics Counts: Report of the Commission of Inquiry into the Teaching of Mathematics in Schools**, Her Majesty's Stationery Office, London.

van de Heuvel G, 1992, *Fashion fashionable,* (Dutch), **Nieuwe Wiskrant**, October 1992, pp 12-16.

de Lange J, 1994, *Curriculum Change: an American-Dutch Perspective*, in Robitaille D *et al.*, (eds.) **Selected Lectures from the 7th International Congres on Mathematical Education**, Les Presses de l'Université Laval, Sainte-Foy, pp 229-248.

van der Linden J *et al.*, 1993, **Mathematics and Work: Teaching Materials from Professional Practice**, (Dutch), Utrecht.

Miow, 1994, **Basic education in the Netherlands: the attainment targets**, Zoetermeer.

National Centre on Education and the Economy, 1995, **Performance Standards, English Language Arts, Mathematics, Science, Applied Learning** (consultation draft). New Standards, University of Pittsburg, Pennsylvania

van der Zwaart P, (ed.), 1996a, **A New Kitchen** (Dutch), in press, SLO, Enschede.

van der Zwaart P, (ed.), 1996b, **Behind the Scenes of the Hospital** (Dutch), in press, SLO, Enschede.

APPENDIX: KITCHEN PLAN

The following is an extract from van der Zwaart (1996a)

Marita and Tim take their kitchen plan to the shop.
You can see the plan below.

1.   Where do you see the outside door in their plan? Write that in the plan.

2.   Two constructional changes need to be made to renovate the kitchen. Which are
        they?

3.   Where did they plan the sink? Write that in the plan.

The wall cupboards are indicated by a cross. Marita and Tim want the cupboards side by side
along the entire wall.

4.   Marita and Tim did not draw in all the cupboards. Draw the missing cupboards.

5.   They drew the microwave in the corner.
        What height would be best for a microwave: a base unit, on the working top or in a
        wall cupboard?

As a seller of kitchens you always explain to people that there are three basic rules they have
to bear in mind if they wish to design the new kitchen themselves.

> **Rule 1**
> For right-handed people it is best to have the cooker on the right side of the sink.
> Tim and Marita are both right-handed.
>
> **Rule 2**
> There is a working top of some 50 cm between the sink and the cooker.
>
> **Rule 3**
> There is at least 30 cm between the cooker and the window.

6.a   Did Tim and Marita put the sink and the cooker in the right place?
  b   Why?

There is also a fourth rule, namely:

Rule 4
In front of the working top you need at least 75 cm for moving around easily. If you often cook together, this should be 100 cm.

10.   Can Marita and Tim easily work in their kitchen at the same time?

11.   Make a rough draft of the kitchen the way they eventually want it to be.

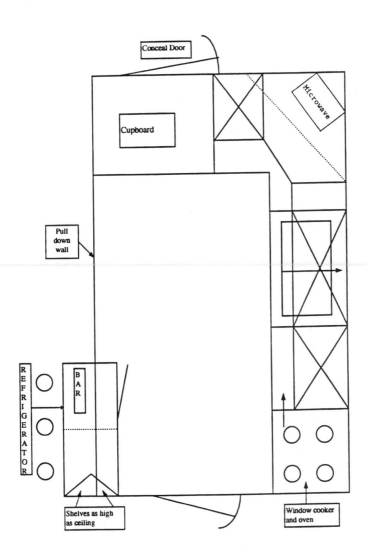

18

Mathematical Modelling for 16-19 Vocational Courses

Julian Williams and Geoff Wake
Centre for Mathematics Education, University of Manchester, Manchester
e-mail: julian.williams@man.ac.uk

ABSTRACT

Mathematics education within vocational programmes poses a special problem related to the conflict between education and training. This is explored in relation to the current crisis in assessment and testing of mathematics in UK vocational qualifications. The needs of the student for a mathematical education and for progression into further education and training suggest a focus on mathematical ideas, comprehension and structure. The training needs, and the students' own personal motivation, suggest a focus on competence in practical situations.

An example of a general mathematical competence in a course being developed by the Mechanics in Action Project in collaboration with the Nuffield Science in Practice project is called 'Handling Experimental Data Graphically'. Students' experimental skills are seen to be enhanced by their increased effectiveness with the mathematical skills involved in collecting, handling and interpreting data. The treatments of statistics, graphs and errors are clearly motivated and integrated.

This curriculum development leads us to work with colleges on materials and teaching methodology, and with industry to identify case studies which demonstrate the validity of the mathematics to the students and teachers.

INTRODUCTION

The UK has seen the introduction of new general vocational courses and qualifications (GNVQs) for the increasing number of 16 year olds seeking further education and aspiring to enter the job market. These are strongly influenced by the assessment practices of our national vocational qualifications, now called NVQs, and their language of 'competence'. In contrast to some continental systems, (see Smithers, 1993), NVQs are training qualification courses which have little or no general educational component.

The general national vocational qualifications (GNVQs) on the other hand serve a general educational function but also prepare students for a field of employment and further vocational training; a subsidiary aim is to prepare students for higher education. The qualifications are relatively new, struggling to achieve public recognition and respectability, but there are signs that many students entering these courses in fact aspire to higher education (Wolf, 1994).

The mathematical component of these courses has been widely criticised, both in delivery (see the Ofsted report of 1994 which found most college systems inadequate), and in specification. On the latter, particularly the engineering industry and the universities have been expressing concern (see Sutherland and Pozzi, 1994, and the IMA report of 1995). The uneasy place of mathematics in a general vocational course lies in the dual view of mathematics both as an academic discipline and a practical problem-solving tool. A good way to examine this dichotomy is to look at assessment and testing.

ASSESSMENT AND TESTING MATHEMATICS IN THE VOCATIONAL COURSE

Teachers from schools and universities outside the vocational training sector find mathematics in these courses described in new and unfamiliar 'competency' terms. The language of competence requires that mathematics is described as an ability to perform a specific task, using a range of techniques, facts and skills. An example from Advanced GNVQ Engineering Mathematics as follows.

Performance Criterion: Use functions and graphs to model engineering situations
Range: Functions: linear, quadratic, trigonometric, exponential, logarithmic.

Although this language is unfamiliar in academic circles, in fact it can be used to specify mathematical knowledge traditionally described by a syllabus. This has been done by using competency statements, intended to be criterion referenced and supported by the range statements, to specify a syllabus. Furthermore, new amplification statements and examples of good practice in assessment have provided more detail and guidance than teachers normally expect for their academic courses. Only in one respect are current general vocational assessment practices falling below those of the General Certificate in Education (GCE) and General Certificate in Secondary Education (GCSE) boards at present, and that is in test construction, provision of suitable test papers and marking guidelines. It seems that the use of 'competency' based assessment borrowed from the NVQ culture has provoked a crisis. But why? Compare two questions asked of students in the Intermediate GNVQ Engineering Unit, Science and Mathematics for Engineering. ('Intermediate' refers to the level of GNVQ qualification prior to the Advanced level, and is considered equivalent to the GCSE.)

Test focus: Identifying devices suitable for measurement of physical quantities.
(91.21% of candidates successful)

Which instrument is used to measure current, voltage and resistance?
   A   *manometer*
   B   *megger*
   C   *thermocouple*
   D   *multimeter*

Test focus: Calculating numerical values. (7.01% of candidates successful).

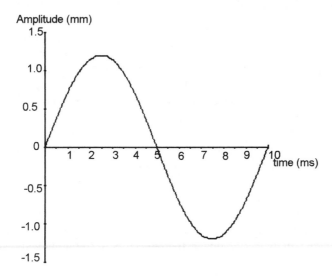

The frequency of the vibration waveform shown is:
   A   *0.1 Hz,*
   B   *10 Hz,*
   C   *12 Hz,*
   D   *100 Hz*

These questions show a huge disparity in facility level. The multistep question makes considerable academic demands on the student beyond simple factual recall. The student gains no benefit here from the placing of the mathematics within a familiar, practical, vocationally-based activity. Indeed the reverse is the case - the 'practical' context here makes the problem more challenging than a purely numerical calculation might be. In fact it is an attempt to be an application without being practical. Here is a less convoluted item, but still with only a 50% facility, from the same test.

A small motor requires 50W of electrical input in order to produce 45W of mechanical power output. What is the motor's efficiency?

A	*5%*
B	*45%*
C	*90%*
D	*95%*

The pass criterion becomes all important - it is normally assumed that the student will achieve mastery of competence, and pass marks are expected to be 70%. This often seems to be ignored when academics criticise the small mathematics syllabuses, comparing them with an A-level course on which a student has only to master part of the material to acquire a good pass grade. Hoever, it does lead us to question the mastery pass rate for what is essentially an academic test - can and should mastery criteria based on NVQ practices be applied to an academic test in mathematics - and can a traditional academic test actually assess real competence? Clearly these questions do not test competence, and the SMP 16-19 experience of trying to test modelling and application skills is that this can only be very limited, and cannot be a valid measure of modelling competence as a whole (see Kitchen and Williams, 1993).

One is reminded of the research in situated cognition by Lave (1988), Saxe (1991) and many others which demonstrates that people are generally nearly 100% effective in mathematical work in familiar practical settings, even when their mathematics test results on superficially similar items are very low. This is because the practical setting lends much intuitive and tacit support to the mathematical thinking, rather than placing added demands on an overloaded, consciously-guided, working memory of propositional knowledge. I am drawing on the notion of a dual architecture of cognition here, which is now believed to be essential for understanding problem solving in real non-academic settings - see Reber (1989), Grunwald (1992) and Broadbent (1993).

The implication might be that if mastery at this level is required, then we either need to assess the mathematics entirely through practical work or we must reduce the test items to the trivial, such as the rote-recall science item above.

The academic test may be a fair place to assess understanding and skill with mathematics but, if so there seems little point in trying to adopt the mastery criterion and the multichoice test. What we really need is a strategy for assessing genuine mathematical competence through practical vocational work. In this context it seems fair to demand mastery, without being trivial and unchallenging.

Up to a point this may be a short-term crisis, but it can be argued that it is a surface eruption of a deeper unresolved problem - the clash of cultures between education and training, university and industry, between mathematics for academic understanding and mathematics for vocational competence. The clash of aims, language and perceptions is a fundamental one, recognisable in the education of 16-19 year olds across national boundaries - see, for example, the discussion in Blum and Strasser (1992).

The target group at age 16-19

The group of students targeted for vocational courses is thought to include many who are capable of formal academic study, but focuses on those who see the vocational route as more suitable and achievable. This includes many who need to enhance their work-related skills to enter a competitive job market, but who feel they are ready to work and have a notion of a vocational field if not a specific job.

This target includes the aspiring technicians identified as the cause of the UK's competitive problems *vis a vis* continental Europe by Prais and Wagner (1985). The system of GNVQs is expected to address this target, and expectations run high - the system is soon expected to qualify about as many students as the academic sector. It is widely believed that mathematics is a crucial area to be addressed, as it is generally considered to be many students' area of weakness and also the area which limits their progress in the scientific disciplines. Clearly a major problem is the gap between the achievement and motivation of the target group and the expressed needs of higher education and industry for the highly mathematically qualified and competent product alluded to above. (See Lord, Wake and Williams, 1995: Sutherland and Pozzi, 1995).

The improvements in the relevance and accessibility of some of the new A-level courses have extended the range of students studying mathematics (see Dolan, 1990, and Kitchen and Williams, 1993). For instance, our research shows that the SMP 16-19 course now succeeds with many who have achieved only a grade C at GCSE level - rare in the past and even with some current A-level courses. However, it is clear that the range of students entering GNVQ will be much wider still (in previous achievement at GCSE and also in motivation) and that a qualitatively new approach to the mathematics curriculum will be required to support these students.

Motivation and accessibility are key issues. Previous experience in dealing with this problem at A-level has shown some success is achievable through coursework, practical approaches and high quality materials and texts - for example the work with SMP Mathematics described in Kitchen and Williams (1992) and Kitchen (1993). Indeed, even among the most able sixth formers, our evaluations have revealed dislike of final examinations. Kyeleve and Williams (1995) have found significant differences in attitude favourable to those following modern modelling courses, too. However, for our target group we expect at best:

1. little or no intrinsic motivation to study mathematics itself,
2. a potential appreciation for mathematical models in so far as they are useful in practice,
3. a willingness to invest time and energy in acquiring general mathematical competence in so far as it enhances technical capability.

The mathematical aims of industry and higher education

Another problem is the perceived dichotomy between training (for industry and applications) and education (for higher education and understanding). Our analysis of the respective needs of industry and those of higher education is that a synthesis is achievable in the debate about aims for mathematics. Our research with higher education indicates that the education community requires attention to the internal coherence of subject content and development of mathematical understanding. This is most often expressed in terms of syllabus content, though recently there has been a growing emphasis on skills such as modelling and communication (see Houston, 1993). On the other hand, the industrial influence calls for specific problem-solving competences in vocational contexts; where mathematics content is concerned industrialists have less to say, and are less coherent and homogeneous.

Our view of competences regularly practised in most vocational contexts is that they require very little mathematical understanding. It is usually the learning of new skills or the adaptation to new techniques and processes which demand mathematical depth, and it is more likely to be in the retraining, upgrading or adaptation that mathematical understanding pays off.

The nature of situated cognition is such that effective functioning *in situ* may not require depth of understanding - the often mentioned use of the 3-4-5 triangle in building construction is a good example. In fact the fluency of effective use of mathematical skills can be undermined by the user's attention being drawn to the underpinning concepts and ideas - the important thing is to count the knots in the string correctly, not to recall the principle of squares on sides. The need for a depth of mathematical understanding therefore arises less from competence in a specific problem in a vocational situation, and more from:

a)   the need to be able to adapt and transfer skills into a new area, recognising the limitations of the usual methods and skills and adapting them to the new situation,

b)   the ability to learn new mathematical skills which build on existing understanding rather than existing skills.

A synthesis requires that a mathematics curriculum should be driven by competences (that is, applicable skills) and assessed through vocational application, but with due concern for an educationally sound subject approach which is coherent, progressive and aims to develop depth of understanding as well as skill. One cannot overemphasise the problem of rote learning of a mass of ill-digested mathematics, applied ritualistically to a few repetitive applications which in no way mirrors the realities of mathematical work in practice (Clegg, 1992).

Mathematical models and modelling

Our vision for the new mathematics courses is based on a synthesis of the vocational need for competence in practice and the educational need for understanding of mathematics. The key concept which allows a synthesis rather than a compromise is that of 'mathematical modelling' in the sense of Blum (1991) - the process of applying mathematics to a real, relevant problem. We propose that vocational mathematics should be viewed as a collection of models; using these models involves the adaptation and application of mathematical knowledge and skill, and leads to mathematical 'competence'. Learning when and how to use these models in a range of applications therefore involves developing an understanding of mathematics as a subject; the understanding of the subject is motivated by the powerful applications it may facilitate.

This approach emphasises the motivation of the mathematics curriculum through problems of interest to the student, and the modelling skills needed to marshal the mathematics in an appropriate way to solve them.

Even at the simplest mathematical level there is a need for the student to learn about mathematics as a modelling process, to appreciate that mathematical models have a time and a place, that they may need some creative adaptation, and that they have limited general validity.

A central judgement in practice in the general mathematical competence described below, for example, is how much you can believe in the relationship established between two experimental variables. The range of possible linear graphs and the range of possible gradients when two variables are proportional is clearly as important in practice as one numerical result.

One of the short activities written to support students when calculating volumes also illustrates this. The students are asked to estimate the volume of brain matter in their head. They are shown how to model the head as a cuboid, prism or hemisphere, collect the relevant measurements and perform the relevant calculations; the differences in the results are noted and validity discussed.

Note that in both cases the idea of models and modelling lends a richness both to the extrinsic motivation of the activity and also to the intrinsic mathematics. The students are thereby encouraged to take the enquiry into mathematics itself so as to deepen their understanding. It is a short step from enquiring into mathematical models to enquiring into mathematics itself.

General mathematical competence

Our strategy involves focusing mathematics on 'general mathematical competences', which we use to organise a field of relevant applications around a corpus of relevant mathematics (models, concepts and skills). The student learns how, when and where the mathematical model applies and this requires attention to its underlying assumptions. This also ensures that the mathematical work can be assessed in its entirety in a project which is both significant, practically and vocationally relevant.

The principle is that general mathematical competence serves a dual function for the student and the teacher.

1.   It is understood immediately to relate to a range of vocational tasks, although it will need minor adaptation to circumstances.
2.   It organises a body of mathematics - a range of concepts, skills, facts and models which are motivated by, understood, practised and assessed in the context of general mathematical competence

In GNVQ Science an obvious example is the handling of experimental data relating two measured variables: *Handling experimental data graphically*. This is one of three chapters being produced for the Nuffield Science in Practice project for trial, see below. We think that the key features of teaching and learning a general mathematical competence are:

1.   the student's motivation is aroused by the immediate applicability, by the fact that the maths is needed to improve their vocational work, and by the obvious future value to other related applications of interest;
2   there is a substantial body of mathematical knowledge wrapped up in the work which will take a student some time to complete. This will involve practice where necessary in exercise form (for example, through helpsheets cross referenced to the module), but which is never too distant from the purpose of the module and which is not allowed to interrupt its flow;
3   the emphasis on adaptation of models and a critical approach to models encourages the students to think mathematically, and to become interested in deepening their work in mathematics itself.

This philosophy is being realised through a collaboration with Nuffield Science in Practice. The intention is to explore this approach across a range of levels and vocational fields through the development of learning materials for colleges. Important elements will be the integration of case studies of industrial practices and exemplary work by students, showing the value added to their vocational competence by their mathematical work.

Materials supporting curriculum delivery

In practice we have developed this strategy by writing mathematical modules (*Handling Experimental Data Graphically, Models of Direct and Inverse Proportion* and *Interpreting a Large Data Set*) in support of Advanced GNVQ Science during 1995. These are at present on trial in schools and colleges and will be refined during the autumn of 1995.

Each module is focused on a 'general mathematical competence' having a structure which is outlined below. This structure is exemplified by reference to the module *Handling Experimental Data Graphically*.

Each module will:

(a) have an introduction using case studies, illustrations and/or practicals from the GNVQ Science course which can clearly be seen to be generalisable to other relevant applications;

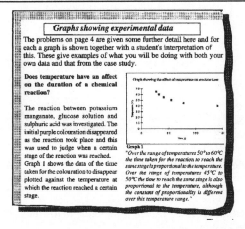

Figure 1: in the example module all three (case study, practical work and illustrations from all areas of science) are used to show the wide range of applications of the mathematics within the module.

(b)   teach an identifiable body of mathematical knowledge (facts, concepts and skills) from the application of number syllabus and, where necessary, beyond;

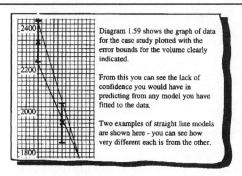

Figure 2. this module, for example, teaches about variables, measurement, errors, proportion, graphs and gradients. The example here shows how a graphical interpretation of error can inform model development.

(c)   explicitly teach the mathematics as a set of adaptable models, wherein the
            assumptions, interpretation and validation issues are critical.

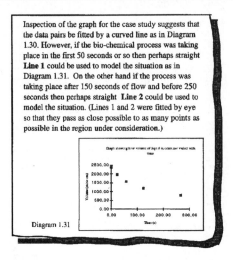

Figure 3: here, for example, the idea of being able to model a restricted range of data
using a straight line is introduced.

(d)   identify potential points where remedial help may be needed, and cross
            reference help sheets;

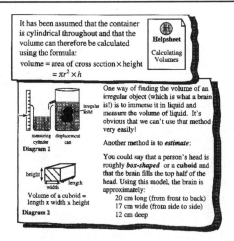

Figure 4: in this module, for example, students are required to find the volumes of
cylinders. This is subsidiary to the 'general mathematical competence' so it is covered by
a Helpsheet. The Helpsheet itself is based in science and has a modelling flavour.

(e)   identify starting points for interesting mathematical inquiries which may take the student beyond the minimal requirements of 'competence'.

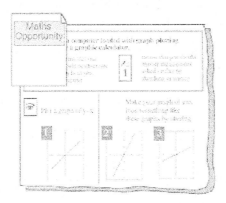

Figure 5: an opportunity for students to study straight lines and their equations in the form y=mx+c is developed in this unit. An accompanying activity allows students to explore this.

Discussion

The strategy we are advocating will not ignore considerations of internal coherence or overlook the desire of many students to progress in academic mathematics. We argue that the motivation requires a more practical approach to mathematics motivated by real problems. An open question is the degree to which we must accept lack of mathematical coherence in the implementation as the price for relevance and 'competence'.

Links across vocational fields are desirable for reasons of efficiency and transferability. We seek a compromise here in the range of different vocational fields which a general mathematical competence can address. Evidently the *Handling of Experimental Data* will find ready applications in science, engineering, construction and, perhaps, health and medicine. On the other hand, other general mathematical competences may be less relevant in their transfer across vocational fields. The kinds of statistical processes required may be more strongly vocationally structured, for instance. The degree of linkage possible remains as yet another open question.

A most significant outcome of the research will be case studies of implementation, describing ways in which vocational teachers and mathematics teachers work, and ways in which mathematical competence is demonstrated in training and work-based settings.

CONCLUSION

Our analysis suggests the need for the development of a mathematics curriculum motivated by vocational problems which demand mathematical competence - that is the adaptation and application of mathematical models.

However, this mathematics must be coherent, progressive and educationally sound - that is, based on understanding which underpins mathematical concepts.

Curriculum units and teaching approaches are being developed by The Mechanics in Action Project which attempt to implement a curriculum based on modelling with 'general mathematical competences'. These will be trialled and evaluated before publication.

REFERENCES

Blum W, 1992, *Applications and Modelling in Mathematics Teaching* in Niss M, Blum W and Huntley I, (eds.), **Teaching Mathematical Modelling and Applications.** Ellis Horwood, Chichester, pp 10-29.

Blum W and Strasser R, 1992, *Mathematics Teaching in Technical and Vocational Colleges: Professional Training versus General Education*, **Zentralblatt fur Didaktik der Mathematik: International Reviews on Mathematical Education, 92**, 7, pp 92-97.

Broadbent D, 1993, *Planning and Opportunism*, **The Psychologist, 6,** pp 54-60.

Clegg A, 1992, **A Case Study Of Mathematics in A level and BTEC Science and Engineering**. Unpublished M Ed Thesis, University of Manchester, Manchester.

Dolan S, 1990, *16-19 Mathematics*, **Teaching Mathematics and its Applications, 9**, 4, pp 145-149.

Greenwald A, 1992, *Unconcious Cognition Reclaimed*, **American Psychologist, 47**, pp 766-779.

Halsey AH, Postlethwaite N, Prais SJ, Smithers A and Steedman H, 1991, **Every Child in Britain**, Report of the Channel Four Commission on Education.

IMA, 1995, **Mathematics Matters in Engineering**, Institute of Mathematics and its Applications, Southend-on-Sea.

Kitchen A, 1993, *The Mechanics in Action Project*, in Houston SK, (ed.), **Developments in Curriculum and Assessment in Mathematics**, University of Ulster, Coleraine, pp 57-66.

Kitchen A and Williams JS, 1994, *Implementing and Assessing Mathematical Modelling in the Academic 16-19 Curriculum,* in Breiteig T *et al.*, (eds.), **Teaching and Learning Mathematics in Context**, Ellis-Horwood, Chichester, pp 138-150.

Kyeleve J and Williams JS, 1995, *Gender, Courses and Curricular Effects on Students' Attitudes to Mathematical Modelling in 16-19 Mathematics Programmes*, **Proceedings of BSRLM,** University of Birmingham, Birmingham, pp 18-22.

Lave J, 1988, **Cognition in Practice: Mind, Mathematics and Culture in Everyday life,** Cambridge University Press, Cambridge.

Lord K, Wake GD and Williams JS, 1995, **Mathematics for Progression from Advanced GNVQs to Higher Education**, UCAS, Cheltenham.

Ofsted, 1994, **GNVQs in Schools 1993/4**, HMSO, London.

Prais SJ and Wagner K, 1985, *Schooling Standards in England and Germany: Some Summary Comparisons Bearing on Economic Performance*, **National Institute Economic Review,** pp 53-72.

Reber A, 1989, *Implicit Learning and Tacit Knowledge*, **Journal of Experimental Psychology: General, 118**, pp 219-35.

Saxe GB, 1991, **Culture and Cognitive Development: Studies in Mathematical Understanding**, Lawrence Erlbaum, New Jersey.

Smithers A *et al.*, 1993, **All Our Futures. Britains's Education Revolution**, A Dispatches Report on Education, Channel Four Television, London.

Sutherland R and Pozzi S, 1995, **The Changing Mathematical Background of Undergraduate Engineers: a Review of the Issues**, The Engineering Council, London.

Williams JS and Kitchen A, 1989, **Modelling with Functions and Graphs**, Mechanics in Action Project, Department of Education, University of Manchester, Manchester.

Wolf A, 1994, **GNVQs 1993-94 : A National Survey Report : An Interim Report of a Joint Project**, Published jointly: Further Education Unit and Institute of Education, University of London, The Nuffield Foundation, London.

19

Mathematical Modelling and Children's Development of Science Concepts

Brian Doig
The Australian Council for Educational Research
Private Bag 55
Camberwell 3124
Australia
e-mail: doig@acer.edu.au

Susie Groves
Deakin University
221 Burwood Highway
Burwood, 3125
Australia
e-mail: grovesac@deakin.edu.au

Julian Williams
University of Manchester
Oxford Road
Manchester, M13 9PL
U.K.
e-mail: jwilliams@fs1.ed.man.ac.uk

ABSTRACT

This paper reports on an Australian Research Council funded project, 'Practical Mechanics in Primary Mathematics', and some collaborative work with primary and secondary school children in the United Kingdom by the 'Mechanics in Action Project'. The research examines the potential of practical mechanics activities to challenge children's thinking about, and explanations of, a variety of situations involving force and motion, and to develop these through the mathematical modelling of such situations. One of the sets of activities used with the children is discussed, together with the purpose and nature of the mathematical modelling, and some of the strengths and weaknesses of this approach with young children.

INTRODUCTION

Newtonian mechanics is the classical paradigm for the mathematical modelling of natural phenomena. The extent to which a person's view of a real life situation can be influenced by the presence or absence of a working knowledge of Newtonian mechanics is illustrated by a United States campaign, *Falling Bullets Kill*, which attempts to stop people from shooting into the air on New Year's Eve. A leading campaigner, who had his chest pierced by a falling bullet a few minutes before the beginning of 1994, described people who shoot in the air as good people who "don't understand or don't remember that everything falls at the same speed" (*Campaigner Gets a Shot*, 1996). While it is not clear whether this is intended to mean 'the same speed as it leaves the gun' or 'all things fall at the same speed' - perhaps terminal velocity - in either case it illustrates a much better understanding of the real situation than that shown by the shooters.

As with the New Year shooters, children's everyday experiences of situations that can be explained through the use of mechanics are extensive, but these experiences unfortunately often lead to intuitions which clash with scientific explanations (Williams, 1985). Students' conceptual models in various domains have been the focus of research for over two decades. This research indicates that students come to science classes with spontaneous conceptions that differ from accepted scientific conceptions (Osborne and Freyberg, 1985), are remarkably resistant to change, particularly in the area of mechanics (Gil-Perez and Carrascosa, 1990) and, at least for force and motion, are already deep–seated by age 10 and are not completely eradicated even by formal university study (Eckstein and Shemesh, 1989). Teaching experiments in mechanics have, however, emphasised secondary or tertiary students rather than primary pupils.

The Australian Research Council funded project *Practical Mechanics in Primary Mathematics* and the *Mechanics in Action Project* in Manchester, UK (Savage and Williams, 1990) are collaborating to investigate ways in which practical mechanics activities can be used to link upper primary and lower secondary students' spontaneous concepts with Newtonian mechanics.

The initial phase of the research attempts to identify the features of practical activities that are attended to by the children, namely, investigate the role of mathematical modelling in children's recording and representation of their experiences and identify opportunities to generalise children's notions and legitimise the formal language of force and motion. There is no formal attempt to teach a Newtonian model. Rather the research, which is based on a constructivist view of learning, examines the potential of the practical activities to challenge children's thinking about, and explanations of, a variety of situations involving force and motion.

The special, practical, interactive and intuitive nature of mechanics makes it most suitable to study with young children. The possibility of recording observations mathematically to obtain data also makes it an ideal field for the examination of the role of mathematical modelling in the process of learning science. Typical activities involve students obtaining real data that can be used to generate graphs or tables, from which predictions can be made and hypotheses tested - for example the students may use a 'timer ball' to measure the time taken for a ball to fall to the ground when dropped. Clearly in such situations the use of mathematics and mathematical modelling are crucial.

Using mathematics to model real-world situations is often viewed as suitable only as a secondary school activity. However, from their earliest days at school, children are modelling their environment in many mathematical ways using number, measurement and pictorial representation as modelling tools (Groves and Stacey, 1990).

While no attempt is made to teach the modelling process formally, the recording and representation of the data from the activities and the subsequent interpretations and explanations given by the children involve them in various aspects of the modelling process, albeit mostly at an *ad hoc* level (see Williams, 1989, for a discussion of intuitive, *ad hoc*, and scientific levels of modelling in practical mechanics tasks).

The modelling serves a number of purposes. It encourages children to focus on relevant features - quantities and variables to be considered in the situation, such as the distance travelled in each second by a ball rolling on a track. It also provides a framework within which to discuss and think about a situation. For example, when investigating the motion of the ball on the track, the graph assists children in their attempts to describe and explain the phenomena observed. In particular it focuses children's attention on the speed of the ball and changes in its speed. This is a crucial point in shifting explanations towards a more scientific point of view. The model (perhaps at this stage more accurately described as the mathematical representation) can then be manipulated, allowing children to make and test predictions and hypotheses. The children are able to consider the graphical model as an object separate from the motion it describes, mentally manipulate this object, and then consider its relation to a hypothetical motion. This process of reification of the motion through the mathematical model may allow a cognitive break from situated, intuitive thinking which may signal the beginning of formal, scientific thinking based on hypothesising and testing. There is a parallel here with the process of reification within mathematics, which has been noted by many authors and is known to have pitfalls as well as gains (Sfard and Linchevski, 1994).

In our research, children usually work in groups of five or six, and class discussion of the results obtained and children's explanations for the phenomena observed are crucial features of the lessons. While the focus is not on the complete modelling process, children are also being exposed to, and practising different aspects of, mathematical modelling in familiar, intrinsically motivating, contexts.

This paper will discuss one of the sets of activities used with the children, the purpose and nature of the mathematical modelling, and some of the strengths and weaknesses of this approach. Data for this paper were obtained through the use of video tapes and children's written work.

ROLLING BALLS

The activities

One of the sets of activities involves children recording the distance travelled in successive seconds by a ball rolling on a horizontal, upward or downward sloping track, or one covered with felt. The data are then displayed in graphical form. Different activities result for example in approximately uniform motion, acceleration and deceleration due to gravity, and deceleration due to friction.

A metronome is used to time one–second intervals, with different children marking positions of the moving object by placing small blocks on the table beside the track. Measurements are taken, or paper streamers used, to record distances between the blocks. Graphs showing the distance travelled in each second are then either drawn or constructed from the paper streamers. Children are asked to interpret their graphs and hence attempt to explain the motion of the ball.

The use of streamer graphs is designed to facilitate the mathematical modelling of situations using the data obtained, without the need to resort to complicated calculations or accurate graphing with pencil and paper. However some classes moved directly to measuring and drawing scale graphs.

By modelling the motion with graphs children can observe, describe and attempt to explain the motion of the ball. As part of the activities children are also asked to predict what the graph would look like for different situations, such as when the ball is released from different heights along the 'launcher', if the track were sloping, or if felt were placed on the track.

Children's Interpretations of the Graphs

When attempting to interpret the graphs the children exhibited different degrees of mathematical and scientific sophistication, these being consistent with varying degrees of use of Newtonian ideas of force causing changes in motion. The following examples of graphs and their interpretations are taken from the written conclusions of 10 to 12 year–old children. Examples have been chosen to exemplify increasingly sophisticated responses to the activity.

1. In some responses the children made no connection between the conclusion and the mathematisation - for example, "the ball goes faster the higher up the launcher you let it go". This is a sound observation but is disconnected from the mathematisation in the activity and is irrelevant to the rest of the activity.

2. A more mathematical response involves a description related to the graph, but with no causal explanation for the motion - for example, "the distance between the seconds got smaller every time" or "it went further on the first second than it did on the last second, it went less distance every second". Such responses show that the children are using the mathematical model and interpreting the graph, but as yet show no evidence of addressing underlying causes of the phenomena.

3. Some explanations offered a mixture of causes, including force and surface features.

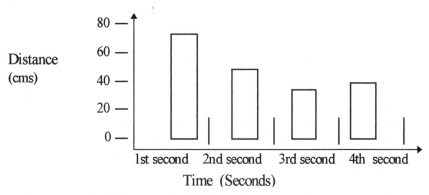

Figure 1: graph of distance travelled in each second by ball on flat track

For example, commenting on the graph in Fig. 1, one child said "I think the ball would have gone further if the track was all one. It slowed down because of friction and it was losing energy from the slope and because of the Sellotape and it made it go faster. Number 1 was the furthest because of the energy from the ball (slope). The further it went it lost power."

Here there is a mixture of surface feature explanations (*eg* the Sellotape, the slope) and appeals to scientific concepts (friction and losing power). The latter flips between force and energy and between impetus and possible Newtonian explanation. However, there is no evidence of any connection between the causes and the mathematical model of the motion.

4. In some cases one begins to observe Newtonian causation, showing an understanding of a force acting to change motion.

Dista

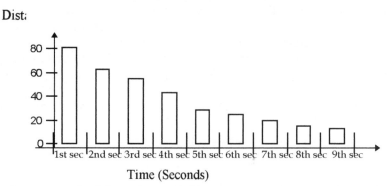

Time (Seconds)

Figure 2: graph of distance travelled in each second for ball rolling on felt.

For example, commenting on the graph in Fig. 2, a child said "The distance wasn't getting as far as the one before because of the friction slowing the ball down. After a while it stopped". This reveals a scientific explanation using the friction force. However at this point it does not necessarily indicate an understanding of friction as a force - it may just be that friction is the name given to something that slows things down.

Another child comments "It got smaller because the friction was a little greater than air pressure each time until the friction finally slowed it down to a stop". Although the effect of air is called 'pressure', there are signs here of an appreciation of friction as one of a number of forces, which combine and cancel, and whose result may cause a change in the motion. Note that in this case the forces are being considered as entities that explain the change in speed.

Some strengths and weaknesses of the modelling approach

The examples above illustrate the power of the graphical model of the motion in facilitating discussion and providing teachers with the opportunity to probe children's understandings of the scientific concepts involved (*ie* speed, changes in speed, and causes of changes in speed).

Work with the children, however, also reveals some of the inherent pitfalls in relying too much on the graphical model.

In order to interpret the graphs sensibly, children need to understand that a longer strip represents faster motion. Many children, however, intuitively believe that the faster the ball travels, the shorter the strip (recall that a strip represents the distance travelled in one second). This suggests a limited understanding of the relationship between speed, distance and time. The language of the children often reveals their confusion: something happens quickly if it takes a short time, and therefore should involve a short distance.

Despite the fact that a child may understand clearly that a faster car or runner goes further than a slower one in the same time, in this context the strip length, a distance, is not obviously related to speed.

The issue here is the co-ordination of speed with the concepts of distance and time, rather than the intuition of speed as such. Piaget (1970) claims that the full development of this relationship awaits formal thinking, and his results on the development of the qualitative concept of speed have found some support from Trowbridge and McDermott (1980) and Perry and Obernauf (1987). Crucial elements in these studies are the features of the motion attended to and the ability of the children to ignore certain features while focusing on others - some critical and some irrelevant. In this case it is important that the children are able to keep in mind the act of timing, the motion of the ball, the marking of the points for each second and also that the streamers are a measure of the distance between marks.

Children's confusion regarding which strip represents faster motion raises an important point related to the process of modelling. It is not our intention to teach children ready-made models, but rather to encourage children to use *ad hoc* models as an aid to learning. However there is a real danger that, because we believe a particular model is self–evident and requires little or no explanation, we fail to recognise the complex nature of the translation from the real situation to the model and *vice versa*. This is a problem which we believe is not unique to young children.

In most cases this confusion between speed, length of streamer and time was clarified through discussion and, where necessary, by reverting to the simpler task of recording the motions of two children walking at a steady speed - one walking slowly and one walking quickly. One child insisted however that, while longer streamers represented the faster walking speed, this was not true for a rolling ball. Several other children appeared convinced but, as might be expected, had reverted to their earlier intuitions by the time of the next lesson.

The resolution of this conflict, though sometimes protracted, was usually followed by the ability to discuss the hypothetical shape of the graph for bodies that speed up or move steadily. The graphs also focused attention on the need for a vocabulary that adequately describes acceleration and deceleration. When merely observing the motion, children often described the ball as 'going faster'. It was often impossible to determine whether children meant 'faster than last time' or 'getting faster and faster' (*ie* accelerating). Although using the graph did not necessarily resolve this problem, it did provide opportunities for some discussion of what was actually meant. Thus the strip graph as a mathematical model seems to focus discussion, enabling children to generalise about motion and allowing new predictions in hypothetical situations to be formalised.

The errors involved in the data frequently provoked children to formulate spurious explanations for changes in speed. Thus a graph that goes slightly up and down might provoke a description of a ball slowing down and then speeding up because "it hits a bump and speeds up". The mathematical representation (*ie* the graph) which was supposed to allow the children to abstract and hence generalise from the immediate reality can, it seems, allow the children to detach their explanations from the reality they have correctly perceived and formulate erroneous descriptions. The dilemma for a teacher then becomes whether to accept the children's data and their inferences from them or to 'correct' one or both of these.

While these inaccurate data provide opportunities for discussion of several features of practical work - such as the need for repeated trials, accuracy in measuring and the use of some form of averaging for data smoothing - many children were fixated on even quite small differences in strip lengths. They often constructed quite complex explanations for the ball's differing motion in each of the seconds, regardless of whether their own observations, theoretical explanations or predictions supported this or not.

Although the project seeks to use the collection and representation of data as a vehicle to promote reflection and hypothesising, it was disturbing to note that the reverse sometimes seemed to occur. Children could agree that the data were suspect, but at the same time use them to provide an apparently precise description of the motion of the ball. Often the children commented that "it doesn't really make sense" or, worse still, "what I have learnt today is that whatever I think will happen won't". This is one of the difficulties with the reification of the model referred to earlier: the fact that it facilitates the production of ideas that may be disconnected from children's perceptions and reality. The graphs were seen as reality while the motion of the ball was adjusted to suit. Alternative solutions, such as noting that the differences were (frequently, but not always) small when viewed as a proportion of the total lengths, or of averaging the entire set of lengths, were rarely raised by the children.

This raises another important issue. Making observations is an active process, influenced by the observer's conceptual framework, with observers often 'seeing what they want to see' and hence focusing on irrelevant factors or neglecting relevant features (Driver, 1983; Duschl and Gitomer, 1991). Osborne and Wittrock (1985) stress the active role of the learner in constructing meaning, which requires links to be generated between stimuli and existing knowledge. It soon became evident that 'experts' approached the graphs in quite a different way from the children. For example, for motion along a relatively smooth, flat track, we viewed the graph with a pre-conceived model that dictated either constant speed or a slight deceleration. We were then able to 'explain' all but the most blatant irregularities in the data in terms of inaccuracy or by appealing to order of magnitude arguments. This bore little resemblance to the children's view, as described above. This highlights the fact that data alone cannot provide a mathematical model for a situation - some means of validation, acceptable to the children, is also essential. Moreover, the data cannot describe a single 'correct' mathematical model because one has to select from a range of 'common sense' models and the range of these that is considered sensible is itself an expression of our pre-conceptions about the situation.

CONCLUSION

Williams (1991) suggests a range of *ad hoc* mathematical models that apply to empirical data gathered in practical mechanics situations. These range from 'concrete' graphs (*eg* the streamer graphs in this case), through qualitative graphs, quantitative graphs and rules, up to algebraic and generalised algebraic functions. The models at the intuitive end of this spectrum are largely descriptive, while at the formal analytic end they are usually predictive.

Children engaged in the rolling ball activities used their practical streamer graphs to make predictions by means of a mental model, which, in this case, is a qualitative graph (*ie* a line or curve which is intuitively extendible). The mathematical model appears to reify the motion, allowing children to detach their thinking from the particular situation and begin to manipulate the graph as a model, thus allowing predictions to be made and hypotheses to be tested.

The mathematisation which occurs can sometimes confront children's intuitive models and encourage them to reflect on the outcomes of the activities, thus encouraging more formal thinking about force and motion. Nevertheless, the use of modelling is often problematic. The examples discussed earlier illustrate the gulf in understanding which often exists between a person who devises (or has long standing acquaintance with) a particular model and someone for whom it is new.

Other difficulties occur due to errors and inaccuracies in data. The technology of data collection could become an issue here and the possibilities for data logging with high technology need to be explored, although caution is needed as we know that technology can have a dramatic impact on the nature of the activity and children's responses.

ACKNOWLEDGMENTS

The *Practical Mechanics in Primary Mathematics* project is funded by the Australian Research Council. The project is being carried out by Susie Groves and Brian Doig, in collaboration with Julian Williams, Director of the *Mechanics in Action Project* at the University of Manchester.

REFERENCES

Campaigner Gets a Shot at Making Festivities Safer (1996, January 1) **The Age,** Melbourne, p 6.

Driver R, 1983, **The Pupil as Scientist?**, Open University Press, Milton Keynes.

Duschl R and Gitomer DH, 1991, *Epistemological Perspectives on Conceptual Change: Implications forEeducational Practice*, **Journal of Research in Science Teaching**, **28**, 9, pp 839–858.

Eckstein SG and Shemesh M, 1989, *Development of Children's Ideas on Motion: Intuition vs. Logical Thinking.* **International Journal of Science Education**, **11**, 3, pp 327–336.

Gil–Perez D and Carrascosa J, 1990, *What to do about Science "Misconceptions"*, **Science Education**, **74**, 5, pp 531–40.

Groves S and Stacey K, 1990, *Problem Solving — A way of Linking Mathematics to Young Children's Reality.* **Australian Journal of Early Childhood**, **15**, 1, pp 5–11.

Osborne R and Freyberg P, 1985, **Learning in Science: The Implications of Children's Science,** Heinemann, Auckland.

Osborne R and Wittrock M, 1985, *The Generative Learning Model and its Implications for Science Education*, **Studies in Science Education**, **12**, pp 59–87.

Perry B and Obernauf P, 1987, *The Acquisition of Notions of Qualitative Speed: The Importance of Spatial and Temporal Alignment*, **Journal of Research in Science Teaching**, **24**, 6, pp 553–565.

Piaget J, 1970, **The Child's Conception of Movement and Speed** translated by Holloway GET and Mckensie MJ, Basic Books, New York

Savage M and Williams J, 1990, **Mechanics in Action**: Cambridge University Press, Cambridge.

Sfard A and Linchevski L, 1994, *The Gains and Pitfalls of Reification: The case of Algebra*, **Educational Studies in Mathematics**, **15**, pp 379–395.

Trowbridge DE and McDermott LC, 1980, *Investigation of Student Understanding of the Concept of Velocity in One Dimension*, **American Journal of Physics**, **48**, 12, pp 1020–1028.

Williams JS, 1985, *Using Equipment in Teaching Mechanics*, in Orton A, (ed.), **Studies in Mechanics Learning**, Centre for Studies in Science and Mathematics Education, University of Leeds, pp 55–86.

Williams JS, 1989, *Real Problem Solving in Mechanics: The Role of Practical Work in the Teaching of Mathematical Modelling*, in Blum W, Niss M and Huntley I, (eds.), **Modelling, Applications and Applied Problem Solving**, Ellis-Horwood, Chichester, pp 158–167.

Williams JS, 1991, *Modelling in Mechanics: A Cross Curricular and Problem Solving Approach to Learning Mathematics*, in Niss M, Blum W and Huntley I, (eds.), **Teaching of Mathematical Modelling and its Applications**,. Ellis-Horwood, Chichester. pp 271–279.

20

The Development of a Secondary-School Course in Probability, Statistics and Modelling that Attracts and Empowers Students

Thomas L Schroeder and Barry E Shealy
State University of New York at Buffalo, Buffalo, NY.
e-mail: schroede@ubvms.cc.buffalo.edu **or** *bshealy@ubvms.cc.buffalo.edu*

ABSTRACT

A course in probability, statistics and mathematical modelling is being developed by high school and community college mathematics teachers and university mathematics educators in Western New York. The course is designed to build on secondary school mathematics students' geometry and algebra skills developed in the New York State Regents courses and promote the development of higher-order thinking, mathematical power and critical attitudes. The course centres on investigations of real situations and data, particularly related to current events, with implications for the students' lives.

INTRODUCTION

New York State currently requires secondary school students to take two years of mathematics. Many students take only this minimum thus leaving them ill-prepared for post-secondary education and vocational training. Colleges and universities in the State devote substantial amounts of their resources to the teaching of so-called developmental mathematics courses *ie* secondary school level mathematics courses needed as prerequisites to college level courses. Recognition of this problem led the State University of New York system to fund the Mathematics Alert Program (MAP) in cooperation with New York schools. The ultimate goal of MAP is to encourage more high school students to take more mathematics while they are still in high school, so that they can:

- avoid having to take developmental mathematics in college,
- move smoothly into the college mathematics sequence for their major,
- maximise their choices among college majors.

MAP pursues these goals by giving students in the penultimate year of secondary schooling a mathematics achievement test and a questionnaire about their educational plans. In return, students are given individualised feedback concerning the mathematical requirements of two post-secondary programmes of their choice, and an indication of what developmental courses, if any, they would be required to take in their first year of college should their mathematics skills remain at the level tested. School counsellors use the scores to advise students about what mathematics courses they should take in their senior year to prepare for the major(s) they might want to pursue in college.

As a follow-up to the MAP, we undertook a critical investigation of the existing secondary mathematics curriculum and considered whether there was a need to develop additional, alternative mathematics courses that would be attractive to students facing the question of whether to take a third or fourth high school mathematics course. The current New York State Regents Sequential Mathematics Courses I, II, and III focus on algebra, geometry and trigonometry, thus preparing students to take calculus. These courses give much less attention to such mathematical topics as probability, statistics, mathematical modelling and linear algebra, topics that would be relevant to a wider range of vocational and educational goals. These findings were not unexpected as they are characteristic of most curricula used until recent years in the United States. Students' opportunity to learn these topics varies greatly and, when addressed, the topics are generally considered with little depth and intensity (McKnight *et al.*, 1987). At the same time, many recognise these concepts as increasingly relevant, even essential, to today's work force (National Council of Teachers of Mathematics [NCTM], 1989).

Furthermore, the Sequential Mathematics curriculum, the textbooks with which it is taught, and the Regents mathematics examinations have been criticised as 'sterile'. Thus, like students in many jurisdictions, New York students tend to view mathematics as a disconnected series of techniques and procedures that is largely irrelevant to 'real life' (Schoenfeld, 1988 and 1989). We have proposed, and are now developing, a course designed to attract students who might not otherwise take a third or fourth mathematics course in high school, and to provide them with mathematical experiences relevant to their current life and future careers.

DISTINCTIVE FEATURES OF THE COURSE

The overall goal of the course is to provide students with an opportunity to learn the mathematical content and processes necessary for pursuing their vocational and educational goals, while empowering them to use the mathematics they have previously studied. The content of the course is in line with current mathematics curriculum reform efforts both nationally (*eg* NCTM, 1989) and internationally. We believe the strength and innovation of the course comes from the method of its development and the context it provides for the mathematical concepts and processes that are studied through its activities.

Content

The content of the course spans three general themes that emphasise data analysis, statistics, probability and modelling. The three themes are:

- describing and presenting data,
- dealing with uncertainty,
- inferring and analysing relationships in data.

The first theme includes both descriptive statistics and various graphical means for presenting data. Many of the procedures (*eg* histograms and computing descriptive statistics) are developed superficially as abstract concepts earlier in the Sequential Mathematics curriculum. In our course we extend these techniques with new topics (*eg* by introducing stem-and-leaf diagrams and box plots). Furthermore, the students use the techniques as methods of argument or interpretation and analyse the assumptions, the appropriateness, and possible abuses of the various descriptive and graphical display methods. In the second theme the students build on their understanding of theoretical probability - they connect theoretical and experimental probability, design and interpret probability-based simulations and analyse and construct probabilistic arguments. The final theme includes inferential statistics (*eg* sampling and inferences from confidence intervals) and mathematical modelling of both linear and nonlinear relationships through curve fitting. The content is selected to build on the substantial knowledge and skills in algebra and geometry, which the students have developed in the first two Sequential Mathematics courses, while challenging students who have attained only instrumental understanding of these topics.

Course development

A particular strength of our course development process is its collaborative nature. The process involved high school teachers, community college mathematics professors, and teacher educators from a college and a university. Over the course of the past year the team met periodically to discuss potential activities, to learn content not all were familiar with, and to share ideas and experiences from piloting initial materials. In addition, we provided a graduate-level mathematics education course for teachers to devote more intense efforts to course development. This process of learning by investigating, piloting, and sharing experiences was crucial to the conceptualisation and development of the course materials.

Guiding principles

Out of the early meetings, several guiding principles arose for the development of activities for the course. A basic assumption is that learning is an adaptive activity. We believe students make sense of their environment by bringing to bear their existing knowledge, assimilating and accommodating new information when existing structures are inadequate and constructing new meanings (von Glasersfeld, 1990; NCTM, 1989). This assumption implies that the most powerful contexts for learning would be situations relevant to the students' lives and goals.   Ideally the students will then develop a need to solve the problems. Where we do not challenge students, or expect them to make sense of the solutions, they often fail to do so, developing only isolated and meaningless techniques. Thus we constructed the course so that it develops mathematical ideas through problem-based investigations. The problem situations involve real data and are initiated both by the teacher and the students. To facilitate working with real data the students make extensive use of advanced calculators, computers and electronic measurement devices. With the concepts relevantly situated, the students bring richer constructs to the solution of the problem and more power for the development of new constructs (Lambert *et al.*, 1989).

If we take seriously the assumption that learning is adaptive, the selection of problem situations is crucial. Our selections were guided by the three criteria of

- significant mathematical content,
- potential for developing mathematical power (as expressed by the NCTM, 1989) and higher-order thinking (Resnick, 1987),
- potential for addressing critical issues (de Lange, 1987; Skovsmose, 1990).

These criteria may be seen as integrally related and mutually supportive. Once we define the content, the ideas of mathematical power and higher-order thinking provide guidance for presenting and investigating the content and describing the nature of the understanding students should develop. The criterion of critical issues then emphasises the importance of mathematics as an empowering mode of reasoning.

By significant mathematics we mean that the content of the course should focus on the study of data analysis, statistics and probability as outlined in the NCTM Curriculum Standards (1989) and build strongly on the algebraic and geometric ideas developed in the first two Sequential Mathematics courses. Since the Sequential Mathematics courses include basic concepts of probability, combinatorics and descriptive statistics, the new course should reinforce the students' understanding of these concepts and their ability to interpret and apply them in new situations; it should not simply revisit those concepts. The content should bridge the gap between these basic ideas and the level of abstraction required in a university-level course in mathematical methods for a particular major. For example in modelling relationships it is important to involve extensive consideration of nonlinear functions. Thus we define significance in terms both of a foundation in previously developed concepts and also movement toward a higher level of abstraction and more complex conceptualisation.

A concern for the notions of mathematical power and higher-order thinking extends and strengthens the foundation of significant mathematics. The mathematical methods and techniques in the course are developed not merely for their own sake but as vehicles for solving meaningful problems thus conveying the message that it is important for students to be able to invent methods, to choose among plans and procedures and to understand computations. We endorse the NCTM Standards definition of mathematical power as the ability to "investigate, to make sense of and to construct meanings from new situations; to make and provide arguments for conjectures; and to use a flexible set of strategies to solve problems from both within and outside mathematics" (NCTM, 1989). The activities of the course encourage students to develop conjectures and arguments for their conjectures as well as to evaluate the mathematically oriented arguments of others.

A focus of the course is on higher-order thinking. Rather than developing a cookbook of algorithms and formulaic approaches to well-defined problems the students are placed in situations where higher-order thinking is encouraged. According to Resnick (1987), higher order thinking:

- is *nonalgorithmic* - the path of action is not fully specified in advance,
- tends to be *complex* - the total path is not "visible" (mentally speaking) from any single vantage point,
- often yields *multiple solutions*, each with costs and benefits, rather than unique solutions,
- involves *nuanced judgment* and interpretation,
- involves the application of *multiple criteria*, which sometimes conflict with one another,
- often involves *uncertainty* - not everything that bears on the task at hand is known,
- involves *self-regulation* of the thinking process - we do not recognise higher order thinking in an individual when someone else 'calls the plays' at every step,
- involves *imposing meaning* - finding structure in apparent disorder,
- requires *effort* - there is considerable mental work involved in the kinds of elaborations and judgments required.

In the development of the course we take these characteristics seriously. They reflect the adaptive nature of learning and provide a guide for the nature of the students' investigations of significant mathematical concepts and procedures.

A consequence of higher-order thinking as described by Resnick is the development of critical attitudes, *ie* the ability to judge mathematically oriented presentations (de Lange, 1987). This emphasis is an aspect of higher-order thinking but we consider it separately in order to emphasise the students' ability to deal with mathematics that has implications— social, political, and vocational. As Berlinski has said, "mathematical descriptions, when applied, tend to drive out all others" (quoted in Davis, 1991), but by emphasising critical attitudes we want students to develop a mindset to see mathematics as problematic rather than cut-and-dried, and thus to question mathematical descriptions. Contrary to what is often public perception, guiding interests and underlying assumptions do affect mathematical arguments. Moving toward this realisation the students judge others' presentations and interpretations of data through considering excerpts from current newspapers and magazines. They then develop their own presentations and interpretations. We challenge them to evaluate not only the mathematical processes and reasoning but also the assumptions, guiding interests, and implications inherent in the arguments. When the students develop their own interpretations, they evaluate critically their own presuppositions and guiding interests, both from inside and outside mathematics, that may affect the solution of the problem and its interpretation (Skovsmose, 1990). Such considerations provide the students with practice in using mathematical reasoning to make real-life decisions and thus empower them to make decisions in the future.

EXAMPLES OF COURSE ACTIVITIES

In this section we outline three examples of activities we have developed for the course and discuss them in terms of the principles and criteria mentioned above. The first two examples are related to the mathematics of linear equations in two variables, content that is presented in Courses I and II. They meet our criterion of mathematical significance to the extent that they cause students to deepen their understanding of these concepts, extend students' ability to interpret and apply them in new situations and move students toward a higher level of abstraction and more complex conceptualisation. The third example highlights the issues of critical attitudes and empowering students to use mathematical reasoning to make real-life decisions.

Cost of sequestering the O J Simpson jury

Quite a few of our activities are based on items from current newspapers and magazines. One such item appeared on the front page of USA TODAY for January 19, 1995. Under the heading 'OJ jury: An expensive stay' it gives estimates of the cost of sequestering the jury (including alternates) for the trial of OJ Simpson, the former American football star accused of murdering his wife. The problem we ask students to investigate is 'where do these numbers come from?'. Our decision to explore this topic was based partly on the fact that the Simpson case was such a prominent topic of everyday conversation that there could be no question of its impact on students.

On the surface there is no way of determining how the authors/designers arrived at the estimates but we have seen that, after carrying out a few calculations, students can make educated guesses and hypotheses thereby demonstrating mathematical power. Although we do not want teachers to "call the plays at every step" (Resnick, 1987), we have prepared a series of leading questions that teachers can use as hints to foster and guide students' thinking within the problem's context (thinking about numbers of days, costs per day or over a number of days, *etc*) and other questions designed to evoke applications of skills related to linear equations and their graphs. These hints in the form of questions should be used sparingly by the teacher; our hope is that the students themselves will pose these or similar questions.

Students can easily find out how many days into the future each projection is, counting from the date of publication. They may then find that dividing each estimated total cost by the corresponding number of days gives three different estimates of cost per day (*ie* \$4409.50 to June 30, \$4376.18 to July 31, \$4352.07 to August 31). If students do not think of doing so themselves we suggest that they calculate the costs per day for the periods of time between the three future dates given. The costs per day in these periods turn out to be the same, namely \$4202.00 per day. If they then divide each of the given cost estimates by this cost per day (*eg* \$714 340 ÷ \$4202/day = 170 days) the students find that the calculated number of days does not match the actual number of days (*eg* June 30 is only 162 days from the date of publication not 170 days). With such facts as these before them, the students are faced with the problem of accounting for discrepancies, a problem that clearly demands 'nuanced judgment and interpretation' (Resnick, 1987).

Teachers may also suggest that students think of the situation in terms of linear equations by asking them to express the three estimates as ordered pairs with x = number of days and y = total cost. Following on from this hint we expect students to find that the three points are collinear, that the equation of the line passing through them is $y = 4202x + 33616$, that the x-intercept is -8 (days) and the y-intercept is 33 616 (dollars). We think this situation provides an excellent opportunity for students to relate their pre-understanding of the situation and their formal, abstract knowledge of algebra, experiencing what the NCTM Curriculum Standards calls mathematical connections and demonstrating mathematical power (NCTM, 1989). However we recognise that these connections may be difficult for students to come up with on their own since there is little in the surface appearance of the bar graph to suggest an equation in the form $y = mx + b$ or its graph. Furthermore the coefficients of this linear equation are much larger than the coefficients typically found in equations in American high school mathematics textbooks and the *x*-intercept at (-8) days may seem strange.

Whether they work in the context of the real world situation or in the context of its representation as an algebraic equation, students should reason that 8 days prior to the date of publication the cost of sequestering the jury was zero and they should interpret this as meaning that the sequestration of the jury must have begun then (a fact that can be verified by reference to other articles in the same newspaper). We also expect students to identify and analyse the assumptions underlying the cost projections. Whoever developed them probably did so by taking the amount spent per day so far ($4202) and projecting into the future at that constant rate and we expect students to question whether it is reasonable to assume that the cost per day will remain constant (considering the impact of dismissed jurors, seasonal variations, *etc*). Finally we hope that students will put the issue of the 'expensive stay' into perspective by asking such questions as:

- What social values need to be considered in deciding whether the cost is 'reasonable' or 'too expensive'?
- What interests and whose interests are served by spending public funds on jury sequestration?
- How should mathematical evidence be used in answering these questions?

Cancer death rate and exposure to radioactive contamination

Many of our activities involving linear equations are for the purpose of modelling existing data rather than projecting into the future. We introduce students to line of best fit procedures by working through examples using the median-median method also known as the three-point line (Tukey, 1977) because we think it is less complicated and more intuitively understandable than classical least squares linear regression. Later we encourage students to use TI-82 calculators for both median-median and least-squares linear models and also for higher-order-polynomial, logarithmic and exponential regression analyses.

We introduce the median-median fit line with an example using data from a 1965 article in the Journal of Environmental Health (as presented in North Carolina School of Science and Mathematics, 1988). That article investigated cancer death rates per 100,000 population in several communities along the Columbia River near Portland, Oregon and an index of their exposure to water-borne radioactive contamination from the atomic energy plant at Hanford, Washington which has produced plutonium since World War II. Although the article is a bit dated the issue of industrial pollution continues to be a major public concern (since we live within 50 km of the infamous Love Canal site) and the media have recently reported new problems at other local chemical waste disposal sites. We think these circumstances increase the students' ownership of this problem and heighten their awareness of the need to consider social and political implications and the guiding interests of the various groups involved in dealing with newly-discovered problems.

Educational data analysis

Perhaps, not surprisingly, we have considered many activities involving educational data ranging from marks on mathematics assignments to large data sets available on the Internet. In many cases students learn from these activities that there is not a very strong association between one variable and another. While this might be expected in such cases as the midterm test and final exam for a course it is also found in such cases as comparisons of college entrance exam scores with college grade point averages. For many students this raises the critical issue of whether the use of such test scores in college admissions is justified or fair.

Faced with decisions about which post-secondary institutions to apply to, many students turn to college guidebooks which provide general information about the institutions and data describing the students admitted to them. For example one popular guidebook gives the range of Scholastic Aptitude Test (SAT) scores for the middle 50% of students admitted to each college. While this may appear to be a straightforward factual statistic, how it could be or should be used by an individual student deciding whether or not to apply to an particular college is far from clear. The difficulties were compounded this spring with the publication of articles in Time magazine and The Wall Street Journal (Stecklow, 1995) showing that some colleges provided different information to the guidebooks than they did to the National Collegiate Athletic Association (NCAA) or to bond-rating agencies. These articles detail such practices as excluding the scores of students admitted under special programmes or excluding the verbal (but not mathematical) scores of international applicants and they also present various issues concerning the meaning of scores or score ranges for different groups of students . We hope that students will not just become cynical and conclude that 'figures lie and liars figure.' Instead we hope they will consider issues carefully and critically and make judgments about whether these constitute legitimate differences of opinion and interpretation or out-and-out misrepresentations.

CONCLUSION

The examples presented above show our work in progress in relation to our goals and values. All those involved in the project - secondary school mathematics teachers and mathematics educators based in colleges or the university - have been challenged to ensure that the mathematics we engage students in is sound and significant and that our activities will develop students' abilities to think critically and use mathematics thoughtfully. Although it is unreasonable to expect each and every activity to meet all of our criteria, we hope they will each contribute to the creation of a course that is both attractive and empowering.

REFERENCES

Davis PJ, 1991, *Applied Mathematics as a Social Instrument*, in Niss M, Blum W and Huntley I, (eds.), **Teaching of Mathematical Modelling and Applications**, Ellis Horwood, Chichester, UK.

de Lange J, 1987, **Mathematics, Insight and Meaning**, OW & OC, Utrecht.

Lambert P, Steward AP, Manklelow KI and Robson EH, 1989, *A Cognitive Psychology Approach to Model Formulation in Mathematical Modelling*, in Blum W, Berry JS, Biehler R, Huntley ID, Kaiser-Messmer G and Profke L, (eds.), **Applications and Modelling in Learning and Teaching Mathematics**, pp 92-97, Ellis Horwood, Chichester, UK.

McKnight CC, Crosswhite FJ, Dossey JA, Kifer E, Swafford JO, Travers KJ and Cooney TJ, 1987, **The Underachieving Curriculum**, Stipes, Champaign, Illinois.

National Council of Teachers of Mathematics (NCTM), 1989, **Curriculum and Evaluation Standards for School Mathematics**, NCTM, Reston, Virginia.

North Carolina School of Science and Mathematics Department of Mathematics and Computer Science, 1988, **Data Analysis**, NCTM, Reston, Virginia.

Resnick L, 1987, *Education and Learning to Think*, **National Academy Press**, Washington, DC.

Schoenfeld A, 1988, *When Good Teaching Leads to Bad Results: The Disasters of 'Well Taught' Mathematical Courses*, **The Educational Psychologist**, **23**, pp 145-166.

Schoenfeld A, 1989, *Explorations of Students' Mathematical Beliefs and Behavior*, **Journal for Research in Mathematics Education**, **20**, pp 338-350.

Skovsmose O, 1990, *Reflective Knowledge: Its Relation to the Mathematical Modeling Process*, **International Journal of Mathematical Education in Science and Technology**, **21**, pp 765-779.

Stecklow S, 1995, *Cheat Sheets: Colleges Inflate SATs and Graduation Rates in Popular Guidebooks*, **The Wall Street Journal**, **76**, April 5, pp A1, A4.

Tukey JW, 1977, *Exploratory Data Analysis*, Addison-Wesley, Reading, MA.

Section D

Tertiary Case Studies

21

A Unified Approach to the Mathematical Modelling of Mechanical Systems

Krzysztof Arczewski
Warsaw University of Technology
Warsaw, Poland.
e-mail: krisarcz@grape.meil.pw.edu.pl

Wojciech Blajer
Technical University of Radom
Radom, Poland.
e-mail:wsiraom@atos.warman.com.pl

ABSTRACT

The paper presents an innovative approach to teaching mathematical modelling of complex mechanical systems, introduced in the advanced courses of mechanics (analytical dynamics) at technical universities. Both holonomic and nonholonomic systems are treated in same way, and no attention is paid to the equations of motion the derived in generalised velocities or quasi-velocities. Compared with the traditional methods of classical mechanics, the present approach turns out to be extraordinarily short, elementary and general. Due to its compact matrix formulation, it is also perfectly suitable for numerical applications and computer symbolic manipulation. Here we describe the method very briefly and concentrate on showing its convenience by way of examples. Reflections and experiences on teaching and learning with this approach are also included.

INTRODUCTION

The mathematical modelling of complex mechanical systems is the subject of lectures on advanced mechanics (analytical dynamics) delivered to students, in the senior years of technical universities, specialising in mechanical engineering. The traditional teaching process of the courses includes the classical methods of Lagrangian mechanics, supplemented by various methods for nonholonomic systems. The way of presenting the methods is strongly influenced by various historical approaches such as the virtual principles, and the mathematical language used is obsolete and difficult to follow. Moreover, substantially different methodologies are used for holonomic (H) or noholonomic (NH) systems and/or for the dynamic equations derived in generalised velocities or quasi-velocities. The fact that the methods are introduced for systems of particles may also cause some confusion. It is the opinion of the authors that all this practice is pointless, sometimes even misleading.

Many of the available methods, with their usual complexity and impenetrability (what are they really saying, cause many students (and, we believe, some of their teachers) to be unable to apply the methods to the modelling of real (multibody) systems. In this paper we would like to present an alternative approach to the solution of mechanics problems. This innovative method, called the 'projection method' (Blajer, 1992a, b), is conceptually simple, of an intuitive nature, compact and general. Both H and NH systems are treated in the same way, and no attention is paid to the dynamic equations expressed in generalised velocities or quasi-velocities. Due to its compact matrix formulation, it is also applicable to numerical applications and computer symbolic manipulations. The experience of the authors shows that the method is much more understandable and easy to use for students than the other methods available in the literature. Actually, after some training, they are able to model quite complex mechanical systems, achieving much more with less effort.

In this paper a compact matrix formulation of the projection method is presented, and its generality and simplicity emphasised. We then focus on showing its convenience by way of examples. Reflections and experiences on teaching and learning with the use of the approach are also included.

The projection method

The starting point is an n-degree-of-freedom unconstrained mechanical system, whose equations of motion can be written in the following general matrix form:

$$\dot{\mathbf{x}} = \mathbf{A}(\mathbf{x})\,\mathbf{v} \tag{1}$$

$$\mathbf{M}(\mathbf{x})\,\dot{\mathbf{v}} = \mathbf{h}(\mathbf{x},\mathbf{v},t) \tag{2}$$

where $\mathbf{x} = [x_1,\ldots,x_n]^T$ and $\mathbf{v} = [v_1,\ldots,v_n]^T$ are the system position and velocity variables, $\mathbf{A}$ is an $n{\times}n$ matrix of transformation between $\dot{\mathbf{x}}$ and $\mathbf{v}$ velocity components, $\mathbf{M}$ is an $n{\times}n$ generalised mass matrix, $\mathbf{h}$ is an n-column representing the applied and gyroscopic forces on the system, and t is time. The said unconstrained system can be either:

- a collection of unconstrained particles,
- a collection of unconstrained rigid bodies,
- a Lagrangian (open-loop, tree structure) system,
- a combination of subsystems as above (whichever unconstrained system).

Let the system defined in Equations. (1) and (2) be subject to m_h H and m_{nh} NH constraints, $m_h + m_{nh} = m < n$, whose original equations and their differentiated forms are:

$$\Phi_h(\mathbf{x},t) = 0$$
$$\mathbf{C}_{nh}(\mathbf{x},t)\,\mathbf{v} + \eta_{nh}(\mathbf{x},t) = 0 \qquad \xrightarrow{\;(\textit{differentiation})\;} \qquad \mathbf{C}(\mathbf{x},t)\,\dot{\mathbf{v}} + \xi(\mathbf{x},\mathbf{v},t) = 0 \qquad (3)$$

where the $m{\times}n$ constraint matrix $\mathbf{C} = [\mathbf{C}_h^T \;\; \mathbf{C}_{nh}^T]^T$ is of maximal rank, $\mathbf{C}_h = (\partial\Phi_h/\partial\mathbf{x})\,\mathbf{A}$, the m-vector $\xi = \dot{\mathbf{C}}\mathbf{v} + \dot{\eta}$, and the m-vector $\eta = [(\partial\Phi_h/\partial t)^T, \eta_{nh}^T]^T$ vanish for scleronomic constraints. Assuming that the initial values of $\mathbf{x}$ and $\mathbf{v}$ satisfy the lower-order constraint conditions, ie $\Phi_h(\mathbf{x}_0,t_0) = 0$ and $\mathbf{C}(\mathbf{x}_0,t_0)\,\mathbf{v}_0 + \eta(\mathbf{x}_0,t_0) = 0$, the governing equations of the constrained system can then be written in the following form of differential-algebraic equations (DAEs), often referred to as Lagrange's equations of type one:

$$\dot{\mathbf{x}} = \mathbf{A}\mathbf{v} \qquad (4a)$$
$$\mathbf{M}\dot{\mathbf{v}} = \mathbf{h} + \mathbf{C}^T\lambda \qquad (4b)$$
$$0 = \mathbf{C}\dot{\mathbf{v}} + \xi \qquad (4c)$$

where $\lambda = [\lambda_1,\dots,\lambda_m]^T$ are the Lagrange multipliers corresponding to the respective constraints (3). An advantage of formulation (4) is that it holds good for both H and NH systems, ie for a dynamic system described in Equations. (1) and (2) and then subject to any combination of H and NH constraints (3). The only difference between a H and a NH system is the number of initial value conditions imposed on $\mathbf{x}_0$ - equal to m for a H system and equal to $m - m_{nh}$ for a NH system (NH constraints do not restrict the system position).

Without going into details, the crux of the projection method lies in the partition of the linear n-dimensional space N, referred to the configuration space of the unconstrained system, into an m-dimensional constrained subspace C, and a k-dimensional tangent subspace D, $k = n - m$, such that $C \cup D = N$ and $C \cap D = 0$. As C is spanned by m constraint vectors contained in $\mathbf{C}$ as rows, in order to define D it suffices to determine an $n{\times}k$ maximal rank matrix $\mathbf{D}(\mathbf{x},t)$ being an orthogonal complement to $\mathbf{C}$, ie

$$\mathbf{D}^T\mathbf{C}^T = 0 \quad \Leftrightarrow \quad \mathbf{C}\mathbf{D} = 0 \qquad (5)$$

and the vectors that span D are represented as columns in $\mathbf{D}$. Then, k independent tangent speeds $\mathbf{u} = [u_1,\dots,u_k]^T$ specifying the system velocity in D can be defined (Blajer, 1992b):

$$\begin{bmatrix} \mathbf{D}^T\mathbf{M}\mathbf{D}\mathbf{u} \\ 0 \end{bmatrix} = \begin{bmatrix} \mathbf{D}^T\mathbf{M}\mathbf{v} \\ \mathbf{C}\mathbf{v} + \eta \end{bmatrix} \equiv \mathbf{T}^T\mathbf{M}\mathbf{v} + \begin{bmatrix} 0 \\ \eta \end{bmatrix} \qquad (6)$$

where $\mathbf{T} = [\mathbf{D} \quad \mathbf{M}^{-1}\mathbf{C}^T]$. From Equation (6) it follows directly that:

$$\mathbf{u} = (\mathbf{D}^T\mathbf{M}\,\mathbf{D})^{-1}\mathbf{D}^T\mathbf{M}\,\mathbf{v} \qquad\qquad (7)$$

$$\mathbf{v} = \mathbf{D}\mathbf{u} + \gamma \qquad\qquad (8)$$

where $\gamma(\mathbf{x},t) = -\mathbf{M}^{-1}\mathbf{C}^T(\mathbf{C}\mathbf{M}^{-1}\,\mathbf{C}^T)^{-1}\eta$ vanishes for scleronomic constraints.

It may be worth noting that the tangent speeds **u** are notionally equivalent to the independent 'kinematic parameters' used in the Maggi and Gibbs/Appell methods (see, for instance, Neimark and Fufaev, 1967), and to the independent 'generalised speeds' exploited in Kane's method (Kane and Levinson, 1985). In view of the latter method, **D** can also be identified with the partial velocity matrix. None of the methods mentioned, however, provides the user with an explicit formula for the determination of **u** in terms of **v** and **x**, which is of paramount importance for the successful initialisation of the integration process of the purely kinetic equations of motion that follow. The definition (6) and the consequent relation (7) fill this void in the literature.

Projecting Equation (4b) into D and C, *ie* premultiplying the dynamic equations by $\mathbf{T}^T$, and then substituting Equation (8) and its time-derivative for **v** and $\dot{\mathbf{v}}$, respectively, the tangential projection leads to the following pure differential equations of motion:

$$
\begin{array}{ccc}
\dot{\mathbf{x}} = \mathbf{A}\mathbf{D}\mathbf{u} + \mathbf{A}\gamma & & \dot{\mathbf{x}} = \hat{\mathbf{A}}(\mathbf{x},t)\mathbf{u} + \hat{\mathbf{a}}(\mathbf{x},t) \\
\mathbf{D}^T\mathbf{M}\mathbf{D}\dot{\mathbf{u}} = \mathbf{D}^T[\mathbf{h} - \mathbf{M}(\dot{\mathbf{D}}\mathbf{u} + \dot{\gamma})] & \Leftrightarrow & \hat{\mathbf{M}}(\mathbf{x},t)\dot{\mathbf{u}} = \hat{\mathbf{h}}(\mathbf{x},\mathbf{u},t)
\end{array}
\qquad (9)
$$

whereas the orthogonal projection (into C) enables one to determine **l**,

$$\lambda(\mathbf{x},\mathbf{u},t) = (\mathbf{C}\mathbf{M}^{-1}\mathbf{C}^T)^{-1}\mathbf{C}(\dot{\mathbf{D}}\mathbf{u} + \dot{\gamma} - \mathbf{M}^{-1}\mathbf{h}) \qquad\qquad (10)$$

Equations (9) form $n+k$ first-order ordinary differential equations (ODEs) in **x** and **u**, and correspond to Maggi's equations of motion, see also Blajer (1992b). Based on the solutions $\mathbf{x}(t)$ and $\mathbf{u}(t)$ to the ODEs, **l**(t) can be synthesized from Eq. (10). For carefully modelled constraint equations (3), **l** may denote the physical reaction forces of the constraints.

A successful application of the projection method is strongly dependent on the effective determination of **D**, and then $\dot{\mathbf{D}}$ and $\dot{\gamma}$ in order to formulate the minimal-order governing equations (9). In fact, the task can be accomplished in many ways. For many simple problems, $\mathbf{D}(\mathbf{x}, t)$ that satisfies the condition (5) can simply be guessed or, after a while, found by inspection. Then $\dot{\mathbf{D}}$ and $\dot{\gamma}$ can be derived symbolically. This is the most popular approach to solving problems of academic interest (see Example 2). The other natural method, for H systems only, is to introduce k independent (generalised) coordinates $\mathbf{q} = [q_1, \ldots, q_k]^T$. Then, the H constraints can be introduced implicitly by way of the relationship.

$$\mathbf{x} = \mathbf{g}(\mathbf{q}, t) \quad \xrightarrow{\ (differentiation)\ } \quad \mathbf{v} = \mathbf{D}(\mathbf{q}, t)\,\dot{\mathbf{q}} + \boldsymbol{\gamma}(\mathbf{q}, t) \tag{11}$$

where $\dot{\mathbf{q}}$ can be identified with $\mathbf{u}$ as in Equation (8), and ODEs (9) transform to

$$\mathbf{D}^T \mathbf{M} \mathbf{D}\,\ddot{\mathbf{q}} = \mathbf{D}^T[\mathbf{h} - \mathbf{M}(\dot{\mathbf{D}}\dot{\mathbf{q}} + \dot{\boldsymbol{\gamma}})] \quad \Leftrightarrow \quad \hat{\mathbf{M}}(\mathbf{q}, t)\,\ddot{\mathbf{q}} = \hat{\mathbf{h}}(\mathbf{q}, \dot{\mathbf{q}}, t) \tag{12}$$

The same result can also be obtained from Lagrange's equations of type two (Blajer, 1995), Kane's equations and the Gibbs-Appell equations (Blajer, 1992a).

Finally, let us point out that the projection method can also be used for changing from one reference frame to another. In such a case $\mathbf{D}$ should be replaced by a transformation matrix between velocity components (see Example 2 for an illustration).

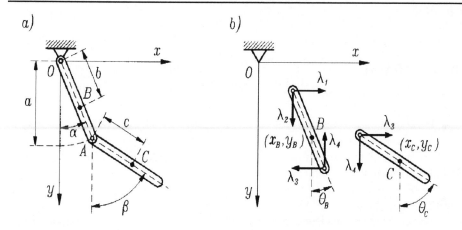

Figure 1: double physical pendulum

CASE STUDIES

Example 1: double physical pendulum (Fig. 1).
This is a classical academic demonstration, usually solved by using the Lagrange equations of type two (Arczewski and Pietrucha, 1993). Setting $\mathbf{x} = [x_B, y_B, \theta_B, x_C, y_C, \theta_C]^T$ for the unconstrained system shown in Fig. 1b, Equation (1) simplifies to $\dot{\mathbf{x}} = \mathbf{v}$, while $\mathbf{M}$ and $\mathbf{h}$ as in Equation (2) are:

$$\mathbf{M} = diag(m_B, m_B, I_B, m_C, m_C, I_C); \qquad \mathbf{h} = [0, m_Bg, 0, 0, m_Cg, 0]^T$$

where m_i and I_i ($i = B, C$) are, respectively, the rod masses and mass moments of inertia about their mass centres B and C, and g is the gravity acceleration. Denoting $\mathbf{q} = [\alpha, \beta]^T$, the constraints on the rods due to the kinematical joints can be introduced implicitly by

$$\mathbf{x} \equiv \begin{bmatrix} x_B \\ y_B \\ \theta_B \\ x_C \\ y_C \\ \theta_C \end{bmatrix} = \begin{bmatrix} b\sin\alpha \\ b\cos\alpha \\ \alpha \\ a\sin\alpha + c\sin\beta \\ a\cos\alpha + c\cos\beta \\ \beta \end{bmatrix} \equiv \mathbf{g}(\mathbf{q}) ; \qquad \dot{\mathbf{x}} = \begin{bmatrix} b\cos\alpha & 0 \\ -b\sin\alpha & 0 \\ 1 & 0 \\ a\cos\alpha & c\cos\beta \\ -a\sin\alpha & -c\sin\beta \\ 0 & 1 \end{bmatrix} \begin{bmatrix} \dot\alpha \\ \dot\beta \end{bmatrix} \equiv \mathbf{D}(\mathbf{q})\,\dot{\mathbf{q}}$$

Then the dynamic equations can easily be determined in the form $\hat{\mathbf{M}}(\mathbf{q})\,\ddot{\mathbf{q}} = \hat{\mathbf{h}}(\mathbf{q},\dot{\mathbf{q}})$ as in Equation (12), where:

$$\hat{\mathbf{M}} \equiv \mathbf{D}^T \mathbf{M}\,\mathbf{D} = \begin{bmatrix} I_O + m_C a^2 & m_C ac\cos\varphi \\ m_C ac\cos\varphi & I_A \end{bmatrix}$$

$$\hat{\mathbf{h}} \equiv \mathbf{D}^T(\mathbf{h} - \mathbf{M}\dot{\mathbf{D}}\,\dot{\mathbf{q}}) = \begin{bmatrix} m_C ac\dot\beta^2 \sin\varphi - (m_B b + m_C a)g\sin\alpha \\ m_C ac\dot\alpha^2 \sin\varphi - m_C cg\sin\beta \end{bmatrix}$$

$I_O = I_B + m_B b^2$, $I_A = I_C + m_C c^2$, and $\varphi = \beta - \alpha$. Note that the above equations were obtained by following elementary matrix transformations. The same result could be obtained by using clearly traditional Lagrange equations (Arczewski and Pietrucha, 1993), but in a much more complicated way.

Example 2: the rolling disc. (Fig. 2)
This is another classical academic demonstration, relating a NH system. Reported in many textbooks (see Neimark and Fufaev, 1967; Desloge, 1982; Kane and Levinson, 1985), the case is seldom solved in a unified way as a system subject to both H and NH constraints.

Consider a sharp-edged homogeneous disc of radius r and mass m that rolls without sliding on the horizontal plane Oxy (Fig. 2a). Let the generalised coordinates of the unconstrained disc be $\mathbf{x} = [x_G, y_G, z_G, \varphi, \theta, \psi]^T$, where x_G, y_G and z_G are the coordinates of the mass centre G with respect to the inertial frame $Oxyz$, and the angles φ, θ and ψ are as indicated. Let us then choose $\mathbf{v} = [w_1, w_2, w_3, \omega_1, \omega_2, \omega_3]^T$, where v_i and ω_i ($i = 1,2,3$) are the components of the absolute linear velocity of G and the absolute angular velocity of the disc, respectively, both with respect to the moving reference frame $G123$ (the components of $\mathbf{v}$ are quasi-velocities), chosen so that the $G12$ plane is that of the disc, and the $G1$ axis remains parallel to Oxy plane. The kinematic equations, corresponding to Equation (1), are

$$\mathbf{\dot{x}} \equiv \begin{bmatrix} \dot{x}_G \\ \dot{y}_G \\ \dot{z}_G \\ \dot{\varphi} \\ \dot{\theta} \\ \dot{\psi} \end{bmatrix} = \begin{bmatrix} \cos\varphi & -\sin\theta\sin\varphi & \cos\theta\sin\varphi & 0 & 0 & 0 \\ \sin\varphi & \sin\theta\cos\varphi & -\cos\theta\cos\varphi & 0 & 0 & 0 \\ 0 & \cos\theta & \sin\theta & 0 & 0 & 0 \\ 0 & 0 & 0 & 0 & \cos^{-1}\theta & 0 \\ 0 & 0 & 0 & -1 & 0 & 0 \\ 0 & 0 & 0 & 0 & -\tan\theta & 1 \end{bmatrix} \begin{bmatrix} w_1 \\ w_2 \\ w_3 \\ \omega_1 \\ \omega_2 \\ \omega_3 \end{bmatrix} \equiv \mathbf{A(x)v} \qquad (13)$$

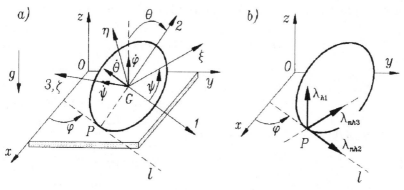

Figure 2: the rolling disc and the constraint reactions

The derivation of the dynamic equations for the unconstrained disc in the $G123$ reference frame requires a little effort. Let us introduce a body-fixed reference frame $\xi\eta\zeta$ (Fig. 2a) that rotates with respect to $G123$ with the angular velocity $\dot{\psi}$ about $\zeta = G3$ axis. The dynamic equations of the disc in $\xi\eta\zeta$ frame (the standard Newton-Euler equations) are

$$\mathbf{\tilde{M}}\,\mathbf{\dot{\tilde{v}}} = \mathbf{h}(\mathbf{x}, \mathbf{\tilde{v}}, t)$$

where

$$\mathbf{\tilde{M}} = diag\big(m, m, m, I, I, I_0\big), \qquad \mathbf{\tilde{h}} = \begin{bmatrix} m(w_\eta\,\omega_\zeta - w_\zeta\,\omega_\eta - g\cos\theta\sin\psi) \\ m(w_\zeta\,\omega_\xi - w_\xi\,\omega_\zeta - g\cos\theta\cos\psi) \\ m(w_\xi\,\omega_\eta - w_\eta\,\omega_\xi - g\sin\theta) \\ -\omega_\eta\omega_\zeta(I_0 - I) \\ \omega_\xi\omega_\zeta(I_0 - I) \\ 0 \end{bmatrix}$$

$\tilde{\mathbf{v}} = [w_\xi, w_\eta, w_\zeta, \omega_\xi, \omega_\eta, \omega_\zeta]^T$, I and I_0 are the disc diametrical and axial moments of inertia, and g is the gravity acceleration. Since the interdependence between $\tilde{\mathbf{v}}$ and $\mathbf{v}$ components is

$$\tilde{\mathbf{v}} \equiv \begin{bmatrix} w_\xi \\ w_\eta \\ w_\zeta \\ \omega_\xi \\ \omega_\eta \\ \omega_\zeta \end{bmatrix} = \begin{bmatrix} \cos\psi & \sin\psi & 0 & 0 & 0 & 0 \\ -\sin\psi & \cos\psi & 0 & 0 & 0 & 0 \\ 0 & 0 & 1 & 0 & 0 & 0 \\ 0 & 0 & 0 & \cos\psi & \sin\psi & 0 \\ 0 & 0 & 0 & -\sin\psi & \cos\psi & 0 \\ 0 & 0 & 0 & 0 & 0 & 1 \end{bmatrix} \begin{bmatrix} w_1 \\ w_2 \\ w_3 \\ \omega_1 \\ \omega_2 \\ \omega_3 \end{bmatrix} \equiv \mathbf{B(x)}\,\mathbf{v}$$

the dynamic equations of the 'flying' disc, expressed in $G123$ and in terms of $\mathbf{x}$ and $\mathbf{v}$, are

$$\mathbf{B}^T \tilde{\mathbf{M}} \mathbf{B} \dot{\mathbf{v}} = \mathbf{B}^T (\tilde{\mathbf{h}} - \tilde{\mathbf{M}} \dot{\mathbf{B}} \mathbf{v}) \qquad \Leftrightarrow \qquad \mathbf{M}\dot{\mathbf{v}} = \mathbf{h}(\mathbf{x}, \mathbf{v}) \tag{14}$$

where $\mathbf{M} = diag(m, m, m,\ I,\ I,\ I_0)$ and $\mathbf{h} = \begin{bmatrix} m(w_2 \tan\theta - w_3)\omega_2 \\ m(w_3 \omega_1 - w_1 \omega_2 \tan\theta - g\cos\theta) \\ m(w_1 \omega_2 - w_2 \omega_1 - g\sin\theta) \\ (I\omega_2 \tan\theta - I_0 \omega_3)\omega_2 \\ (-I\omega_2 \tan\theta + I_0 \omega_3)\omega_1 \\ 0 \end{bmatrix}$

The governing equations composed of Eqs. (13) and (14) are very convenient for further analysis. The equations of constraints on the system, defined as in Eq. (3), are:

$$z_G - \rho\cos\theta = 0$$
$$w_1 - \rho\omega_3 = 0 \tag{15}$$
$$w_2 \sin\theta - w_3 \cos\theta + \omega_1 \rho \tan\theta = 0$$

where the first equation stands for a H constraints, while the two others for NH ones. The constraint matrix $\mathbf{C}$, the matrix $\mathbf{D}$, and the term $\mathbf{x}$ are:

$$\mathbf{C} = \begin{bmatrix} 0 & \cos\theta & \sin\theta & -\rho\sin\theta & 0 & 0 \\ 1 & 0 & 0 & 0 & 0 & \rho \\ 0 & \sin\theta & -\cos\theta & \rho\cos\theta & 0 & 0 \end{bmatrix} \xrightarrow{\text{(inspection)}} \mathbf{D}^T = \begin{bmatrix} \rho & 0 & 0 & 0 & 0 & -1 \\ 0 & 0 & 0 & 0 & 1 & 0 \\ 0 & 0 & \rho & 1 & 0 & 0 \end{bmatrix}$$

$$\xi = \begin{bmatrix} (w_2 \sin\theta - w_3 \cos\theta)\,\omega_1 + \omega_1^2 \rho \cos\theta \\ 0 \\ -(w_2 \cos\theta + w_3 \sin\theta)\,\omega_1 + \omega_1^2 \rho \sin\theta \end{bmatrix}$$

and the minimal-dimension governing equations of the disk, corresponding to Equations (9), are:

$$\dot{x}_G = \rho u_1 \cos\varphi + \rho u_3 \cos\theta \sin\varphi,$$
$$\dot{y}_G = \rho u_1 \sin\varphi - \rho u_3 \cos\theta \cos\varphi$$
$$\dot{z}_G = \rho u_3 \sin\theta$$
$$\dot{\varphi} = u_2 \cos^{-1}\theta$$
$$\dot{\theta} = -u_3$$
$$\dot{\psi} = -u_1 - u_2 \tan\theta$$

$$(I_0 + m\rho^2)\dot{u}_1 = -m\rho^2 u_1 u_2$$
$$I \dot{u}_2 = -(I_0 u_1 + I u_2 \tan\theta) u_3$$
$$(I + m\rho^2)\dot{u}_3 = m\rho^2 u_1 u_2 + (I_0 u_1 + I u_2 \tan\theta) u_2 - m g \rho \sin\theta$$

The above equations form nine ODEs in **x** and **u**. Given $\mathbf{x}_0$ and $\mathbf{v}_0$, which satisfy the conditions of constraints (15), the initial values of tangent speeds **u** can be determined from Equation (7) as:

$$u_{10} = \frac{m\rho}{I_0 + m\rho^2}\, w_{10} - \frac{I_0}{I_0 + m\rho^2}\,\omega_{30}$$

$$u_{20} = \omega_{20}$$

$$u_{30} = \frac{m\rho}{I + m\rho^2}\, w_{30} - \frac{I}{I + m\rho^2}\,\omega_{10}$$

Finally, Equation (10) enables one to determine the constraint reactions (Fig. 2b):

$$\lambda_{n1} = m\rho\left(\frac{I}{I+m\rho^2} u_2^2 \tan\theta \sin\theta - u_3^2 \cos\theta + \frac{I_0 + m\rho^2}{I + m\rho^2} u_1 u_2 \sin\theta\right) + \frac{I + m\rho^2 \cos^2\theta}{I + m\rho^2} mg$$

$$\lambda_{nh2} = \frac{I_0\, m\rho}{I_0 + m\rho^2}\, u_2 u_3$$

$$\lambda_{nh3} = m\rho \left(\frac{I}{I+m\rho^2} u_2^2 \sin\theta - u_3^2 \sin\theta + \frac{I\cos^{-1}\theta + m\rho^2 \tan\theta\sin\theta - I_0\cos\theta}{I+m\rho^2} u_1 u_2 \right)$$

$$+ \frac{m\rho^2}{I+m\rho^2} mg\sin\theta\cos\theta$$

CONCLUSIONS

The projection method provides one with a general purpose and convenient tool for effective mathematical modelling of complex mechanical systems. Its advantages can be summarised as follows.

- Ease of derivation, simplicity, and intuitive in nature.
- Compact matrix formulation, useful in computer applications.
- Diversity of applications (not mentioned in this paper).
- Unified treatment of holonomic and nonholonomic systems.
- No attention need be paid to the components of **u** and **v** which are generalised velocities and/or quasi-velocities (in many other methods the concerned problems are very subtle and lead to enormous complexity).
- No scalar function (*eg* kinetic energy) needs to be introduced and then differentiated, that usually is a laborious and difficult to computerize task.
- Constraint reactions can be determined.
- The method is a unification of many other known methods of mechanics. One general procedure for solving constrained system problems is given

The authors had delivered various courses on mathematical modelling of mechanical systems following traditional analytical mechanics, and started to promote the projection method only recently. Our experiences to date have been very positive for a number of reasons:

- The theoretical background given to the students can be much shorter, and we can devote more time to case studies and engineering applications.
- The proposed matrix procedure is very automated and easy to use. Most of our students, after some elementary training, are able to follow the algorithm and solve quite complex problems, doing much more with less effort.
- The students are very happy with having only one general method for the modelling of mechanical systems.
- The proposed formulation is perfectly suitable for numerical applications and/or computer symbolic manipulations. The students can apply their knowledge gained during the courses on numerical methods and computer algebra.

-     Experienced in applying the projection method, students are well prepared to use (and understand) diverse multibody software systems, which they usually encounter in their engineering courses.

Having experimented with many sample problems taken from standard books on classical mechanics, the authors have found the projection method especially useful in teaching analytical mechanics. We encourage all lecturers to judge the merits of the method for themselves by trying it out on some problems of their own and then, eventually, to spread the projective approach amongst their students. Apart from its tutorial qualities, the projection method, being well suitable for computational calculations, may also eventually become a generally accepted method in engineering applications.

REFERENCES

Arczewski K and Pietrucha J, 1993, **Mathematical Modelling of Complex Mechanical Systems, Vol. 1: Discrete Models**, Ellis Horwood, Chichester.

Blajer W, 1992a, *A Projection Method Approach to Constrained Dynamic Analysis*, **ASME Journal of Applied Mechanics**, **59**, pp 643-649.

Blajer W, 1992b, *Projective Formulation of Maggi's Method for Nonholonomic Systems Analysis*, **AIAA Journal of Guidance, Control, and Dynamics**, **15**, pp 522-525.

Blajer W, 1995, *Projective Formulation af Lagrange's and Boltzmann-Hamel Equations for Multibody Systems*, **ZAMM (Zeitschrift für Angewandte Mathematik und Mechanik)**, **75 SI**, pp S107-S108.

Desloge, EA, 1982, **Classical Mechanics**, Wiley, New York.

Kane TR and Levinson DA, 1985, **Dynamics: Theory and Applications**, McGraw-Hill, New York.

Neimark JI and Fufaev NA, 1967, **Dynamics of Nonholonomic Systems**, (Russian), Nauka, Moscow (Translations of Mathematical Monographs, Vol. 33, American Mathematical Society, Providence, RI, 1972).

22

Performance Modelling of Parallel Algorithms

D B Clegg
Liverpool John Moores University, Byrom Street, Liverpool L3 3AF
e-mail: d.b.clegg@livjm.ac.uk

ABSTRACT

Speed-up models are developed for simple calculations on transputer arrays. The outcomes of student attempts to determine results for the calculation of standard deviation are reported and issues raised in the modelling process are assessed. For some students this type of problem has provided an initial experience with practical aspects of modelling. A model for optimising performance on a tree network is derived and shown to compare well with practical values. Some student feedback is reported.

INTRODUCTION

A popular metric for parallel algorithms is speed up, the ratio of the times of execution of a sequential algorithm to that of a parallel algorithm. We shall show that it is possible to develop accurate performance models of parallel algorithms on transputers at the design stage. Hence, at the outset it can be determined whether a particular parallel algorithm is worthwhile implementing before any codes are developed.

Our attention is confined to simple examples, as these allow students new to parallel computing to concentrate on the parallel aspects of the algorithms. The same principles are applicable to more complex situations. Readers interested in the technical detail of the issues for parallel program design are referred to Foster(1995).

TRANSPUTER TECHNOLOGY

A transputer (Inmos, 1988) is a microprocessor with its own arithmetic units, memory and links for connecting to other transputers. Each processor is programmable as a separate unit, and a parallel computer can be constructed by connecting processors through links.

The main features of the technology are as follows:
- processors have their own non-shared memory,
- pransputers have 4 links and 8 channels,
- link speeds are 10 and 20 mega-bits/s,
- point-to-point communication is possible between processors,
- data sharing between processors is by message passing,
- message passing is unidirectional on a single channel
- there are various types / families known as T400, T800 and T9000.

A schematic diagram of a T800-type transputer illustrating links, channels and arithmetic units is illustrated below.

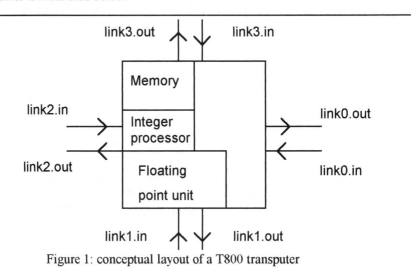

Figure 1: conceptual layout of a T800 transputer

The simplest arrangement is for a 'host' transputer to be located in a PC, which acts as a file server for the transputer board. Links to other transputers are then made from the 'host' transputer board. For example, with two transputers data from the 'host' would be transmitted to the second transputer through the connected links. Each processor would be separately programmed with instructions that included the transmission of data. The first programming language available to do this simply was occam (see Burns, 1988), a language which was developed in parallel with the transputer hardware.

Parallel features of a T800-type transputer

To develop performance models, a basic understanding of parallel features is required. Full details are available in the Inmos transputer reference manual. The parallel features on each T800-type transputer are:
- communication along all 8 channels,
- integer arithmetic unit,
- floating point arithmetic unit.

In total a maximum of 10 parallel operations is possible on a single processor. In other words, it is possible to achieve parallel operations using both arithmetic units and the four input and four output channels. In addition, parallel operations can be made across a network using linked processors. In the diagram below we illustrate a linear chain of p processors, with a master and $p - 1$ slave processors. A single slave can be programmed and the actions replicated across all slave processors.

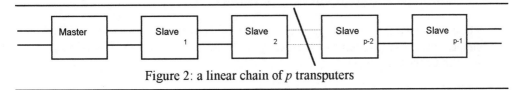

Figure 2: a linear chain of p transputers

Below is an example of a tree network of four transputers, for which each of the branches could contain virtually identical coding.

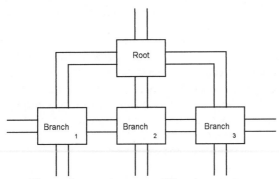

Figure 3: a tertiary tree of four transputers

Performance
We obtain speed up through the time estimation of both the sequential and parallel algorithms. For this purpose, binary floating point operation times between array elements are required. A typical set of values is shown below.

Floating point operation	$a_i + b_i$	$a_i - b_i$	$a_i * b_i$	a_i / b_i	Average
Time $(10^{-6}\,\text{s})$	2.00	2.06	2.30	2.60	2.24

Table 1. Typical times for binary floating point operations $\times 10^6$ seconds

Floating point operation	$a_i * x_y + y_i$
Time $(10^{-6}\,\text{s})$	2.80

Table 2. Typical time for saxpy floating point operation $\times 10^6$ seconds

From the above we note the following:

1.   the time for a single floating point binary operation is nearly constant,
2.   chained operation times are not additive,
3.   average floating point operation times are application dependent,
4.   predicting time (performance) accurately may not be possible if algorithms contain a great degree of chained arithmetic operations.

The communication time for passing information between processors is described by a simple model. To communicate a packet of n like items between two processors, the time is given by the expression

$$T = T_s + nT_c$$

where T_s the start up time, is topology and data type dependent, T_c is data type dependent and is directly proportional to link speed. For practical purposes T_s and T_c are constants in any application but these vary slightly with the degree of parallelism in operation. Arevalillo (1994) has provided extensive information on this effect.

CASE STUDY

We consider an elementary application which has been presented to final-year students on a parallel programming course. Students were requested to design, implement and evaluate a parallel algorithm to calculate the mean and standard deviation of a sample of n values, $\{x_i, i = 1, 2, \ldots n\}$ on a linear chain network of transputers. The evaluation stage is concerned with contrasting the performance of both the sequential and corresponding parallel algorithms.

The specified formulae were:

$$\text{Mean} \quad \bar{x} = \frac{1}{n} \sum_{i=1}^{n} x_i \, ,$$

$$\text{Standard deviation} \quad \sigma = \sqrt{\frac{1}{n} \sum_{i=1}^{n} (x_i - \bar{x})^2}$$

The problem specification includes the following algorithmic features:
*   all data are initially on the master processor,
*   data sets are to be broadcast to the network from the master processor,
*   parallel in and out transmission of data should be made by all the slave processors,
*   the calculation is to be equally sub-divided between all processors.

To determine an estimate of performance, we shall make the following additional assumptions:
*   number of processors is p,
*   for simplicity, $m = n/p$, an exact integer,
*   time for a single floating point operation is T_f,
* T_s and T_c are the times to start up and communicate with a single real32.

The work is requested in three stages as follows:
1. the sequential solution in which T_f is investigated,
2. the development of a pseudo-parallel solution in which the logical aspects of the coded solution are verified,
3. the implementation of the parallel solution and the presentation of a model.

Performance Model

If we let T_1 be the time of execution for the sequential algorithm on a single processor and T_p be the time of execution for the parallel algorithm on p processors, then the speed up is defined by:

$$S_p = \frac{T_1}{T_p}$$

Sequential Algorithm

The operation count for the standard deviation is simply established. For the mean $\bar{x}$, n numbers are to be summed and this is followed by a single division by n. If the summation is programmed, the following algorithm is usual:

```
s:=0
FOR i:= 1 to n DO
s:=s+x[i]
```

We therefore assume that for all summations of the form $\sum_{i=1}^{n} ?$, n additions are required.

The mean $\bar{x}$ therefore requires n additions and one division, a total of $n+1$ floating point operations. For the standard deviation, there is one subtraction for each $(x_i - \bar{x})$ one multiplication to obtain the square of each term, and a total of n additions for the summation with a final division by n. These $4n+2$ operations are then followed by a square root calculation.

Hence the total number of operations is $4n+2+n_{sqrt}$, where n_{sqrt} is the number of floating point operations to calculate the square root of a real number. (Al-Jumeily et al. (1994) provide details of operation counts for other algorithms).

Assuming that all floating point operations take time T_f, the required time estimate is $(4n+2+n_{sqrt})T_f$.

Parallel algorithm

This is readily derived through the various stages of the algorithm as specified below:

1.   broadcast p-1 data packets to the network, $(p$-$1)(T_s+mT_c)$,
2.   summation of m locally stored values mT_f,
3.   master processor gathers partial sums $(p$-$1)(T_s+T_c)$,
4.   master calculates the mean $(p$-$1)T_f$,
5.   mean is broadcast throughout the network $(p$-$1)(T_s+T_c)$,
6.   local summation of $(x_i - \bar{x})^2$ $3mT_f$,
7.   master gathers partial sums $(p$-$1)(T_s+T_c)$,
8.   calculate $(p+1+n_{sqrt})T_f$.

The total time estimate

$$T_p = (4m+2p+2+n_{sqrt})T_f + (p-1)(m+3)T_c + 4(p-1)T_s$$

Hence the theoretical speed up is

$$S_p = \frac{(4n+2+n_{sqrt})T_f}{(4m+2p+2+n_{sqrt})T_f + (p-1)(m+3)T_c + 4(p-1)T_s}$$

$$S_p = \frac{p+\dfrac{2}{4m}+\dfrac{n_{sqrt}}{4m}}{1+\dfrac{(2p+2)}{4m}+\dfrac{n_{sqrt}}{4m}+\dfrac{(p-1)}{4}(1+\dfrac{3}{m})(\dfrac{T_c}{T_f})+\dfrac{(p-1)}{m}\dfrac{T_s}{T_f}}$$

In the limit as $n \rightarrow \infty$, $S_p \rightarrow \dfrac{p}{1+\dfrac{(p-1)}{4}(\dfrac{T_c}{T_f})}$

We therefore observe the following:

1.   speed up is dependent upon the compute to communicate ratio T_c/T_f. As these characteristic values change, algorithms which were effective on yesterday's technology may not be worthwhile on today's technology and vice versa.

2. $S_p > 1$ if $T_c/T_f < 4$.

3.   as $T_c/T_f \rightarrow 0$, $S_p \rightarrow p$ as expected.

For T800 technology, the value of T_c/T_f has a minimum value of about 1.25 whereas for T400-type transputers, the best value is about 0.16. Hence effective parallel algorithms developed for T400-type technology may not be worthwhile on T800-type technology!

The ideal speed up would be p, which is clearly not the case unless the ratio T_c / T_f is sufficiently small, as illustrated in the graphs below.

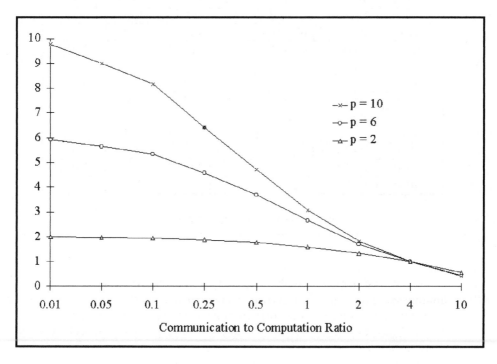

Figure 4: variation of limiting speed up with the ratio T_c / T_f

Practical results

We present results for both the sequential and the parallel algorithm on 5 transputers. The rate of computation is conveniently expressed in terms of mflops (million of floating point operations per second). Values of interest are as follows.

$T_f = 1.611E\text{-}06$ $\quad n_{sqrt} = 48$ $\quad\quad\quad\quad T_s = 1.20E\text{-}05$

$T_c = 4.40E\text{-}06$ $\quad$ Limiting speed up $= 1.340$

Sample size	Mflops	Measured T_f		Practical speed up	Theoretical speed up	Simple theory
100	0.5208	1.645E-06		0.857	0.899	1.136
200	0.5682	1.624E-06		1.048	1.064	1.343
300	0.5859	1.616E-06		1.103	1.139	1.437
400	0.5952	1.612E-06		1.167	1.182	1.490
600	0.6048	1.608E-06		1.216	1.230	1.549
800	0.6098	1.606E-06		1.242	1.255	1.581
1000	0.6127	1.605E-06		1.259	1.271	1.601
2000	0.6188	1.603E-06		1.287	1.304	1.642
4000	0.6188	1.609E-06		1.312	1.322	1.664
8000	0.6196	1.611E-06		1.321	1.331	1.675
16000	0.6203	1.610E-06		1.326	1.335	1.681
32000	0.6205	1.611E-06		1.329	1.338	1.684
64000	0.6207	1.611E-06		1.329	1.339	1.685

Table 3. Results for standard deviation

We can make the following conclusions:
1.   performance estimation is highly correlated with the observed values,
2.   limiting speed up value is achieved as sample size increases,
3.   the simple time estimates (last column) provide an upper bound for performance! This will generally be the case as chaining of floating point operations reduces the average value T_f.

Student algorithms

It was a surprise to discover that students used four basic coding structures for the standard deviation. The variations were presented in the calculation of $\sum (x_i - \bar{x})^2$. If we assume that the data values are held in the array x, and mean is the calculated mean, the four occam structures used are shown below.

```
1.              SEQ i = 0 FOR n
                SEQ
                 difference := x[i] - mean
                 s := s + (difference*difference)
                 standard.deviation := SQRT(s/(REAL32 ROUND n))

2.              SEQ i = 0 FOR n
                 s := s + ((x[i] - mean)*(x[i] - mean))
                 standard.deviation := SQRT(s/(REAL32 ROUND n))
```

3.
```
SEQ i = 0 FOR n
  s := s + POWER(ABS(x[i] - mean), 2.0(REAL32))
standard.deviation := SQRT(s/(REAL32 ROUND n))
```

4.
```
SEQ i = 0 FOR n
  SEQ
    difference := x[i] - mean
    difference.squared := difference * difference
    s := s + difference.squared
standard.deviation := SQRT(s/(REAL32 ROUND n))
```

The first method is the expected solution and the third method is very inefficient. Each of these structures can be effectively modelled. Students often exclude n_{sqrt} in the operation count and thus typical reported operation counts for the above codes, excluding the value $n+1$ for the mean are:

Method 1	$3n+1$
Method 2	$4n+1$
Method 3	$(3n+1)!$ The calculation required for each application of the function is ignored.
Method 4	$3n+1$

It is feasible to choose effective floating point operation counts by simply considering the algebraic formulation and relating the mflop rates to the real world number of algebraic operations. Experience has shown that students find the justification for such an approach too intellectually challenging. The operation count for the full calculation would then be obtained through the addition of an additional $n+1+n_{sqrt}$ floating point operations.

In the table below are mflops for the sequential solution.

Method	Mflops	Time (seconds)	Operation count
1	0.621	0.0644	$4n$
2	0.726	0.0688	$5n$
3	0.0483	0.8281	$4n$
4	0.589	0.0679	$4n$

Table 4. Reported Mflop rates for sequential solution with $n=10,000$

The absolute time is also shown as this indicates that, although a higher rate of mflops is achieved by Method 2, Method 1 is achieved in minimum time. Students often find it difficult to understand that inefficiencies of the coding have to be taken into account in this manner.

Another aspect of the results are the values for speed up which have been observed. Typical limiting values are shown in Table 5.

Method	Speed up	Mflops	Op count	Parallel time
1	1.30	0.84	$4n$	0.0598
2	1.37	0.89	$5n$	0.0709
3	4.18	0.19	$4n$	0.8964
4	1.42	0.82	$4n$	0.0672

Table 5. Reported speed up for all methods

Summary

- The best speed up is obtained with the most inefficient method,
- The worst speed up is found with the most efficient coding,
- The least time is obtained when the coding is most efficient.

Experience has also confirmed that good correlation is possible for every solution which a student produces. Consequently all students have a well-defined task, namely to produce an applicable performance model.

If poor correlation exists between theory and practice then:

- the model could be incorrect,
- the coding could be incorrect.

In either case, the variation can be investigated until a satisfactory explanation is obtained or the source of error discovered. Common errors in students work are as follows:

- the use of inconsistent operation counts,
- failure to implement the same coding structures in both the sequential and parallel solutions, which results in the incorrect use of the average time for a floating point operation,
- failure to use parallel input/output by each of the slave processors and incorrect modelling of this feature,
- failure to recognise that measured times for small sample sizes could contain significant error.

STUDENT FEEDBACK

The exercise described forms part of a module on parallel programming in an integrated scheme for students studying computing, mathematics and statistics. Students specialising in computing had no previous experience of mathematical modelling. At the end of the course a commonly expressed view was that the modelling was useful and worthwhile. In developing their models, students found the algebra very demanding. This was particularly evident in those cases in which students were asked to model the parallel algorithm which they had constructed and this did not always conform to the specification, the main departure being with respect to data transmission through the network and non-parallel input and output data transmission by slave processors.

Typical student comments are the following:

- I never imagined that mathematics could be so useful in developing parallel algorithms,
- finding a limiting value for large sample sizes was beyond me, but I obtained a reasonable approximation using a spreadsheet,
- developing the model reinforced my understanding of the program codes I had written,
- I found the model helped me to understand the complexity of the hardware,
- the exercise was useful and not as difficult as I had initially conceived. The experience has given me the confidence to think of applying mathematical ideas to other situations,
- I had to think about how to obtain meaningful graphs for the whole range of sample sizes.

The disappointing performance which can be achieved with the specified algorithm has not been a worry to the students. There has been little evidence of undue co-operation and every evidence that work presented is a genuine personal effort.

CONCLUSIONS

1. We have demonstrated that effective models of performance can be constructed for parallel algorithms on transputer networks.

2. Simple performance estimates can be obtained without the need to perform any coding by using a crude estimate of the average time to compute a floating point binary operation. Generally this will provide an upper bound for the speed up.

3. Accurate performance models can be determined if realistic estimates of the average floating point times are available.

4. We have considered the details of practical work for a simple statistical calculation undertaken by students and confirmed that good correlation between theory and practice is readily achieved using a variety of algorithms.

5. We have advocated that the development of performance models for parallel algorithms can be undertaken before any coding is attempted.

6. If performance models do not correlate with measured performance, this can attributed to either the inapplicability of the model or errors in the parallel algorithm. This is a useful check for the programmer.

7. Distributed parallel computing is a fertile area for exploiting performance model developments, with experimentation allowing students to verify results.

REFERENCES

Al-Jumeily DM, Clegg DB, Pountney DC and Harris P, 1994, **Optimising Simple Statistical Calculations on Distributed Memory Computers**, Report 5, School of Computing and Mathematical Sciences, John Moores University, Liverpool.

Burns A, 1988, **Programming in occam2**, Addison-Wesley, Wokingham England.

Arevalillo M, 1994, **On Communication Costs for Transputer Arrays**, Student Project Report, School of Computing and Mathematical Sciences, John Moores University, Liverpool.

Freeman TL and Phillips C, 1992, **Parallel Numerical Algorithms**, Chapter 1, Prentice Hall, London.

Foster I, 1995 **Designing and Building Parallel Programs**, Addison Wesley, USA.

Inmos, 1988 **Transputer Reference Manual**, Prentice Hall, London.

23

Modelling Cancer Chemotherapy

JR Usher and D Henderson
The Robert Gordon University, Aberdeen, Scotland, AB1 1HG
e-mail : jru@scms.rgu.ac.uk **or** *dh@scms.rgu.ac.uk*

ABSTRACT

Mathematical modelling provides an ideal opportunity for students at all levels in their education to learn how to use mathematics, and how to develop a deeper understanding of mathematical ideas. The requirement for a suitable application area is of vital importance if the student's interest is to be sufficiently aroused so that they can appreciate and understand the characteristics of the modelling process as well as practice the mathematical techniques taught to them through direct instruction. The authors' current area of research involves modelling the effect of toxic drugs on cancer tumours and seeking effective treatment strategies. The intention of this paper is to describe briefly the application area and associated aspects of mathematical modelling, before presenting some mathematical models that have been developed over the last 15 years or so, via three case studies. The models, analyses and results discussed provide motivating case-study material for undergraduate mathematical modelling courses as well as giving students the opportunity to appreciate the nature of mathematical modelling and the importance of differential equations for gaining an understanding of practical problems. It will also enable us to demonstrate the usefulness of this application area for students wishing to enhance their mathematical ability and understanding of a variety of techniques and ideas.

INTRODUCTION

Students undertaking a mathematical course must learn how to utilise and develop the mathematical ideas and techniques they are taught. In order to help achieve this they need to practice applying mathematical techniques to solve many different types of practical problem. A mathematical modelling programme provides an ideal vehicle for achieving these aims, providing insight both into the modelling process itself and the process of applied problem solving. Such a programme also provides opportunities for group work, an essential ingredient in any modern-day undergraduate course (Crouch *et al.*,1996).

In the study of biological problems mathematical modelling is playing a significant and ever increasing role. In particular biologists are continually attempting to improve their clinical insight, by whatever means available, in order to aid their treatment decision-making process. The treatment of cancer by chemotherapy is no exception. For this situation numerous mathematical models exist that detail a variety of characteristics associated with the problem, and results generated by these models can often provide valuable clinical insight. A phenomenon of particular interest, when treating a cancer patient with chemotherapy, is the occurrence of drug resistance. It is within this clinical problem area that the authors' main research interests currently lie. We hope to demonstrate the crucial role played by mathematical modelling in ascertaining tumour response to various drug administration regimes, in the presence of drug resistance.

Over the last 15 years or so several models have been generated that attempt to deal with the problem of drug resistance occurring when treating cancer with chemotherapy. Goldie and Coldman's models (1979, 1982) form the foundation for all subsequent models. Usher and Henderson (1996) utilised previous work by Usher (1980, 1994) and Wheldon (1988) in generating their most recent model, which reveals the significant role played by drug resistance and increased drug sensitivity mechanisms when selecting treatment strategies. We intend to demonstrate how this recent research work provides suitable case study material for an undergraduate mathematical modelling programme.

BACKGROUND

Normal cells reproduce themselves by division under the control of the regulating system in the body, the body's so-called homeostatic control mechanism. One model for this birth process is the cell cycle which can be seen as a sequence of events (G_1, S, G_2, M) after which daughter cells can continue in the cell cycle or enter a resting phase G_0. This process is represented in Fig. 1.

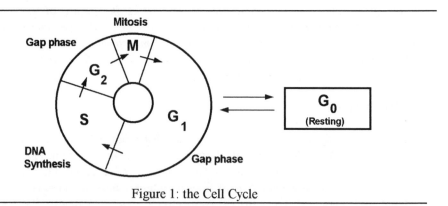

Figure 1: the Cell Cycle

The S phase is the phase in which DNA (Deoxyribonucleic acid) synthesis takes place; this is the chemical material which has a binary structure possessing a helical twist, the famous double helix discovered by Watson and Crick (1953), and contains the genetic code of the cell. At mitosis (the M phase) the two strands of DNA separate, thus passing the genetic information of the mother cell onto the two daughter cells. G_1 and G_2 are two intermediate resting phases.

During the cell cycle normal cells can mutate, thus altering the genetic code of such cells, and (after an appropriate number of such mutations) transform into cancer cells; this process is called carcinogenesis. Anything that contributes to carcinogenesis can be termed a carcinogen - examples include smoking, dietary components or exposure to chemicals. Cancer cells may be thought of as cells which grow outwith the body's homeostatic control mechanism.

The three main treatment possibilities for cancer are surgery, radiotherapy and chemotherapy, or a combination of these; attention here will be focused on chemotherapy. Chemotherapeutic drugs can be usefully classified into two categories-cycle-specific (drugs which affect cells in the proliferating phases G_1, S, G_2, M of the cell cycle but not those in the resting phase G_0) and cycle-non-specific (drugs which affect cells both in the proliferating phases and the resting phase).

Drug resistance is a serious problem faced by oncologists. If a cell in a cancer tumour becomes resistant to the drug or group of drugs that it is being treated with, that cell will continue to divide and eventually cause treatment failure. Various mathematical models exist for describing this phenomenon, most notably those of Goldie and Coldman (1979, 1982). A recent analysis by Usher and Henderson (1996) introduces the idea of an increased drug sensitivity mechanism which competes with the increased drug resistance mechanism. This analysis is investigated in Case Study 3.

ASSOCIATED ASPECTS OF MATHEMATICAL MODELLING

All mathematical models must be underpinned by a set of assumptions; in a practical situation these assumptions must be physically realistic. As stated by Wheldon (1988) the only real purpose of a mathematical model is "to allow unambiguous derivation of the implications of the set of assumptions adopted, these providing the sharp description of the dominant causal influences operative in the field concerned".

The modelling process should begin with users in the field (oncologists in the current area of study) identifying key requirements and what factors influence the outcome of interest. These influential factors then have to be specified in the form of model assumptions which will allow mathematicians to formulate an appropriate mathematical model. The behaviour of the model should then be analysed utilising whatever mathematical techniques are considered appropriate. The results from the mathematical analysis should finally be applied to the physical problem, hopefully satisfying the initial requirements as stipulated by the user in the field and so performing the model validation process. At this stage (particularly within the complex area of mathematical biology) there are often new factors that need to be incorporated into the model. The entire process then starts again, this time utilising the latest model based on further assumptions which the user in the field feels are important.

The following case studies detail the build up to the Generalised-Ex model (Usher and Henderson, 1996) for describing the affects of chemotherapy on tumour growth when drug resistance occurs. The case studies display all the characteristics of the modelling process described above, and thus students undertaking them will gain an appreciation of the process as well as an opportunity of applying various mathematical techniques.

TUMOUR GROWTH AND CHEMOTHERAPY MODELS

Tumour growth in humans is difficult to measure accurately. Numerous studies have, however, been undertaken in an attempt to determine patterns of tumour growth, some of which have been summarised by Steel (1977). The first case study below investigates some mathematical models of tumour growth. The second case study investigates a model which describes the affects of toxic drugs on tumour cells. The third case study investigates a model for dealing with the phenomenon of drug resistance.

Case study 1

We shall focus attention on the following points concerning tumour growth:
- tumours first become clinically observable when there are approximately 10^{10} tumour cells. (N.B. Because of technological improvements this value is lower than the estimate given in Usher and Abercrombie, 1981),
- there is experimental evidence (Wheldon, 1988) that many tumours grow exponentially during the initial 'latent' period and then subsequently follow an asymptotic limiting behaviour as illustrated in Fig. 2.

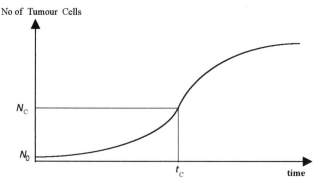

N_o : Initial size of tumour cell population (in reality, unity)
N_C : Size of tumour cell population at transition between two phases of growth
t_C : transition time

Figure 2: two phase growth kinetics

Usher (1980) introduced the following generalised growth model to describe tumour growth, as it incorporated some previous models as special cases.

$$\frac{dN}{dt} = \frac{\lambda N}{\alpha}\left[1-\left(\frac{N}{\theta}\right)^{\alpha}\right]$$

where N denotes the number of tumour cells at time t and
        α, λ, θ are determined from the growth characteristics of the tumour.

In any modelling situation it is crucial that students:
- appreciate the need to relate any new model to previous models,
- investigate limiting cases to gain a better understanding of the model,
- relate results from the model back to the physical situation.

Returning to the generalised growth model, the following limiting forms are of particular interest for tumour growth kinetics:

- the Gompertzian growth model, often used by oncologists when investigating clinical data in the situation when the tumour growth rate is slowing down

$$\alpha \to 0 \quad : \qquad \frac{dN}{dt} = -\lambda N \ln\left(\frac{N}{\theta}\right)$$

- the Verhulst growth model

$$\alpha \to 1 \quad : \qquad \frac{dN}{dt} = \lambda N - \frac{\lambda}{\theta}N^2$$

(this model is rarely used for tumour growth)

- the exponential growth model

$$\alpha \to 0, \theta \to \infty \quad : \qquad \frac{dN}{dt} = \lambda N$$

(this model is employed by oncologists for the growth phase when the doubling time of the tumour is constant.)

A student, once familiar with methods for solving first-order ordinary differential equations, will be able to derive the following solution for the generalised growth model.

$$N(t) = N_0\left\{\left[\frac{N_0}{\theta}\right]^{\alpha} + \exp(-\lambda t)\left(1-\left[\frac{N_0}{\theta}\right]^{\alpha}\right)\right\}^{\frac{-1}{\alpha}}$$

where N_0 is the initial size of the tumour cell population (and will in reality take the value unity).

Typical growth curves are illustrated in Fig. 3 where for convenience N_0 has been taken as 10^{10}, the number of cells present when the tumour first becomes clinically observable. They provide a good example of the need for standard forms and log-linear graph paper.

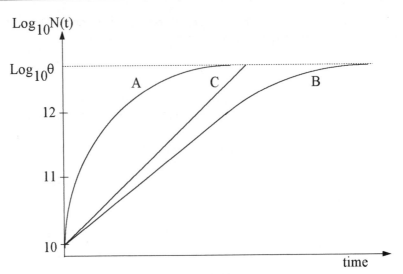

Figure 2: typical growth curves associated with the generalised growth model.
A : Gompertz , B : Verhulst , C : Exponential

Subsequently Wheldon (1988) developed the following 'Gomp-Ex' model to deal explicitly with two-phase growth, consisting of an initial exponential growth phase followed by a Gompertz growth phase:

$$\frac{1}{N(t)}\frac{dN(t)}{dt} = \begin{cases} \lambda & N \le N_C \\ \lambda - \beta \ln\left(\dfrac{N(t)}{N_C}\right) & N \ge N_C \end{cases}$$

where N denotes the number of tumour cells at time t,
λ, β denote parameters describing the two-phase growth characteristics of the tumour,
N_C is the size of the tumour cell population at the transition between the phases of growth (from experimental data N_C is approximately 10^9).

If students have investigated the generalised growth model and its limiting forms they will be able to proceed with solving the Gomp-Ex model to obtain

$$N(t) = \begin{cases} N_0 \exp(\lambda t) & N \le N_C \\ N_C \exp\left(\dfrac{\lambda}{\beta}\left\{1 - \exp\left[-\beta(t - t_C)\right]\right\}\right) & N \ge N_C \end{cases}$$

where N_0 is the size of the tumour cell population (again in reality $N_0=1$),

t_C is the time the tumour takes to reach size N_C, $\qquad ie \ t_C = \dfrac{1}{\lambda}\ln\left(\dfrac{N_C}{N_0}\right)$

This model further demonstrates the importance of relating a mathematical model to experimental evidence.

Case study 2

Now that we have established a hierarchy of models for tumour growth we are in a position to investigate the affects of toxic drugs on tumour cells. As mentioned previously, drugs may be classified as cycle-specific or cycle-non-specific, depending on whether cells are only killed in the proliferating phases of the cell cycle or in all phases including the resting phase. We shall focus attention on the use of cycle-non-specific drugs. It should be reasonable to consider developing a model which combines the features of the models investigated in case study 1 to incorporate a cell kill term representing the effect of the cycle-non-specific drug being used. Usher (1994) has already developed such a model, which is referred to as the 'Generalised-Ex' model. This model is represented by the following equations:

$$\frac{1}{N}\frac{dN}{dt} = \begin{cases} \lambda - \rho C(t) & N \le N_C \\[2mm] \lambda + \dfrac{\lambda_1}{\alpha}\left[1-\left(\dfrac{N}{N_C}\right)^{\alpha}\right] - \rho C(t) & N \ge N_C \end{cases}$$

where $\alpha, \lambda, \lambda_1$ are the tumour growth parameters,

N_C is the size of the tumour cell population at the transition between the phases of growth,

ρ is the dose response parameter of the tumour cells,

$C(t)$ is the concentration of the cycle-non-specific drug administered.

This model assumes an exponential dose-response relationship with parameter r. If the concentration is constant, analytic solutions can be obtained dependent upon various parameter constraints. Details of all such analytic solutions are given by Usher (1994). This work provides ideal case study material for helping students to gain an appreciation of the importance of analytical solutions in the field of mathematical modelling. The behaviour of these analytic solutions can also be investigated by the use of spreadsheets. If however the drug concentration is non-constant then numerical techniques such as Runge Kutta (James, 1996) will be required.

Thus this case study shows students the need for a variety of tools, including software tools, when solving a complex problem; in particular it illustrates the complementary roles played by analytic and numerical solutions of differential equations.

Case study 3

Let us now consider the situation when drug resistance occurs. There appears to be a number of mechanisms causing drug resistance, not all of which are understood by oncologists. There also appear to be mechanisms causing the reverse process, *ie* increased drug sensitivity (i.d.s) but such mechanisms are not yet well understood by oncologists. We shall assume that after treatment has started the total cell population consists of two sub-populations; one sub-population consists of cells sensitive to the drug (drug sensitive (d.s) cells) whilst the other sub-population consists of cells resistant to the drug (drug resistant (d.r) cells). We assume that the d.r mechanisms cited above result in migration of cells from the d.s sub-population to the d.r sub-population while the competing i.d.s mechanisms result in migrations in the reverse direction. The effects of these competing mechanisms are illustrated in Fig. 4.

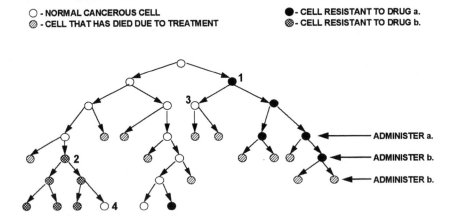

○ - NORMAL CANCEROUS CELL	● - CELL RESISTANT TO DRUG a.
◉ - CELL THAT HAS DIED DUE TO TREATMENT	⊛ - CELL RESISTANT TO DRUG b.

1   Inherent clonal resistance - drug resistance which occurs as a result of mutations before, during or after treatment.

2   Acquired drug resistance - occurs as a result of the application of a particular drug.

3   Inherent loss of resistance - gain in drug sensitivity arising from back-mutations which can occur before, during or after treatment.

4   Acquired loss of resistance - gain in drug sensitivity resulting from the administration of a particular drug.

Figure 4: competing drug resistance and increased drug sensitivity mechanisms

Goldie and Coldman (1979) only considered clonal resistance (one particular d.r mechanism) whereby cells randomly migrate from the d.s sub-population to the d.r sub-population. However the recent analysis of Usher and Henderson (1996) incorporates both d.r and i.d.s mechanisms into Usher's (1994) Generalised-Ex growth model investigated in case study 2.

Usher and Henderson's (1996) model is given symbolically in Fig. 5, with the corresponding differential equations given below.

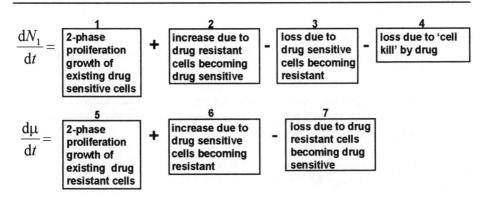

Figure 5. the symbolic representation of the mathematical model which incorporates the various drug resistant and increased drug sensitivity mechanisms. Here $N_1(t)$ and m(t) denote the number of drug sensitive and drug resistant cells respectively at time t

$$\frac{dN_1}{dt} = \begin{cases} \{\lambda_1^* N_1\} & \quad +\{\beta\gamma_2\mu\} - \{\psi\gamma_1 N_1\} - \{\rho N_1 C(t)\} \quad , \quad N_1 \le N_C \\ \left\{\left(\lambda_1^* + \frac{\lambda_1}{\alpha_1}\left[1-\left(\frac{N_1}{N_C}\right)^{\alpha_1}\right]\right)N_1\right\} + \{\beta\gamma_2\mu\} - \{\psi\gamma_1 N_1\} - \{\rho N_1 C(t)\} \quad , \quad N_1 \ge N_C \end{cases}$$

(1a,b)

$$\frac{d\mu}{dt} = \begin{cases} \{\lambda_2^* \mu\} & \quad +\{\psi\gamma_1 N_1\} - \{\beta\gamma_2\mu\} \quad , \quad \mu \le \mu_C \\ \left\{\left(\lambda_2^* + \frac{\lambda_2}{\alpha_2}\left[1-\left(\frac{\mu}{\mu_C}\right)^{\alpha_2}\right]\right)\mu\right\} + \{\psi\gamma_1 N_1\} - \{\beta\gamma_2\mu\} \quad , \quad \mu \ge \mu_C \end{cases}$$

(2a,b)

where N_1 and μ denote the number of d.s cells and d.r cells, respectively, at time t,
N_C and μ_C represent the transition points corresponding to the two phase growth of the Generalised-Ex model (1a,b),
$\lambda_1^*, \lambda_1, \alpha_1$, are associated with the growth characteristics of the d.s cells.
$\lambda_2^*, \lambda_2, \alpha_2$ are associated with the growth characteristics of the d.r cells.
γ_1 and γ_2 are the proportions of d.s and d.r cells respectively engaged in proliferation.

ψ represents the gain in drug resistance (ie migration rate of d.s cells to d.r cells),

β represents the gain in drug sensitivity (ie migration rate of d.r cells to d.s cells).

The analysis of Usher and Henderson (1996) considers the following two situations of clinical interest.

Case A. Treatment yielding either a steady 'log cell kill' decline in d.r cells together with a decaying d.s cell sub-population or constant size sub-populations of d.r and d.s cells.

Case B. Treatment yielding either a steady 'log cell kill' decline in d.s cells together with a decaying d.r cell sub-population, or constant size sub-populations of both d.s and d.r cells.

As an illustrative example of the results generated by the analysis of Usher and Henderson (1996), attention here will be focused on Case B.

We require
$$N_1 = N_{10} \exp(-\omega t) \tag{3}$$
where N_{10} represents the number of d.s cells at the onset of treatment
Substitution of (3) into (1a,b) and (2a,b) yields analytic results for treatment schedules and the corresponding behaviour of the d.r cells for the following cases:-

1i) $\mu_0 \leq \mu_C$, $N_{10} \leq N_C$

1ii) $\mu_0 \leq \mu_C$, $N_{10} \geq N_C$

where μ_0 represents the number of d.r cells at the onset of treatment.

As an example of the results generated by the analysis of Usher and Henderson (1996), attention here will be focused on subcase 1ii). The analysis gives rise to restrictions on the values of the decay rate ($-\omega$) of the d.s cells that can be sustained by an appropriate choice of $C(t)$ which also gives rise to a decaying d.r sub-population. These restrictions are illustrated in the ω-β plane in Fig. 6. Thus, from Fig. 6, for a given migration rate β, ω must lie in the shaded region if the required result is to be obtained

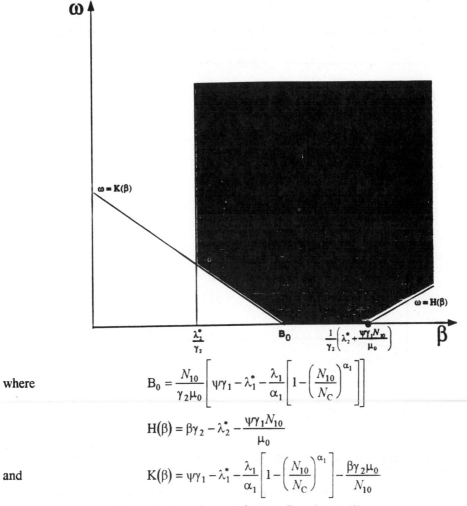

where

$$B_0 = \frac{N_{10}}{\gamma_2 \mu_0}\left[\psi\gamma_1 - \lambda_1^* - \frac{\lambda_1}{\alpha_1}\left[1 - \left(\frac{N_{10}}{N_C}\right)^{\alpha_1}\right]\right]$$

$$H(\beta) = \beta\gamma_2 - \lambda_2^* - \frac{\psi\gamma_1 N_{10}}{\mu_0}$$

and

$$K(\beta) = \psi\gamma_1 - \lambda_1^* - \frac{\lambda_1}{\alpha_1}\left[1 - \left(\frac{N_{10}}{N_C}\right)^{\alpha_1}\right] - \frac{\beta\gamma_2\mu_0}{N_{10}}$$

Figure 6: the ω-β plane for case B, subcase 1ii)

The various threshold values illustrated in Fig. 6 are determined from the following physical considerations.

- The d.r and d.s sub-populations must decay in size or remain of constant size.
- The results must be biologically meaningful. This involves the following;

the drug concentration must be positive throughout treatment,
just before the onset of treatment both sub-populations must be growing in size,
the numbers of d.s and d.r cells during and after treatment must be non-negative.

With suitable restrictions on ω and β Fig. 7 illustrates various drug concentrations which yield both exponentially decaying d.s sub-populations and different types of associated behaviour for the d.r sub-population.

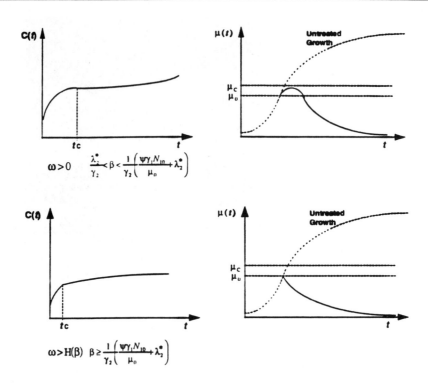

Figure 7: various drug regimes which give the required sub-population behaviour for case B. subcase 1ii) where t_C is the time at which N_1 (the number of d.s cells) = N_C (the sub-population size at the transition point in the two-phase growth)

It is clear from the above analysis that a substantial amount of work is required to ensure that the mathematical analysis represents the real-life situation. Students who are asked to perform the above analysis would therefore not only gain experience of differential equations, their behaviour and solution techniques, but would gain an appreciation of what is required to ensure that the mathematical analysis is useful within the application area.

In order to solve the other two subcases ($\mu_0 \geq \mu_C$, $N_{10} \leq N_C$ and $\mu_0 \geq \mu_C$, $N_{10} \geq N_C$) it is necessary to employ a numerical procedure such as Runge Kutta.

To enable students to readily initiate numerical calculations appropriate parameter estimates are listed below.

Tumour growth can be taken to be in one of three categories:

Fast $\qquad \lambda_1^* = \lambda_2^* = 0.00107$, $\lambda_1 = \lambda_2 = 3.14267*10^{-4}$, $\alpha_1 = \alpha_2 = -0.28287$

Medium $\qquad \lambda_1^* = \lambda_2^* = 6.59956*10^{-4}$, $\lambda_1 = \lambda_2 = 2.48016*10^{-4}$, $\alpha_1 = \alpha_2 = -0.37053$

Slow $\qquad \lambda_1^* = \lambda_2^* = 4.76650*10^{-4}$, $\lambda_1 = \lambda_2 = 1.93755*10^{-4}$, $\alpha_1 = \alpha_2 = -0.40255$

N_C equals 10^9
γ_1 and γ_2 have to be between 0 and 1
ψ and β are in the range 10^{-7} to 10^{-3}
ρ is between 0 and 1
$C(t)$ is between 0 and 5
time is in days

This case study is relatively complex but provides students with an opportunity to follow various pathways.

CONCLUSION

From the case studies investigated it can be seen that mathematical modelling in this area of cancer research provides valuable and much needed clinical insight into both the anticipated behaviour of the drug resistant and drug sensitive sub-populations when seeking effective treatment strategies, and also the various thresholds which determine the success or failure of a specific treatment strategy. These models provide a qualitative representation of the biological problem.

The analysis of Usher and Henderson (1996), briefly described in case study 3, reveals the significance of cells migrating from the d.r sub-population to the d.s sub-population when attempting to ascertain a drug concentration that will result in either constant size or decaying sub-populations of both d.r and d.s sub-populations. There is a threshold value for this migration above which decaying sub-populations can be achieved but below which both sub-populations will grow in size. When attempting to quantify such a mechanism, mathematical modelling has been seen to be an invaluable tool.

Underlying the models discussed in case study 3 are several biological assumptions. These assumptions have been discussed with oncologists and biologists, who feel that the models and the associated parameters are reasonable and provide a good place to start before incorporating more complex biological mechanisms that affect the treatment of patients. For example, we are currently incorporating the idea of drug resistant and drug sensitive cells competing for a common nutrient supply and investigating the affects of toxicity constraints. The analysis of Usher and Henderson (1996) has yielded useful qualitative results; experimental data is currently being sought so that the model parameters can be estimated, thus allowing quantitative as well as qualitative results to be obtained. This will then enable the construction of a practical decision-support system for the oncologists which will help enable them to decide which chemotherapy strategy to adopt for a given tumour type. Thus the incorporation of appropriate biological mechanisms and the validation of the mathematical models provide a better understanding of the biological problem and improve clinical insight.

Although the ultimate beneficiaries of the work being undertaken on chemotherapy and drug resistance will be the cancer patients themselves, it is important to recognise that the mathematical models developed, the analysis undertaken and the results obtained provide case study and project material which will motivate many students undertaking mathematical modelling courses. Mathematics students will gain experience of the various techniques utilised for solving differential equations, and an understanding of the way in which they behave. They will also gain an appreciation of the importance of mathematical modelling when attempting to analyse and understand a particular physical problem or mechanism.

REFERENCES

Crouch R, Davies AJ, Usher JR, Mackenzie IG and Forrest E, 1996, *Working in Groups: Developing Personal and Professional skills*, in Haines C and Dunthorne S, (eds.), **Mathematics Learning and Assessment - Sharing Innovative Practices**, Arnold, London.

Goldie JH and Coldman AJ, 1979, *A Mathematical Model for Relating the Drug Sensitivity of Tumours to their Spontaneous Mutation Rate*, **Cancer Treatment Reports, 63**, pp 1727-1733.

Goldie JH and Coldman AJ, 1982, *Rationale for the Use of Alternating Non-Cross-Resistant Chemotherapy*, **Cancer Treatment Reports, 66**, pp 439-449.

James G, 1996, **Modern Engineering Mathematics 2nd edition**, Addison Wesley, Wokingham.

Steel GG, 1977, **Growth kinetics of tumours**, Clarendon, Oxford.

Usher JR, 1980, *Mathematical Derivation of Optimal Uniform Treatment Schedules for the Fractionated Irradiation of Human Tumors*, **Mathematical Biosciences, 49**, pp 157-184

Usher JR, 1994, *Some Mathematical Models for Cancer Chemotherapy*, **Computers Math Application, 28**, 9, pp 73-80.

Usher JR and Abercrombie DA, 1981, *Case Studies in Cancer and its Treatment by Radiotherapy*, **Int.J.Math.Educ.Sci.Technol, 12**, 6, pp 661-682.

Usher JR and Henderson D, 1996, *Some Drug Resistant Models for Cancer Chemotherapy*, **IMA Journal of Mathematics Applied in Medicine and Biology**, **13**, pp 99-126.

Watson JD and Crick FHC, 1953, *Molecular Structure of Nucleic Acids: a Structure for Deoxypentose Acids,* **Nature, 171**, pp 737-738.

Wheldon T, 1988, **Mathematical Models in Cancer Research**, Adam Hilger, Bristol and Philadelphia.

24

Motoring - Modelling in the Fast Lane

M J Herring
Cheltenham and Gloucester College of Higher Education, P. O. Box 220, The Park, Cheltenham, GL50 2QF, UK.
e-mail: mherring@chelt.ac.uk

ABSTRACT

When considering modelling problems, it is important that such problems are accessible to all students, that they are topical, and that data can be readily obtained for parameter estimation or validation. This paper will review a number of situations that arise from motoring and road traffic, an area that provides a rich source of modelling problems for various levels of mathematics. Initially two problems that are appropriate for an introductory modelling group will be discussed. Further areas for student investigation and project work will then be discussed.

BACKGROUND

Students at Cheltenham and Gloucester first encounter mathematical modelling in a Year 1 - second semester module. These students have a wide variety of backgrounds and mathematical experience, such as GCSE, A-level in Mathematics, Access qualifications, Polymaths or Open University foundation course credit. However, all students are strongly advised to take the first-semester module, Mathematical Toolkit, where they are introduced to the spreadsheet EXCEL and its applications to various mathematical and statistical techniques. The second part of the module is geared to a students' background. Those without A-level Mathematics or equivalent would study an area entitled Functions and Graphs that exposes them to the basic functions, their graphs and the underlying algebra needed in using such functions. The approach is based largely on Polymaths Book B (LeMasurier and Townsend, 1977).

PROBLEMS TO BE MODELLED

The selection of appropriate problems and case studies to be studied by such a diverse group of students needs careful deliberation. The group includes Mathematics students on the modular degree, BEd (Primary and Secondary) students who have chosen mathematics as an elective subject, and a small input of students from other fields who require an additional module for the scheme requisites. Certain facets of the problems need identification. They include

- knowledge required of the specific background of the subject,
- difficulties in collecting and organising data,
- model selection,

Also there is the educational requirement for the selection of problems

- the work to be done must be motivating for the students.

An appeal for topicality in the problems to be considered (Raggert, 1984) is also seen as an important consideration.

An area that lends itself to mathematical modelling at various levels is consideration of motoring and road traffic. With reference to the points above, the prior background mathematical knowledge and expertise are an important consideration since a number of students will have only limited mathematical experience. Therefore, problems for solution need to be tractable to all students. It is to be hoped that the problems are also capable of extension. The problems must still be seen as 'real' and, whenever possible, within the experience of the majority of students. In order that the problems may be solved within the time constraints, data should be readily obtainable. I believe the area under consideration meets these criteria.

DESCRIPTION AND EVALUATION OF SOME EXEMPLARY PROBLEMS

In order that students begin modelling with confidence, they are introduced initially to some empirical models. The first problem is to deduce a functional relationship between the overall stopping distance of a car for a given speed using data from the Highway Code. This allows the student to use a variety of approaches in obtaining the relationship (Burghes, 1991; Stone and Huntley, 1982). Validation of their result may be made by plotting d/v against v, where d is the stopping distance and v the velocity of the vehicle. Another empirical model investigated is the relationship between lighting up times and the time of the year, where data may be obtained from any good diary. We then consider two deterministic models.

Example 1 Flow past a set of traffic lights set on green

The problem is to determine how many cars in a long stationary queue may pass through the lights when they turn green for a set period.

The manner in which traffic behaves at lights can be complex, so the student is encouraged to make simplifying assumptions and hence determine a functional relationship between the number of cars, n, passing the lights and the period, T, in which the lights are on green.

The students are encouraged to consider the case where no traffic turns left or right at the lights and where there is a clear exit across the junction. Another assumption is that the queue is sufficiently long to allow the traffic density to remain constant.

Other typical assumptions made are as follows.

(i)   All vehicles are of the same length.
(ii)   The distance between stationary vehicles is constant.
(iii)   All cars start from rest, and there is a constant (reaction) time between the car in front starting to move and the following car starting to move.
(iv)   The first car moves off the instant the light changes to green.
(v)   Each car moves with a constant acceleration.
(vi)   Each car will accelerate until a maximum permissible velocity is reached (due to the urban setting), when the acceleration becomes zero instantaneously.
(vii)   No car will start to pass through the lights after the instant they change from green.

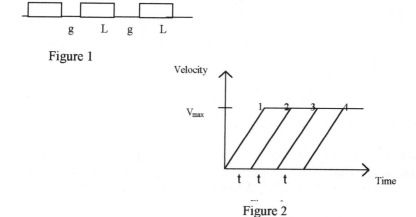

Figure 1

Figure 2

A suitable symbol definition list is then made, for example:

t	time elapsed since light turns green (s),
L	length of each vehicle (m),
g	gap between each stationary vehicle (m),
$v_n(t)$	speed of the nth vehicle t seconds after lights turn green (m/s),
T	time lights are set on green (s),
n	the number of cars in the stationary queue,
v_{max}	the maximum allowed speed of a vehicle (m/s),
$s_n(t)$	distance travelled by the nth car in time t (m),
a	acceleration of each car (m/s^2)
τ	constant reaction time (s).

Students are encouraged to make some illustrative sketches as in Figs. 1 and 2 above to obtain a qualitative feel for the problem and begin a mathematical formulation. Formulation may be obtained as follows.

Velocity attained by the nth car is given by

$$v_n(t) = \begin{cases} 0 & t < (n-1)\tau \\ a(t-(n-1)\tau) & (n-1)\tau \le t < v_{max}/a + (n-1)\tau \\ v_{max} & t \ge v_{max}/a + (n-1)\tau \end{cases}$$

and the distance travelled by this car as

$$s_n(t) = \begin{cases} 0 & t < (n-1)\tau \\ \dfrac{1}{2}a(t-(n-1)\tau)^2 & (n-1)\tau \le t < v_{max}/a + (n-1)\tau \\ \dfrac{v_{max}^2}{2a} + v_{max}(t-(n-1)\tau - v_{max}/a) & t \ge v_{max}/a + (n-1)\tau \end{cases}$$

To determine how many cars pass the lights in a period T then we require the greatest integer, n, such that

$$s_n(T) > (n-1)(L+g)$$

where (n-1)(L+g) is the initial distance of the front of the nth car from the lights.

Data may be obtained simply by observations and given values. Results can be obtained by calculation, or by the use of a spreadsheet to find the speed and distance each car travels in time T and also its position relative to the lights at this time.

From this example, students gain confidence in making simplifying assumptions in order to obtain a model that describes a real situation adequately. The model may be validated easily by simple observations and measurements. It is interesting to note students' reactions to assumptions (ii) and (iii) when they consider their own driving experience. The effect of omission of assumption (i) is also of interest. Once a sufficiently long time T has elapsed, the early cars that pass the traffic lights achieve high speeds.

Another feature of the problem is that it may be the first occasion that students need to define expressions that are not the same for the whole domain. A major criticism of the model is the behaviour of the drivers once the lights change from green, and an extension to incorporate how the drivers will behave as the lights change colour (or after the lights have been green for a sufficiently long time) may be implemented.

Example 2 Restricted view on approaching a bend

The problem is to advise motorists of a safe maximum speed of approach into a bend. This will be affected by means of a traffic-warning sign indicating maximum speed. The road is assumed to be enclosed by high hedges on either side, restricting view forward, and is a single lane with double white lines in the middle (Fig. 3).

This problem may be considered from two aspects: geometrical and mechanical. The approach employed will depend on the background knowledge of the student.

Certain basic assumptions are first made.

(i)   The dimensions are such that the geometry may be regarded in two dimensions.
(ii)   The bend is a circular arc and carriageways are of constant width.
(iii)   As there are double white lines, vehicles are restricted to their carriageway and there is no likelihood of meeting a car 'head on'.
(iv)   The road surface is flat, horizontal and in good condition.
(v)   The car must be able to stop before reaching any obstruction, irrespective of size.
(vi)   The driver's view will be most restricted if the nearside edge of the car is adjacent to the left hand side of the road.
(vii)   The car is in proper condition, and the driver can stop within the limits as laid down by the highway code.

The following is a symbol definition list:

r radius of circular arc measured to double white line (m),
w width of each carriageway (m),
d_1 distance of head of driver from near side (left side) of car (m),
d_2 distance of head of driver from front of car (m),
v speed of car at instant obstruction spotted (m/s),
α angle subtended at the centre of curvature of the circular bend by the driver and point on nearest edge (radians),
a, b parameters associated with the highway code rule (values obtained as discussed earlier).

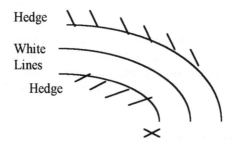

Figure 3

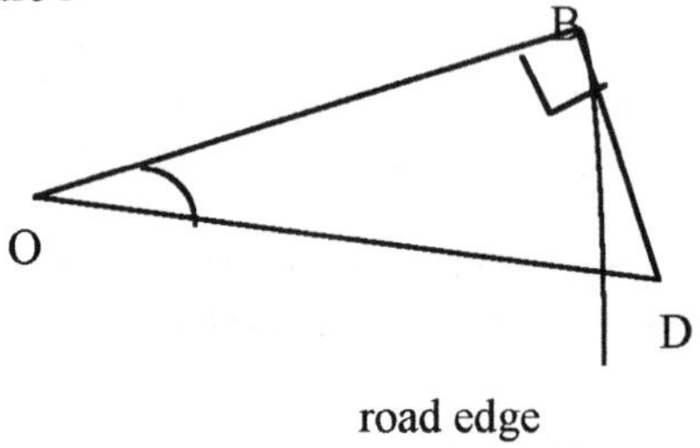

Figure 4

Fig. 4 indicates that the driver's view is restricted to DB. The driver must be able to stop within a distance measured on the arc of road less the distance of the driver's head from the front of the car, so that $(r-w)\alpha - d_2$,

α is then given by $\qquad \cos\alpha = \dfrac{OB}{OD} = \dfrac{r-w}{r-w+d_1}$

and, if the stopping distance (thinking and braking) is taken as $s = av+bv^2$ we require that v must be restricted to

$$av+bv^2 < (r-w)a - d_2$$

ie $\qquad av+bv^2 < (r-w)\cos^{-1}[(r-w)/(r-w+d_1)] - d_2 \qquad (*)$

Data values are not so readily available in this problem, particularly for r, but an interesting discussion may be generated on the determination of this parameter.

A second consideration is the possibility of skidding. Further assumptions are made as follows.

- The car, if driven too rapidly around the bend, will skid sideways rather than overturn.
- The coefficient of friction is such that the limiting case for the frictional force is not reached.

- The centre of mass of the car and driver is in the midpoint of the car, which may therefore be treated as a particle.

Further symbols needed are:

μ coefficient of friction between car and road,
M   mass of car (kg),
g   acceleration due to gravity (m/s^2),
d_3 width of the car (m).

We have the frictional force $< \mu Mg$ and the force towards the centre of the circle is given by

$$\frac{Mv^2}{r - w + \frac{1}{2}d_3} \qquad \text{where} \quad r - w + \frac{1}{2}d_3 \quad \text{is the radius of the circle.}$$

Thus, to avoid skidding $\quad \dfrac{Mv^2}{r - w + \frac{1}{2}d_3} < \mu Mg$

$$\textit{ie} \qquad v < \sqrt{\mu g (r - w + d_3/2)} \qquad\qquad (**)$$

The safe speed for traversing the bend is therefore determined by finding the value of v that satisfies inequalities (*) and (**). The parameter μ is not easy to determine, and some interesting ideas may emerge from discussions on how to obtain a numerical value for it.

The interesting feature of this problem is that two very different approaches may be employed and that the conditions obtained from geometric considerations should not be discounted.

Student feedback

The responses from students when tackling these problems have been encouraging and, in general, they have shown enthusiasm for considering practical problems. Comments such as "at last I see a purpose for studying mathematics, it was not obvious at school" and "these were problems I could relate to" were common place. Students also became aware of a structure they could adopt when undertaking other modelling problems. Statements such as "I often thought about such situations but did not know where to begin to solve them" and "I now feel confident to start to formulate models for my own ideas" were frequently expressed. Another feature commented upon was the different approaches that may be made, particularly with problem 3.2 above. This led to a feeling of ownership of the problem and the belief that individuals were achieving as much as colleagues with a greater knowledge of mathematical techniques. Another issue, relating to gender, was reflected upon. Females were often initially fearful of the subject content as illustrated by one female student, "I thought that this would be like physics in the man's world, but I could understand the tasks and I enjoyed solving the problems." Students were also interested in how the 'real world' solved such problems. Some negative comments were expressed however. Not all students were motivated, stating for example, "why do we bother, these problems have been solved already". It is to be hoped that such opinions do not prevail, particularly by those studying on the BEd (Primary). I often challenge such students to devise their own ideas. Some students who have come through the more traditional routes are initially disconcerted by the absence of a 'correct answer' and uncertain that they are using the 'right method'. Experience usually allays these fears.

The feedback comments clearly indicate the need for the development of real problems which are of interest to all participants of the module. These problems should also be capable of continued development. The ability of students to become independent learners is reflected in their comments. It is to be hoped that the ideas are continued in later studies. A second year module Scope of Mathematics (Herring and Bloomfield, 1995) has continued the ideas, in particular the input of mathematicians in industry into the module has been reinforced and the presentation of problems they introduce has gained greater significance. The development of other modules, such as Simulation and Financial Studies, will now be motivated by modelling approaches and case studies.

FURTHER IDEAS

Example 2 above discussed two basic problems that introduced students to the art of mathematical modelling. As the student acquires further mathematical knowledge, the problems that are able to be modelled become more complex. However, the theme of motoring still offers opportunities for further investigations and project work. Four areas are identified as being amenable to mathematical investigations (Ashton 1976).

(1)   Movement of vehicles at intersections and on the open road, including parking problems.
(2)   Engineering problems such as design of road systems and the investigation of systems of vehicle control including traffic lights and roundabouts.
(3)   Transportation and scheduling problems.
(4)   Accident and safety aspects.

The methods used in these investigations may be deterministic or stochastic. The use of stochastic models develops after students have acquired these more sophisticated ideas and techniques.

Problems that have been investigated in this section include parking problems related to the college campus.

(i)   The demand for the car park throughout the working day was estimated. As the college has various sites, a significant number of students travel between campuses during the day. A mathematical function of the expected demand against time of day is fitted using information collected using a survey; thus this investigation required use of various statistical techniques. The results of this investigation proved to be of interest to the College planning committee.

(ii)   A design of a car park layout between two buildings was proposed that involved the allotment of bays and their angle of alignment. A similar problem was analysed by Hunter and Young (1982).

(iii)   Another investigation undertaken was an evaluation of the traffic flow along a stretch of urban road and the effects on the flow if certain feeder side roads are closed so that all vehicles joining this stretch of road would be forced to join at one junction only, a proposed restriction that a student was to encounter on his route to College.

Design of road systems has proved to be a popular area of study, particularly the traffic dynamics at roundabouts.

Some introductory problems have included the identification of patterns of traffic flow at road junctions and the sequencing of traffic lights in order to maximise the flow. In the case of roundabouts, problems due to high traffic density and the effect of traffic light installation on access roads to a roundabout were considered.

Other design problems have involved the placing and shape of rear window wipers (Herring 1991) and the suitability of mud guards/splashers, particularly on large lorries, in order to reduce the amount of spray being thrown up at a following vehicle.

Another investigation undertaken recently has been the scheduling of the mini bus services introduced onto town routes in Cheltenham. Investigations have been made into the problem of 'bunching' of these buses on a frequent service and how they should be scheduled to reduce this effect. Other studies have been made regarding the routing of the services in the town.

Aspects of safety and causes of accidents provide a number of problems. The Open University course MST204, Mathematical Models and Methods, has set two interesting problems, one concerning the determination of the width between successive yellow lines painted on a carriageway with the purpose of reducing vehicle speed when approaching an obstruction such as a roundabout or road junction (Open University, 1995) and the other concerns the spacing and design of road humps to restrict the speed of traffic (Open University, 1989). The problem of design in the latter problem may be tackled from a purely geometric standpoint or by considerations of the dynamic effects on the driver.

Another interesting aspect is accident investigations (Smith and Hurst, 1990), concerning the determination of the speed of the vehicle before braking prior to an accident, and the skidding of cars (Smith and Thatcher, 1991).

CONCLUSIONS

When students are drawn from a broad spectrum of interests and backgrounds, they need to be motivated and convinced of the 'realness' of the problems they are to tackle. Topics must be chosen which are accessible to all in order that they are not distracted by technicalities. I have attempted to demonstrate that the area of motoring and road traffic has proved to be a suitable area that may lead to further investigations. It is an area where students can readily extend existing problems and devise their own investigations and case studies, an activity that is an important feature of mathematical activity.

The student feedback comments indicate that they wish to have an opportunity for independent thought, and to develop ideas and mathematical models from their own observations. The two examples discussed in section 3 hopefully gave a motivation for this and certain themes have been used by a number of students in later investigations and project work as illustrated in section 4. If students are exposed to carefully devised and motivating modelling problems early in their course then many of the objectives expressed above are achievable.

REFERENCES

Ashton A, 1976, *Stochastic Models for Road Traffic Situations*, in Andrews JG and McLone RR, (eds.), **Mathematical Modelling**, Butterworth, pp 127-142.

Burghes DN, 1981, *The Humber Tunnel Authority*, in James DJG and McDonald JJ, (eds.), **Case Studies in Mathematical Modelling**, ST(P) Cheltenham, pp 101-109.

Herring MJ 1991, *The Use of Mathematical Modelling of a Programme of Integrative Assignments*, in Niss *et al.*, (eds.), **Teaching of Mathematical Modelling and Applications**, pp 306-316.

Herring MJ and Bloomfield A, 1995, *Scope of Mathematics Practitioner Involvement in Undergraduate Mathematics*, in Sloyer C, Blum W and Huntley I, (eds.),**Advances and Perspectives in the Teaching of Mathematical Modelling and Applications**, Water Street Mathematics, Yorklyn, Delaware.

Hunter A and Young B, 1982, *Car Park Layout*, **Solving Real Problems with Mathematics Volume 2**, The Spode Group, Cranfield Press, pp 121-126.

LeMasurier D and Townsend G, (eds.), 1977, **Polymaths Book B**, Functions and Graphs, ST(P) Cheltenham.

Open University MST204 Course Team, 1995, **Assignments Booklet III**, Open University Press, Milton Keynes.

Open University MST204 Course Team, 1989, **Assignments Booklet I**, Open University Press, Milton Keynes.

Raggert G, 1984, *Topicality: A Personal Plea for Mathematical Modelling* in Berry JS *et al.*, (eds.), **Teaching and Applying Mathematical Modelling** , Ellis Horwood, pp 1-10.

Smith R and Hurst J, 1990, *Accident Investigations*, in Huntley ID and James DJG, (eds.), **Mathematical Modelling, A Source Book of Case Studies**, Oxford Science Publications, pp 67-80.

Smith R and Thatcher D, 1991, *Skidding of Cars and Quadratic Formulae*, **Teaching Mathematics and its Applications**, **10**, 2, pp 53-57.

Stone A and Huntley I, 1982, *Easing the Traffic Jam*, **Solving Real Problems with Mathematics Volume 2**, The Spode Group, Cranfield Press, pp 106-112.

25

Computer-based Experiments in Mechanics

D A Lawson and J H Tabor
*BP Mathematics Centre, Coventry University, Priory Street, Coventry,
England, CV1 5FB
e-mail: mtx047@coventry.ac.uk* **or** *mtx041@coventry.ac.uk*

ABSTRACT

*Experiments are an important part of the learning process in science. However students
of mathematics (who often study some very similar topics) are rarely given the
opportunity to carry out experiments. This denies them access to a valuable learning
experience. One of the main reasons for this is that mathematics departments do not
usually have access to the laboratory facilities required to allow their students to carry
out experiments.*

*Recent advances in computer technology now make it possible to use the computer as a
substitute for the laboratory. Realistic experiments can be carried out on screen. This
can give students many of the educational advantages of traditional laboratory-based
experiments but at a much lower cost in terms of time and resources.*

*In this paper we describe some computer-based experiments and give an evaluation of
the benefits, and a warning about some of the possible pitfalls, of making use of such an
approach.*

INTRODUCTION

In most, if not all, branches of pure and applied science, experiments are widely used as
part of the learning process. The superficial onlooker or cost-conscious finance officer
may argue that it is an inefficient use of time and resources to have large numbers of
students carry out the same relatively simple experiment to determine results that are
already well-known. However, such arguments show a lack of understanding of the
educational process. It is perfectly true that it would be much cheaper simply to tell

students the information that it is intended that they will learn through carrying out experiments. However this would be to deny them a valuable learning opportunity.

An ancient Chinese proverb says *"I hear I forget. I see I remember. I do I understand"* It is this philosophy that lies, in part, behind the importance of hands-on experimental experience in the education of scientists. The personal involvement in carrying out an experiment, the resulting ownership of the data produced, and the struggle to understand and analyse this data are all aspects that produce deep rather than surface learning. The use of experiments is more time-consuming and costly than the passing on of information, but the learning process is more efficient overall because the students are more likely to gain (and remember) an understanding of the concepts.

Despite the long-term benefits of experiments to the learning process they are an easy target when financial cutbacks must be made. Even in science and engineering courses the amount of time students spend in laboratories has been reduced in many British universities over recent years. For some time the use of computers has been seen as one possible way of giving students some of the benefits of carrying out experiments without the high cost of extensive use of laboratories (see, for example, Smith and Pollard, 1986).

Some early computer-based experiments, such as those presented by Borghi *et al.* (1987), used crude on-screen representations. They were limited by the technology to which students had easy access. However, in recent years there have been enormous advances in computer hardware and software. It is now possible to perform experiments on the computer screen rather than in the science laboratory. The experiments can occur in real-time or, where appropriate, be slowed down or speeded up. The image on the screen can be highly realistic, being taken from photographs or video, or sensibly idealised to focus on the key concepts under consideration. The attractiveness of computer experiments to students from a visual point of view, and to academic staff from an educational point of view, has increased with the improvements in the technology. More work is now being carried out in this area (Bates *et al.*, 1995). The high level of realism which can now be achieved allows the student to experience many (although not all) of the educational advantages of actually carrying out experiments but at a fraction of the cost.

In the remainder of this paper we describe the particular value of experiments to students of applied mathematics, particularly in view of recent developments in education in Britain. This is followed by a description of two on-screen experiments in mechanics that are suitable for first-year undergraduates. Although these experiments are similar in content they are illustrative of computer-based experiments. Furthermore, within the description of each experiment some design issues are discussed. Following this an evaluation of the benefits of computer-based experiments is given. It must be recognised that computer-based experiments have their limitations and a discussion of the possible pitfalls in the use of on-screen experiments is presented.

Applied mathematics and experiments

In many instances it is hard to determine exactly where the boundary lies between science and applied mathematics - mechanics is one such topic. The precise position of the line dividing mechanics as a topic in applied mathematics from mechanics as a topic in physics is not easily located. However, students studying mathematics are rarely given the opportunity to deepen their understanding by carrying out experiments. The primary reason for this is that few mathematics departments have access to laboratory facilities or even the simplest equipment. So, whilst for example a student in physics may carry out experiments to show that the period of a simple pendulum is independent of the mass of the bob, such an opportunity is almost never given to mathematics students. The latter have to be content with studying simple harmonic motion solely from a theoretical perspective.

This state of affairs is to be regretted. In recent years the mathematical community has successfully promoted the importance of mathematical modelling in many disciplines. However, it is almost inevitably the case that in order to formulate a good mathematical model the mathematician needs some insight into the phenomena under consideration. This insight can often be gained by carrying out experiments.

Reasons peculiar to the situation of applied mathematics in Britain also reinforce the need for students to be given the opportunity to gain insight through experiment and experience. About fifteen years ago, the majority of students reading mathematics in higher education in Britain had studied A-levels in Mathematics, Further Mathematics and Physics. In the two mathematics A-levels, about half the syllabus would have been a study of mechanics. As this was linked to a study of physics in their third A-level these students entered university with a strong understanding of basic principles such as Newton's laws. Since then, however, there have been significant changes. The popularity of Further Mathematics as an A-level subject has decreased enormously. Many schools no longer offer it to their pupils. There has been a widening of the range of subjects offered at A-level, and it is now far from uncommon for students to enter university to read mathematics having studied no science after the age of 16. Furthermore there have been changes within the mathematics A-level. The introduction of modular A-levels has tended to reduce the amount of mechanics studied by students. The combination of these factors has resulted in some students beginning mathematics courses at university with little insight into key physical principles.

As has been mentioned earlier, few mathematics departments have laboratory facilities to allow students to gain 'hands-on' experience through experiments. Computer-based experiments can be used to give students some experience of the experimental process and first-hand knowledge of key principles.

From the point of view of applied mathematics, mechanics is an extremely suitable topic for computer-based experiments. By the very nature of the subject there are a number of conceptually simple experiments that are extremely difficult to set up and carry out. It is much easier to have a 'smooth' (*ie* frictionless) surface in a computer-based experiment than in a laboratory. Firing a projectile into a vacuum presents little difficulty when preparing on-screen experiments but demands some effort in the laboratory (Borghi *et al.*, 1987).

SIMPLE COMPUTER-BASED EXPERIMENTS IN MECHANICS

In this section we describe two similar computer-based experiments centred around projecting a small ball vertically upwards. These experiments illustrate a range of features of real experiments that can be incorporated into computer-based experiments. Experiments such as these, and those reported elsewhere (Lawson *et al.*, 1995) can be found in the Mechanics module of Mathwise (Beilby, 1993), the product of the United Kingdom Courseware Consortium (a Teaching and Learning Technology Programme project).

Experiment 1 - gravity in a vacuum

In these simple experiments the student chooses an initial velocity for the ball. An animation then runs in real-time showing the ball travelling upwards in front of a measuring scale. The experimenter must measure the maximum height that the ball reaches. To help the student do this the ball leaves behind it a trail so that when it has fallen back to its start position its path is clearly marked. The user is then able to read off from the scale the maximum height and record this datum on the screen. This is shown in Fig 1.

When five experiments have been carried out a point plot of the results collected is automatically generated. There then follows a discussion of the relevant theory including a figure of the curve showing the theoretical maximum height against initial velocity. When students have read this information they can choose to superimpose their own experimental results on this theoretical curve.

This is a tightly controlled experiment. The student is given a suitable range of values for the initial velocity (between 6 and 26 m/s) and is prevented from entering values outside this range. In addition, the value of the maximum height entered by the student is checked before being transferred to the data table on the right of the screen and, if it is incorrect, the student is warned and requested to insert another value.

Neither of these two features is essential. The limit on the values of the initial velocity is there purely for practical reasons. Below 6 m/s the ball hardly moves at all whilst above 26 m/s the ball goes too high (*ie* off the measured scale). Instead of having a check on the limits it would be possible to allow a student to reject an experiment as not providing useful data. The check on the entered value of maximum height is to stop the student from entering an incorrect reading and thereby producing data points that do not fit the theory to be expounded shortly.

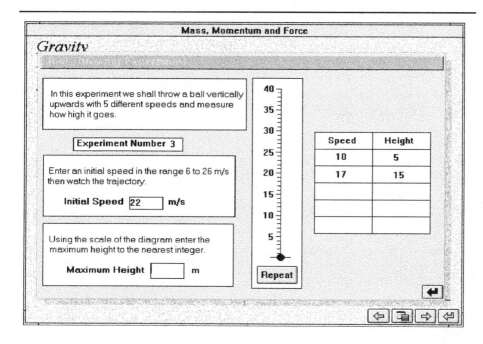

Figure 1: motion in a vacuum under gravity

Experiment 2 - motion in a resisting medium

In this experiment the student is seeking to determine the nature of the resistive force on a ball as it travels through a fluid. The student chooses an initial velocity for the ball and must record the height of the ball at up to twelve times during its motion. The student can treat the simulation as a video of the actual event. The animation can be played in slow motion and paused at any time the student chooses. Each time the motion is paused a measurement is automatically recorded in the data table. This is shown in Fig 2.

When the data have been collected the student is offered tools to analyse them. Velocity and acceleration values may be calculated by using finite differences, and then the student can choose to plot any one of the four columns (time, height, velocity, acceleration) from the data table against any other. A curve-plotting facility is also offered so that the student can try to fit a curve through the data points. As the student is trying to determine the nature of the resistive force (which is a function of velocity) the most useful graph is acceleration against velocity. This is suggested to the student once an incorrect form for the resistive force has been entered. Fig. 3 shows the screen from this experiment with height plotted against time.

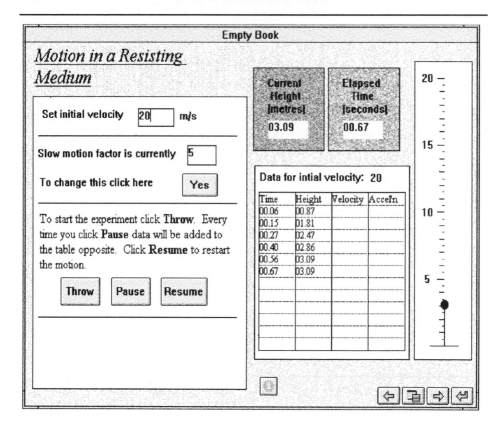

Figure 2: collecting data from resisted motion

In this experiment the student is allowed more freedom than the ones previously described. Any initial velocity is allowed and students can select data at whatever points they like. It is better if the data points can be reasonably equally spaced in time. This is quite hard to achieve when viewing the animation in real time. However this task becomes much easier if the slow-motion option is used in an appropriate manner.

Although there is more freedom for the student in this experiment, there is still a significant amount of control. The values of the height at the various times at which the student pauses the animation are automatically recorded in the data table. The student does not have to make any measurements and so one source of error is eliminated. This is a double-edged feature. On the positive side it prevents students from collecting incorrect data and then wasting considerable time and effort trying to find theories to fit values that are meaningless. On the other hand Magin and Reizes (1990) argue that measurement is part of the experimental process. Therefore, if a computer-experiment is to be as realistic as possible, the students should have as much freedom in measurement as they do in a laboratory.

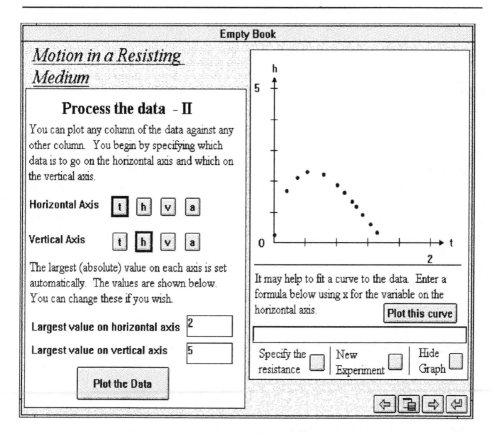

Figure 3: graphs from data of resisted motion

It is certainly possible to allow students considerable freedom in data acquisition. Students can be offered a choice (from a pre-determined list) of what result from an experiment they wish to measure. A poor choice may lead to results from which it is impossible to deduce anything. Measuring instruments can be presented on screen and the student required to use them to acquire data. If a student is not skilled in using these, inaccurate data may be obtained. Whilst these are certainly events that happen to many laboratory experimenters, and certainly to those engaged in research, they can be a hindrance to the educational purpose of simple on-screen experiments. The designer must be clear about the purpose of the computer-experiment, in isolation and as part of a package, when considering how much freedom to give students within a particular experiment.

A design issue common to all computer-based experiments is that of realism. The animation of a blue circle moving up and down a measured scale against a white background is clearly only a characterisation of reality. It would have been a relatively simple matter to achieve greater realism by using a photograph of a wall with a scale painted on it and to move a bit-map of a tennis ball or stone instead of the simple blue circle. For conceptually simple experiments like this one a high level of realism is unnecessary and may even be distracting. However, for more complicated situations, such as the response of a car's suspension system to travelling over bumps, greater realism may be required.

EVALUATION OF COMPUTER-BASED EXPERIMENTS

Benefits

There is much to be gained from using computer-based experiments with students in all scientific disciplines, but particularly students of mathematics. For the vast majority of mathematics students the choice is not between computer-based experiments and those carried out in the laboratory. Rather it is usually a choice between computer-based experiments or no experiments at all. In such circumstances computer-based experiments are the only way to give students access to some of the educational benefits of experiments.

In a direct contest between computer-based and laboratory-based experiments there are some areas in which computer-based experiments have advantages over those in the laboratory.

- Amount of time required - it takes much less time for a student to set up and carry out a computer-based experiment; the student can therefore carry out many more investigations and acquire more data than is possible in a conventional laboratory.
- Ease of repeatability - a student, on analysing the data obtained from a series of experiments, may wish to obtain further results; with laboratory experiments this is often not possible but with computer-based ones it is straightforward.
- Integration of analysis and theory - the tools required to graphically and statistically analyse the experimentally collected data can be built into the same piece of software that is providing the on-screen experiments, and appropriate discussions of the relevant theory can also be available at the click of an on-screen button.
- Impossible experiments - the student can carry out on-screen experiments that would be almost (or indeed actually) impossible in any university laboratory, for example, when considering motion under gravity in a vacuum the mass of the earth can be changed to produce a different acceleration due to gravity.
- Ability to alter time - experiments in the laboratory occur in real-time, whereas experiments on-screen can occur at any time scale. The use of slow motion has already been illustrated, and speeding up time can also be used to advantage, for example in experiments in astronomy where the typical time scales may be years.
- Accessibility - students can carry out computer-based experiments at any time; there is no need for supervision.
- Resources - computer-based experiments require computer technology but nothing else; there is no need for expensive equipment or consumables.

Disadvantages

Despite the impressive list of advantages above, wholesale replacement of conventional laboratory-based experiments by on-screen ones would not be sensible. There are a number of limitations to computer-based experiments.

- No experience of handling equipment - computer-based experiments do not require students to set up their experiment - they gain no feel for how difficult it may be in practice to measure certain quantities, and the on-screen experiments may create the false impression that all experiments are easy to carry out.
- Limited by the author - a student may wish to take an investigation along lines not provided for by the author of the software.
- Not the real thing - the student is experimenting with a model of reality not with reality itself.

This last point indicates the folly of imagining that computer-based experiments could ever totally replace real experiments. The models on which computer-based experiments rest required data from real experiments in their development.

CONCLUSIONS

- Experiments are a valuable part of the educational experience of students studying science. It would be beneficial if similar experience could be given to students of applied mathematics.
- Mathematics departments not having access to laboratories can give their students some experience of experiments by using computer-based experiments.
- Some topics, such as mechanics, are particularly suitable for computer-based experiments.
- Computer-based experiments have a number of advantages (both educational and in terms of resources) over laboratory-based ones.
- Computer-based experiments should never be thought of as potential replacement for all laboratory work.

REFERENCES

Bates B, Leary JJ and Saadat S, 1995, *Virtual Laboratory Experimentation: A Review of the State of the Art and Current Research*, **Proceedings of Hypermedia in Sheffield 1995**, pp 172-181.

Beilby, M, 1993, *Mathwise: A Learning Environment (part 1)*, **Maths & Stats**, 4(4), pp 2-5.

Borghi L, De Ambrosis A, Mascheretti P and Massara CI, 1987, *Computer Simulation and Laboratory Work in the Teaching of Mechanics*, **Physics Education**, **22**, pp 117-121.

Lawson DA, McCabe EM and Tabor JH, 1995, *Virtual Experiments in the Physical Sciences*, submitted to **Computers and Education**.

Magin DJ and Reizes JA, 1990, *Computer Simulation of Laboratory Experiments: An Unrealized Potential*, **Computers and Education**, **14**, 3, pp 263-270.

Smith PR and Pollard D, 1986, *The Role of Computer Simulations in Engineering Education*, **Computers and Education**, **10**, 3, pp 335-340.

26

Modelling Patient Flow through Hospitals

Sally McClean
University of Ulster, Coleraine, Northern Ireland, BT52 1SA.
e-mail: SI.McClean@ulst.ac.uk

ABSTRACT

It is commonly the case that hospital departments engage in two separate forms of clinical activity : acute/rehabilitative and long-stay care. These are organisationally distinct and have very different resource needs. Current hospital-planning models, however, often base performance measures on the implicit assumption that all patients move through the system at the same rate, thus ignoring such inherent heterogeneity in patient behaviour.

A compartmental model has been developed in which the constituent components are the acute/rehabilative, or short-stay, patients and the long-stay patients. Difference equations may then be solved to provide expressions for the average numbers and lengths of stay for short- and long-stay patients thus enabling us to assess the effect of system changes such as converting long-stay to short-stay beds. This basic model has been implemented in a software package for Bed Occupancy, Management and Planning (BOMPS) which uses the daily bed-occupancy census data. It is hoped that such an approach will provide more effective planning for the hospital service.

More recent work extends this approach to consider patients discharged from hospital departments and their subsequent length of stay in the community. We have, therefore, two states of patient behaviour while in hospital. Here patients are first admitted to the acute/rehabilitative state, from which they may die, be discharged back into the community, or admitted to the long-stay state from which they eventually die. The community component has one state, from which patients may be re-admitted into the hospital or go to the fourth state which is death. We are in the process of generalising this four-compartment model to its stochastic analogue and extending the approach to include age-specific transitions between compartments.

INTRODUCTION

The proportion of elderly people in the population is increasing rapidly while the relative proportion of younger people participating in the workforce is declining. The problem of caring for the frail or sick elderly is therefore one requiring urgent attention and careful planning. There are various factors which must be taken into account when planning for an aged population, principally:

- elderly people are more likely to be disabled and to live in single person households,
- they often remain in hospital longer than their younger counterparts,
- diseases present differently in the elderly compared with younger patients.

Geriatric patients may be thought of as progressing through stages of diagnosis, assessment, rehabilitation and long-stay care. Most patients are eventually rehabilitated and discharged. The small proportion who become long-stay remain in hospital for months, or even years. These patients may be very consuming of resources and thereby distort the performance statistics. Simple methods of measuring performance such as the average length of stay for a patient are therefore no longer appropriate.

In collaboration with Professor Peter Millard of St. George's Hospital Medical School, London, a number of mathematical models have been developed which describe the process of geriatric patients moving through hospital and enable us to provide more accurate ways of measuring performance. The basic model, which was developed by Harrison and Millard (1991) and is further discussed in McClean and Millard (1993a), analyses the pattern of occupancy of hospital beds in Departments of Geriatric Medicine.

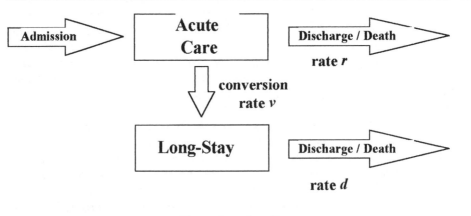

Figure 1: patient flows

The model was developed as a result of the empirical observation that the pattern of bed occupancy is best fitted by a two-term mixed exponential distribution of the form

$T(s)=A\exp(-Bs) + C\exp(-Ds)$

where $T(s)$ is the total number of patients who have been in hospital for s days at a given point in time. Essential features of the model are:

- the concept that movement occurs in streams *ie* groups of fairly homogeneous patients all progressing at much the same rate,
- the idea that discharge rates depend on destination,
- the notion that discharge practices vary from hospital to hospital,
- the observation that the pattern of length of stay in the beds can be described by exponential distributions.

The patient flows within a geriatric hospital are illustrated in Fig. 1, where patients are initially admitted to an acute or rehabilitative state from which they either are discharged or die at a rate r or are converted to a long-stay state at a rate v. Long-stay patients are discharged or die at a rate d. We develop the basic model in terms of linear difference equations, as follows.

$A(s+1)$	$=$	$A(s)$	$- vA(s)$	$- rA(s)$
acute at day s+1		acute at day s	conversions	death / discharge

$L(s+1)$	$=$	$L(s)$	$+ vA(s)$	$- dL(s)$
long-stay at day s+1		long-stay at day s	conversions	death / discharge

The total number of patients with at least s days in hospital $T(s)$ is then given by $T(s) = A(s)+L(s)$ where $A(s)$ and $L(s)$ are found by solving linear difference equations so,

$$T(s) = \frac{(1-k)}{(v+r)} A_0 (1-v-r)^s + \frac{k}{d} A_0 (1-d)^s$$

which is equivalent to the above expressions for $T(s)$.

This is a mixed exponential distribution which has been found to give a good fit to data from hospitals in the SW Thames region.

By thus making a number of plausible assumptions about discharge rates it is possible to find equations giving the numbers of acute and long-stay patients. The model is fitted to occupancy data from the midnight bed returns. This enables us to use the model to assess performance and evaluate the effect of possible changes.

A PC-based software package, known as BOMPS (Bed Occupancy Management and Planning system), has been developed at St George's. We may use it to gain an accurate picture of long-stay patients. The method may in fact be used for specialties other than geriatric medicine which exhibit a similar pattern of short- and long-stay patients. The BOMPS package allows for the possibility of fitting one-two-or three-stage models. An essential feature of the package is the facility to carry out what-if analysis and investigate possible future scenarios. Typical questions which can be answered by using the what-if utility are:

- how many fewer patients might we treat next year if we close 10 beds,
- if we stop admitting patients to a ward, how long will it be before the ward is empty,
- how many more patients would we be able to admit if we increased rehabilitation thus reducing the chances of patients becoming long-stay.

Health Service Managers are currently using the model to assess the impact of hospital closures in London, and the College of Physicians is using it to assess the impact of the Community Care Act on patient throughput in hospitals. In summary, the basic model enables managers to make decisions about the balance of short- and long-stay geriatric patients. Modelling can thus facilitate the operation and planning of geriatric services. With an increasing emphasis on the efficiency and cost-effectiveness of hospital services, a modelling approach must become increasingly important.

THE STOCHASTIC MODEL OF GERIATRIC IN-PATIENT BEHAVIOUR

We now develop a stochastic model for the behaviour of geriatric in-patients. This approach is further described in Irvine, McClean and Millard (1993). The stochastic model is more realistic than the basic deterministic model because it uses probabilities thus enabling means and variances to be calculated. It also has the advantage of being mathematically tractable. We consider two versions of this model, the first of which assumes a constant waiting list which thereby keeps the total number of patients in the unit constant as the beds are always full (Fig. 2). The second version of the model is appropriate to a situation when the admissions arrive at random as patients in the catchment area become ill. We therefore describe such admissions using a Poisson process (Fig. 3). The two models are solved in the same way and yield similar results, which in each case are again a mixture of exponential terms, as in the deterministic case. In each formulation, we describe the movements of patients through hospital in terms of transition probabilities.

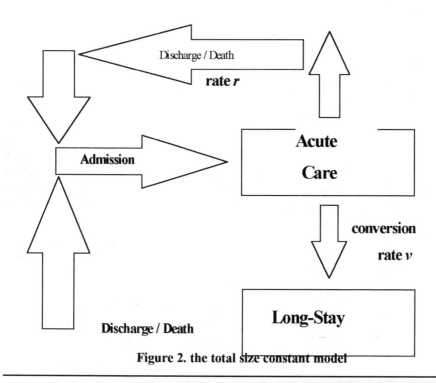

Figure 2. the total size constant model

Let $p_{ij}(t)$ be the probability that a patient who is in state S_i at time 0 is in state S_j at time t where S_1 is acute care, S_2 is long-stay care and S_3 is discharge or death.

Then $\dfrac{d}{dt}P(t) = P(t)A$

where $A = \begin{bmatrix} -(v+r) & v & r \\ 0 & -d & d \\ 0 & 0 & 0 \end{bmatrix}$

Solving, we obtain the probability that a patient admitted to hospital at time 0 is still there at time t is $p_{11}(t) + p_{12}(t) = \dfrac{r-d}{r+v-d}e^{-(r+v)t} + \dfrac{v}{r+v-d}e^{-dt}$

We solve the total size constant model as follows.

$n_1(t)$ and $n_2(t)$ are the number of patients in acute and long-stay respectively.

$G(z_1, z_2 : t) = \displaystyle\sum_{n_1 n_2} z_1^{n_1} z_1^{n_2} p_{n_1, n_2} t$ p.g.f. of n_1 and n_2 at time t.

It follows that

$$\frac{\partial G}{\partial t} = d(z_1 - z_2)\frac{\partial G}{\partial z_2} - v(z_1 - z_2)\frac{\partial G}{\partial z_1}$$

We can solve this for G and hence find the mean and variances of the numbers of acute and long-stay patients at time t namely,

$$E[n_1(t)] = A\left(\frac{d}{v+d} + \frac{v}{v+d}e^{-(v+d)t}\right) + L\left(\frac{d}{v+d} - \frac{d}{v+d}e^{-(v+d)t}\right)$$

$$E[n_2(t)] = A\left(\frac{v}{v+d} - \frac{v}{v+d}e^{-(v+d)t}\right) + L\left(\frac{v}{v+d} + \frac{d}{v+d}e^{-(v+d)t}\right)$$

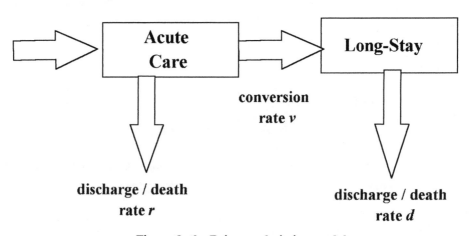

Figure 3. the Poisson admission model

For the Poisson admission model, as before, we obtain a partial differential equation for the generating function.

In this case,

$$\frac{\partial G}{\partial t} = [v(z_1 - z_2) + r(1 - z_1)]\frac{\partial G}{\partial z_1} + d(1 - z_2)\frac{\partial G}{\partial z_2} + G\lambda(z_1 - 1) \quad \text{and}$$

$$E[n_1(t)] = \frac{\lambda}{v+r}(1 - e^{-(v+r)t}) + A_0 e^{-(v+r)t}$$

$$E[n_2(t)] = \left[\frac{\lambda v}{v+r}\left(\frac{1-e^{-dt}}{d} - \frac{e^{-dt}-e^{-(v+r)t}}{v+r-d}\right)\right] + \frac{A_0 v}{v+r-d}\left(e^{-dt}-e^{-(v+r)t}\right) + L_0 e^{-dt}$$

The partial differential equation for the probability generating function is solved to give an expression for G. We then differentiate this expression and put each z equal to 1 to obtain successive moments. In this case, we have presented results for the respective first moments (the mean numbers in acute and long-stay care at time t, $n_1(t)$ and $n_2(t)$ respectively). Of course the second moments, the variances, are of particular interest as they give an idea of how close to the mean our model predictions are likely to be. They are calculated in a similar fashion to the means where, in each case, the solutions are again sums of exponential terms and the expressions for the means are essentially the same as those obtained for the deterministic model.

We now extend the model to take account of the time spent by former geriatric patients in the community. As it is frequently the case that geriatric patients are repeatedly discharged and readmitted, by including the community as a state of the model we may obtain a better understanding of the whole picture.

THE FOUR-COMPARTMENT MODEL

The deterministic model

The flow of patients through the system is illustrated in Fig. 4, where all patients are initially admitted into acute care. From here, they are discharged back into the community at a rate r_1, or die at a rate r_2, or are converted to long-stay care at a rate v. Long-stay patients stay until death, which occurs at a rate d. Ex-patients in the community may be re-admitted back into the department at a rate u_1, or die at a rate u_2. This model is a generalisation of the model of Harrison and Millard (1991) which describes the hospital part of the system. It is further described in Taylor et al.(1996).

We describe the new system by the following equations:

$$A(s+1) = A(s) - r_1 A(s) - v A(s) - r_2 A(s);$$
$$L(s+1) = L(s) + v A(s) - d L(s);$$
$$C(s+1) = C(s) - u_1 C(s) - u_2 C(s).$$

where $A(s)$ is the number of patients who have been in acute care for s days,
            $L(s)$ is the number of patients who have been in long-stay care for s days,
            $C(s)$ is the number of ex-patients who have been in the community for s days.

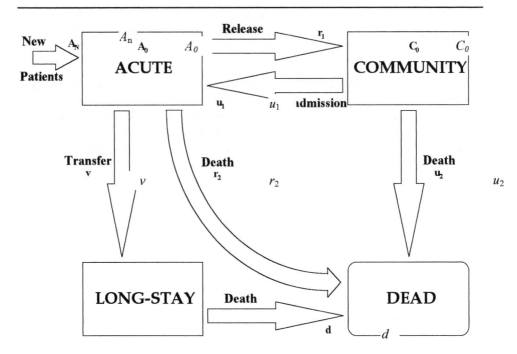

Figure 4: the basic geriatric system model

Solving these equations, we now obtain

$$A(s) = (1 - v - r_1 - r_2)^2 A_0,$$
$$L(s) = k A_0 (1-d)^s - k A_0 (1 - v - r_1 - r_2)^s \quad , \quad k = v / (v + r_1 + r_2 - d);$$
$$C(s) = C_0 (1 - u_1 - u_2)^s.$$

where A_0 is the admission rate into acute care and C_0 is the rate at which patients re-enter the community. These equations are as for the basic model, with the addition of an equation describing the number of ex-patients in the community who are described by a simple exponential equation.

The estimation of parameters was carried out using data consisting of 6994 geriatric patients admitted between 1969 and 1984 to four sites in the health authority containing St. George's Hospital, London (McClean and Millard, 1993). This data set was censused on a chosen date and the model fitted to all patients and ex-patients in the system at that point in time. Such an approach enables relatively easy and available collection of data by the hospital administrators, although obtaining complete data for ex-patients in the community may be more difficult and time consuming. The model may also be fitted to a cohort of patients. However, this would be less practical and more time consuming for a busy hospital manager.

The model was in fact fitted to data for Wednesday 1st December 1976. This date was chosen because, in January 1977, Professor Millard checked the death records for all patients in the data. Therefore the data around this time are statistically complete. A Wednesday was chosen to avoid the phenomenon that patients tend to be admitted towards the beginning of the week and discharged at the end of the week. There may however also be seasonal trends present within the data, and therefore we avoided dates soon after traditional holidays such as Christmas and Easter.

Using the fitted values of the various rates, we may obtain various information which is of use to the hospital managers. Examples are as follows:

$$\text{average time spent in acute care} = \frac{1}{v + r_1 + r_2} = 36.87 \text{ average days};$$

$$\text{time spent in long stay care} = \frac{1}{d} = 1000.5 \text{ days};$$

$$\text{average time spent in geriatric department} = \frac{k}{d} + \frac{1-k}{v + r_1 + r_2} = 72.98 \text{ days};$$

$$\text{average time spent in the community} = \frac{1}{u_1 + u_2} = 714.79 \text{ days}.$$

We may also use the model, as before, to assess the effect of possible changes in the system parameters - for instance changes in the number of beds and changes in the acute care release rate.

The stochastic model

As we saw previously for the hospital in-patient model, we may extend our deterministic model of the whole system of geriatric care to a stochastic compartmental model, which may be regarded as a continuous time Markov Chain.

Then we can show that, in this case,

$$\frac{d}{dt}\mathbf{P}(t) = \mathbf{P}(t)\mathbf{R}, \text{ where } \mathbf{R} = \begin{pmatrix} -(r_{12} + r_{13} + r_{14}) & r_{12} & r_{13} & r_{14} \\ 0 & -r_{24} & 0 & r_{24} \\ r_{31} & 0 & -(r_{31} + r_{34}) & r_{34} \\ 0 & 0 & 0 & 0 \end{pmatrix}.$$

Here **R** is our transition matrix. Solving this equation we are able to calculate the means and variances of the proportion of patients expected to be in either a particular compartment or the hospital system itself at any time t, as before.

The specification for this model, and implementation using maximum likelihood estimates based on exponential movement around the system, is further described in Taylor, McClean and Millard (1995). Such an approach, using census data, may be implemented with relative ease by extracting the required information from the hospital's Patient Administrative System (PAS).

MEASURING PERFORMANCE - A SYSTEMS APPROACH

System performance may either be measured directly or indirectly, using a mathematical model. Mathematical models thus play a vital role in measuring the performance of systems. Typical performance indicators are productivity, responsiveness and utilisation. In terms of our geriatric care system, we may interpret productivity as throughput of patients, responsiveness as length of time on the waiting list and utilisation as bed occupancy. Mathematical models may thus be used to calculate such measures and facilitate better understanding, monitoring and control of the whole system.

Our previous approach to planning services for an elderly population concentrated on improving throughput in the hospital sub-system. However the whole system includes other services such as home-helps, district nursing, sheltered accommodation and residential and nursing homes (Fig. 5). It is only by analysing the whole system of care options and the flows between them that we may understand, evaluate and plan the service. Our existing models consider patient behaviour as consisting of short-and long-stay care. By regarding duration in these two categories as having different distributions we account for a major source of heterogeneity in the data. Other sources of heterogeneity are gender, age, dependency, source of referral and initial diagnosis. By including these variables we are better able to describe patients' spells in hospital and the outcomes ie. whether they are discharged, transferred to another unit or die in hospital. Some preliminary work on the relevance of such additional variables is described in McClean and Millard (1993b) and further work will investigate the possibility of incorporating such additional variables into the models, thus enabling us to better describe the system.

CONCLUSION

A modelling approach enables us to look at the whole problem of describing and monitoring the geriatric care system by considering the system as a whole rather than concentrating on the separate sub-systems. Such an approach to health-care planning can facilitate understanding, performance measurement and evaluation of change. We may thus provide management with tools for monitoring and evaluating care provision.

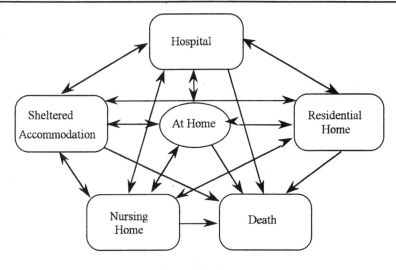

Figure 5: the geriatric care system

REFERENCES

Irvine V, McClean S and Millard P, 1993, *Stochastic Models for Geriatric In-patient Behaviour*, **IMA Journal of Mathematics Applied in Medicine and Biology, 11, pp** 207-216.

Harrison GW and Millard PH, 1991, *Balancing Acute and Long-stay Care: the Mathematics of Throughput in Departments of Geriatric Medicine*, **Methods of Information in Medicine, 30, pp** 221-228.

McClean SI and Millard PH,1993a, *Modelling In-patient Bed Usage Behaviour in a Department of Geriatric Medicine* **Methods of Information in Medicine, 32, pp** 79-81.

McClean SI and Millard PH, 1993b, *Patterns of Length of Stay after Admission in Geriatric Medicine: an Event History Approach*, **The Statistician, 42, pp** 263-274.

Taylor GJ, McClean SI and Millard PH, 1995, *Continuous Time Markov Models for Geriatric Patient Behaviour*, in Janssen JJ and McClean SI, (eds.,) **Proceedings of the Seventh International Symposium on Applied Stochastic Models and Data Analysis**, University of Ulster, pp 582-590.

Taylor GJ, McClean SI and Millard PH, 1996, *Geriatric-Patient Flow-Rate Modelling*, IMA Journal of Mathematics Applied in Medicine and Biology, **13, pp** 297 - 307.

27

Lagging a Pipe or Bandaging a Limb

A J I Riede
Universität Heidelberg, Im Neuenheimer Feld 288, D-69120 Heidelberg, Germany
e-mail: riede@mathi.uni-heidelberg.de

ABSTRACT

In his articles on comprehension tests in mathematics, Houston (1993) pointed out ideas for getting the students to read, write and talk about an article, with understanding. By the example named in the title this paper shows how Houston's ideas were used in teaching mathematical modelling to future teachers. The main example deals with lagging a conical pipe. It leads to a mathematical explanation of a method which is applied in bandaging the legs of persons suffering from heart disease.

INTRODUCTION

In the summer semester of 1995 I gave a course on mathematical modelling for inexperienced students intending to become high school teachers. For the lecture I used the book by Edwards and Hamson (1993) as a general framework. I taught the general methodology and aspects of the main tools of modelling, stochastics and differential equations; individual exercises and group work alternated via the group work I also intended that the students learn to read articles with understanding - in the sense of Houston's (1993a) rationale for a comprehension test. For this reason in my lectures I showed how Houston's list of assessment objectives (see Appendix) can be used as a guide for studying modelling examples from the literature. The students had to read two articles - the first was Handicapping Weightlifters from Burghes *et al.* (1982) formerly used by Houston (1993a) in a comprehension test (see also Berry and Houston, 1995: pages 75-84) the second was an article by Braun (1988) on a model in ecology (see also Riede, 1993: pages 306-313). The study of this article gave new insight into the survival of the fitter of two similar species (see Riede, 1995b).

Furthermore, in accord with the example reported in this article, I had the following aims in mind unchanged:

(a)   Except in courses on Linear Algebra, Analysis (Calculus) and on Introduction to Stochastics, the students very seldom learn subjects during their scientific study that they can use in their later career as school teachers. A lecture course on differential geometry often progresses to a high level too quickly to infinite dimensional manifolds - in order to prepare students for their diploma thesis. My intention was to start with a problem that can really be dealt with at school;

(b)   The second aim was to get the students to learn some of the mathematical background for themselves. I showed them how to use a mathematics dictionary namely the one by Bronstein and Semendjajew (1991). On the basis of background reading I tried to build up a relationship between mathematics at high school and university. In this way I wanted to improve the motivation to attend the more advanced lectures, for example those on differential geometry.

THE PROBLEM AND A FIRST MODEL

The example is taken from the book by Edwards and Hamson, (1993: pages 6, 7 and 37). It deals with the problem of 'properly' lagging a pipe - that means smoothly, without gaps and overlaps, without intermediate space between pipe and tape and without stretching the tape. One of Houston's (1993b) aims in a comprehension test is to demonstrate to students that mathematics is a living subject and is used in contemporary situations. To meet this and the above aim (a), I took with me several tapes and pipes which I found in my home and cellar. I showed them (see Edwards and Hamson, 1993) how, for a cylindrical pipe, this problem can be solved by the theoretical experiment of cutting the lagged cylinder along a line and unwinding the surface onto a plane. As a result the tape is wound in such a way that its boundary forms a helix on the pipe.

At that time my wife told me the striking fact that patients suffering from a heart disease had their legs bandaged in quite a different way. The tape is wrapped downwards towards the thinner part of the leg, becomes horizontal, and is then wrapped upwards again in a way that the tape crosses itself. It is then wrapped horizontally and again downwards and so on.

At this point I could not find a favourable geometrical explanation to explain the way of bandaging just described should be. In order not to undermine the students' satisfaction at having understood some applied mathematical points of lagging and bandaging I did not mention this method.

USING THE LIST OF ASSESSMENT OBJECTIVES

According to the second and third points (see Appendix) of Houston's list I remarked that the assumption that the pipe has a constant cross section was not mentioned. Thus the above solution could be applied to the bandaging of a limb only in so far as a limb could be approximated by a cylinder. Hence we looked for other simple surfaces without constant cross section, and found the cone. It was not difficult to lead the students to the fact that a cone can, like a cylinder, be unwound onto a plane. Perhaps this fact could be useful for solving the problem of properly lagging a cone; this would meet Houston's seventh point. According to Houston's sixth point this was the opportunity to ask for the background material and undertake the task of examining the literature on all surfaces that can be unwound onto a plane. The students did this job quite well and found such surfaces as general cylinders, general cones and the tangential surfaces. I gave a lecture on isometries of surfaces to cover some details and in order to give an overview of differential geometry. There was not enough time to think about which practical applications the tangential surfaces might have. We concentrated on lagging a conical pipe, and kept in mind that a cone should be a better local approximation to the surface of a limb than a cylinder.

LAGGING A CONICAL PIPE

The pitch angle

We study the lagging problem on a cone unwound onto a plane as a circular sector. The first observation is that if you try to lag without gaps the tape has an angle greater than zero once you have lagged round the cone. This occurs as the pitch angle increases if you wrap towards the thicker end of the cone. As we see in Fig. 1, once we are round, the new pitch angle c is the former one, say b, plus the angle a of the circle sector.

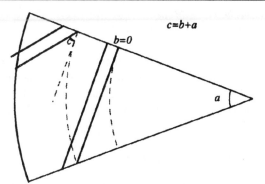

Figure 1: the change of the pitch angle

In Fig. 1 we started with pitch angle $b=0$. Owing to the change of pitch angle, it is not possible to wrap a cone properly in the above sense. The only possibility is to cut the tape after each winding and put a new strip beside the first one and so on, but this is not a good procedure.

Lagging with an elastic tape

Of course you can take an elastic tape and stretch it during the lagging. As a result pressure on the pipe will occur. When bandaging a leg of a patient suffering from heart disease the aim is indeed to provide this pressure. As an easy example to begin with, I dealt with the problem of putting on an elastic cylindrical stocking. It is not difficult to derive from Hooke's law a formula for the pressure as a function of the circumferences of the leg and the stocking.

Let l be the circumference of the unstretched stocking and u the circumference of a certain cross section of the pipe. We approximate a short piece of the pipe by a cylinder of length δb and circumference u. Denote by δK the force which stretches the tape strip of width δb, and by $\delta F = d\,\delta b$ the area of the cross section of this strip, where d is the thickness of the tape. Then by Hooke's law we have $\delta K / \delta F = E\,(u-l)\,/\,l$ with E being Young's elasticity modulus for the tape. It follows that $\delta K = d\,\delta b\,E(u-l)\,/\,l$. The opposite force to δK is the force δK^- acting on the piece of the pipe of length δb and producing the pressure p. Here we assume that the shear stress can be neglected. The approximating cylinder has area $\delta F^- = u\,\delta b$. Hence p is approximately $\delta K^- / \delta F^-$. From the physical law that says force equals opposite force and inserting from above we derive $p \approx \delta K / \delta F^- = \dfrac{d\,\delta B\,E\,(u-l)}{u\,\delta B\,l}$. For $\delta b \rightarrow 0$ the approximation of the piece of pipe by a cylinder gets better and better and we get the exact value of the pressure:
$p = d\,E\,(u-l)/u$. One exercise during the course was to measure the minimal and maximal circumferences of the forearm, to calculate the minimal and maximal pressure and give an estimation of an appropriate circumference of the stocking. Thus the students experienced that mathematics is applicable in everyday life.

In practice an elastic roller is normally used and again my wife told me that it is not good bandaging a leg spirally with constant pitch angle, in the course of which stretching is necessary, because the pressure then differs so much that a leg looks very deformed when it is unbandaged. Along the boundary curve of the tape a deep 'valley' occurs.

Explanation of the special way of bandaging used in hospitals

To my surprise it was exactly the above mentioned observation of the changing pitch angle which proved to be the explanation of the fact that a leg is bandaged not spirally, but in the way described earlier. If you lag a cone from the thicker to the thinner end, without stretching, the pitch angle is decreasing. Call the angle in a plane orthogonal to the symmetry axis of the cone the winding angle ϕ. If we have wound once round then, for the winding angle $\phi = 2\pi$ the constant rate of decrease is a. Thus the pitch angle must reach the value zero after some winding angle ϕ as in Fig. 1. Look now on Fig. 1 in such a way that the lagging starts at the thicker end of the conical pipe.

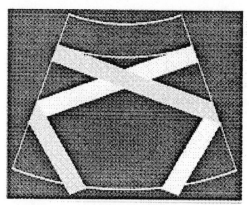

Figure 2: the tape on the circular sector

In Figs. 2 and 3 we see that from this point on the tape automatically winds back to the thicker end of the cone. This can also be done by a uniformly stretched elastic roller thus producing a uniform pressure. Turning to the thicker end of the cone the lagging does not automatically change its direction towards the thinner end. However this can be done by stretching. If you use stretching at the thicker end to change the direction of lagging, and use stretching at the thinner end to produce the required pressure, a fairly uniform pressure is produced. This seems to be the geometric explanation of the special way of bandaging a leg. At this point I told the students of this special way of bandaging. After studying the more theoretical background on isometries of surfaces the students liked these practical geometric considerations. For the sake of completeness I should say that the self-crossings of the tape mentioned can only occur if a < 90°. If this equality holds the number of self crossings - in both directions - of an infinite inelastic tape on an infinite cone is the largest natural number which is less than $\dfrac{90 - a}{a}$ where a is measured in degrees. Of course in later (university) lectures on differential geometry you can use this model to show geodesic lines on a cone, the boundary lines of the tape wrapped onto the cone.

Further thoughts

For bandaging a cone with constant pitch angle and stretching, calculate the boundary curve of the tape after the cone is unwound onto the plane and the difference of the pressure between left and right boundaries of the tape.

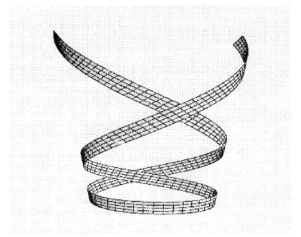

Figure 3: lagging a conical pipe, pipe removed

CONCLUSION

At the end of the course the students completed a questionnaire. They remarked that they had felt that mathematics was sometimes dry but they had found it very excitingwhen applied to real life situations, as they had just done in the course. They stated that they were now better motivated for mathematics and thought that the inclusion of examples from the real world in their later jobs as teachers could be a means of motivating pupils doing mathematics. They found the lagging example very suitable for teaching at high school. The students stated that Houston's test objectives were a very good formulation of how you can read an article with understanding.

REFERENCES

Berry J and Houston K, 1995, **Mathematical Modelling**, Edward Arnold, London.

Braun M, 1988, Das Prinzip der Auslese durch Wettbewerb in der Populationsbiologie, in Braun M, **Differentialgleichungen und ihre Anwendungen**, Ferdinand Springer, Heidelberg, pp 483-492, (second English edition 1978).

Bronstein I and Semendjajew K, 1991, **Taschenbuch der Mathematik**, Teubner, Leipzig.

Burghes DN, Huntley I and McDonald J, 1982, **Applying Mathematics**, Ellis Horwood, London.

Edwards D and Hamson M, 1993, **Guide To Mathematical Modelling**, Macmillan, London.

Houston K, 1993a, *Comprehension Tests in Mathematics*, **Teaching Mathematics and its Applications**, **12**, pp 60-73.

Houston K, 1993b, *Comprehension Tests in Mathematics II*, **Teaching Mathematics and its Applications,12**, pp 113-120.

Riede A, 1993, **Mathematik fuer Biologen**, Vieweg, Braunschweig.

Riede A, 1995a, **Handicapping weight lifters**, preprint.

Riede A,1995b, **Survival of the fittest of two similar species**, preprint.

Riede A,1995c, **Teaching Mathematics to Biologists - Some General Aspects and Modelling Examples**, in Sloyer C, Blum W and Huntley I, (eds.) **Advances and Perspectives in the Teaching of Mathematical Modelling and Applications**, Water Street Mathematics, Yorklyn, pp 301-309.

APPENDIX

Houston's assessment objectives

The following assessment objectives are quoted in Houston (1993).

(i)   explain all statements like 'it can be shown that...' or 'it follows from the above ...' in the article,

(ii)   identify and explain all mathematical modelling assumptions made in the article,

(iii)   make constructive criticism of assumptions made, mathematical analysis and calculations carried out, inferences and deductions made, processes carried out,

(iv)   locate any inconsistencies or incorrect deductions made in the article,

(v)   locate and correct any mathematical or typographical errors in the article,

(vi)   have some wider background knowledge of the situation described in the article,

(vii)   generalise the ideas or apply the ideas to a different situation.

Section E

Tertiary Courses

28

The Relevance of Research in the Development of Undergraduate Courses in Mathematical Modelling

Angela Botham and Jean Crowe
Edge Hill College of Higher Education, St Helens Road, Lancashire, L39 4QP
e-mail thomasb@admin.ehche.ac.uk

ABSTRACT

The mathematics courses at Edge Hill College have recently been supported by the introduction of a Mathematical Modelling Research Group, one of whose purposes is to assess the suitability of material for inclusion in undergraduate areas of study. This paper focuses on one such area, namely the place of mathematical modelling in the BA/BSc (Qualified Teacher Status) course and attempts to examine the influence of a particular case study on the changing provision. The case study is an interpretation of an investigation suggested by the Educational Unit of Ove Arup, originally intended to encourage school children to think of a career in Engineering. The suggested presentation, concerning bridge building, has been adapted for younger children and a piece of classroom teaching analysed.

This paper concentrates on learning mathematical modelling and the inherent problems of assessment at different levels. It is hoped that it will contribute to an on-going consideration of the place of mathematical modelling in the curriculum and the natural influence that this promotes on teaching styles.

INTRODUCTION

Working on a modelling project during the BSc (Qualified Teacher Status) course, children were found to be unwilling to participate or discuss as a group, resulting in initial discussion being heavily teacher directed. The children were concerned with building a bridge constructed mainly of newspaper, and it was only when they started to solve the problem practically that they began to realise that communication as a group and collaboration of ideas was essential for their success (see Non-Statutory Guidance,

1990 : page A1). It was concluded that the initial method of delivery had hindered the children's initiatives by not instilling them with the confidence to work independently. It was found that a wide range of mathematics was being applied and developed throughout the modelling process, although the children were not necessarily aware of this. Reflection on this experience led us to consider whether infant school children can competently tackle real problems - developing skills of mathematical modelling by using and applying the mathematics they know to help them develop their mathematical ability.

Case study - theoretical considerations

A group of Year 1 children (5-6 years) was selected to work with. The initial problem was modelled around characters familiar to them, in the hope of eliminating the element of hindrance previously encountered. Through the medium of story the children were asked to solve a real problem - to design and construct a bridge strong enough to hold the weight of Postman Pat and his bag.

The principle aim of the study was to investigate and analyse the way in which young children, in a problem-solving situation, can apply and extend their mathematical knowledge and understanding. This asked the following questions:

- can young children apply their mathematical ability to the modelling process?
- does this facilitate the further development of their mathematical ability?

As an integral part of the modelling process the project endeavoured to stimulate children into

- applying mathematical knowledge
- discussing thoughts and ideas
- explaining and reasoning
- using 'trial and improvement' methods
- participating and collaborating as a team
- developing autonomy
- communicating and sharing results

To facilitate the fulfilment of these aims each session was assessed and evaluated directly against the modelling sub-skills of Burkhardt *et al.*, (1985: page 14), namely;
(1)   generating variables
(2)   generating relationships
(3)   selecting relationships
(4)   identifying specific questions
(5)   modelling process
(6)   estimating
(7)   validating

To enable the evaluations to be thorough, the very nature of the research activity demanded a small working group. This presented an unrealistic situation since, with a small number of children, such intense work, evaluation and teacher attention is not feasible in everyday teaching situations.

EVALUATION OF LEARNING AND THE MODELLING PROCESS

The children worked through the modelling process using Burkhardt's modelling sub-skills. Through discussion they started *generating variables* by identifying a reasonably sufficient number of key variables. Firstly they listed four properties pertinent to their problem: their bridges were to be strong, safe, carefully built and stable. When considering the variable of size they *selected the relationship* that the size of the bridge would need to be proportional to Postman Pat, generating questions such as 'How long will it need to be?'. By attempting to answer these questions the children started *generating relationships*. One stated that it would need to be "bigger than Postman Pat otherwise he would not fit on it". A second replied it needed to be "wide enough for him". She *selected a relationship* from the *generated relationship* pertaining to size. They then considered the length of the bridge deciding that this would depend on the width of the river. They were beginning to *identify specific questions* by listening to and considering each other's ideas - starting to work co-operatively. At this stage it was important to continue to identify relevant specific questions, and this led to the identification of two questions - what materials will we use and why? To answer these questions the children looped in the modelling process, returning to generate a list of possible materials. In addition to considering the potential strength of the materials, they considered their ease of manipulation, additional tools required and the cost of materials and tools required. From analysis of this first session it was evident that the children were working through the modelling sub-skills and using the modelling process. This became increasingly apparent throughout the project as a whole, with the seven modelling sub-skills providing links between the sessions. The children would appear to be working mainly towards one or two sub-skills each session but, as the study progressed, they started linking the skills together returning to initial sub-skills to help overcome a difficulty or to justify decisions made. This correlates directly with Kapadia and Kyffin (1985) suggesting that modelling is a looping process.

At this point the children were informed of the final part of the problem - the main structure of the bridge (the road) was to be constructed of newspaper. Alex intuitively said "but newspaper isn't strong enough. It won't hold Postman Pat's weight". Nicola retorted saying "we could make the newspaper stronger". Alex had *generated a relationship* between the strength of the bridge and the weight of Postman Pat. The dependence of one quantity on another, the dependence of strength on weight, develops early ideas of a function fostering the child's developing mathematical ability. The links with higher mathematical concepts and knowledge illustrates the importance of the activity at this stage.

Alex also *identified a specific question* requiring an answer before the problem could be considered resolved - how can we make the newspaper stronger? For a few minutes the children investigated various ways of manipulating newspaper. Laura quietly smiled and holding a rolled piece exclaimed "This one feels quite strong". They were working within the fifth sub-skill *modelling*. She had *estimated* that the roll was strong, assuming strength meant it did not bend or break, and then *validated* this by using her hands to apply a simple force at each end, indicating a degree of strength in the centre where it would be most needed. Laura had worked through the whole modelling process in an attempt to find the answer to Alex's identified question illustrating the importance of the fourth sub-skill. The ability to *identify specific questions* breaks the work up into tractable sections and fixes a direction for the modelling process. (Burkhardt *et al.*, 1985). The children generated relationships when considering the relative strength of a variety of examples. They concluded that the thinner the roll the greater its strength.

Building the bridge

The children then organised themselves into two groups to design their bridges, supporting their choices with valid reasons. Their designs were very different - as Cockcroft (1982) reported the same investigation can produce a variety of results which are all valid solutions to a problem.

By drawing their designs the children were beginning to *model* the situation. Williams and Shuard describe mathematics as having three main types of language, the second of these being the use of diagrams saying, that "in the early years this is the most convincing way of expressing the relationships which children perceive." (Williams and Shuard, 1982: 51)

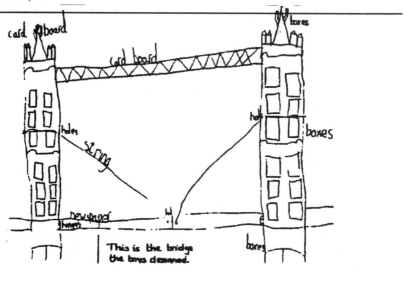

Figure 1: group 1's design

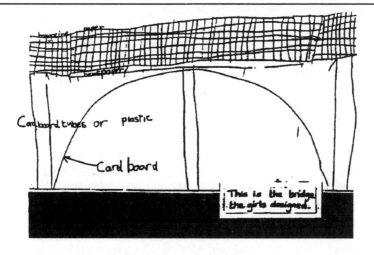

Figure 2: group 2's design

The important property of the sub-skill *modelling* at this level is that it draws on the generated variables, relationships and specific questions identified in the earlier skills hence allowing for the essential features to be simplified and abstracted while other features are ignored. Their drawings are not accurate but illustrate the appropriateness of the sub-skill *modelling*. They contain enough detail to meet the needs of the real problem for these children at their stage of development.

The children identified the materials required, with Group 2 again illustrating the 'looping' property of the modelling process as a whole. They were not sure which material would be more suitable for the arch as their *identified specific question* needed further investigation, so they left the loop open to be completed later while pursuing the more essential features at this stage. This again illustrates the sub-skill *identifying specific questions* - breaking the work up into manageable sections and so providing a direction for the modelling.

Group 1's designs were essentially very similar to those of Group 2. Alex's design (Fig. 1) took this further to include his simplification of how he intended to construct the mechanics of their open bridge. His ingenuity and ability to explain his design concisely and coherently was surprising, since he was described as a child of only average ability, yet his developing mathematical ability is evident in his design. From Alex's drawing and oral explanations, it was evident that he had applied and extended his mathematical knowledge and understanding through the process of mathematical modelling. The modelling project interested and stimulated Alex, providing him with an opportunity to show some of his true potential. Since the two designs are different, involving the application of different mathematical skills and concepts, it seems appropriate to consider how each group progressed.

Different approaches to the same real problem

From their plans Group 1 worked within the sub-skill *modelling*, and began to construct their bridge. When building the road the children (from both groups) were abstracting essential features of the real problem and simplifying them. The group constructed the towers by developing ideas of symmetry and balance related to the nature of their bridge. By raising the road the children discovered the towers needed to be heavy. Consulting the original materials list, looping in the modelling process to reconsider their initial *generated variables*, David suggested using stones. Intuitively he knew that for the towers and ultimately the bridge to be balanced, the stones in each side had to be equal. They *estimated* the required weight which is an important skill for progressing on a modelling problem - "(it is) better to estimate and find out a better value if it proves necessary" (Burkhardt, 1985). They employed 'trial and improvement' methods to decide whether features on their design were necessary to comply with their *generated variables*. Next they attached the string, to complete the mechanics of the bridge, and tested their model. They were *validating* Alex's design, proving that it was an acceptable resolution to the problem. They demonstrated the strength of their bridge - it met the original requirements of supporting a combined weight of 420 grammes, and also its ability to operate as an open bridge thus allowing tall ships to sail through.

Working within the sub-skill *modelling* Group 2 initiated the construction of their bridge. Having built up a supply of thin, strong rolls of newspaper for the road Nicola decided they had the required number. She *estimated* this by considering the width of Postman Pat in relation to the pile of rolls. Laura was unsure. For the *generated relationships* of width there were enough rolls, but she intuitively knew that one roll length would not be long enough for the span of the bridge. Laura *estimated* two rolls were needed to span the set distance. She used the concept of direct comparison to qualify her estimation, again an important skill for progressing with a modelling problem. Joining the rolls together they discovered that the ends were weak so they pushed the ends closer together. After sellotaping the road together, they *validated* their design so far by checking its strength. Kathryn discovered, as Postman Pat's parcel bag fell in the 'river', that their bridge was not safe - one of their *generated variables*. She 'looped' in the modelling process to *identify a specific question* having consulted their design drawings they used small magazine paper rolls to make railings, thus satisfying their design specification by providing more strength, stability and ultimately safety. They 'looped' in the modelling process when *validating* the safety and strength of their bridge. The children were abstracting features of railings on a real bridge to meet their *generated variable* of safety, creating a 'model' situation.

They had worked through the modelling process meeting the main objectives in the real problem. I then asked them to check their original plans to see if their bridge was complete. 'Looping' back to the sub-skill *modelling* they improved their model by adding an arch. Considering a large piece of cardboard they commented that it was too short. After some considerable thought Nicola said, "We could cut it in half that way (*pointing lengthways*) and put the bits together at the end". They *estimated* to find half of the card. When a long ruler was not available Nicola assumed that they could not find the middle of the card. Laura suggested that if the card was folded and cut on the fold they would have two bits the same, meaning the card is cut in half. She had applied her mathematical knowledge to a practical situation conveying her concept of a half. This illustrates the usefulness of modelling activity for assessing children's understanding, providing a context in which children create their own need to use and apply mathematical knowledge.

Using 'trial and improvement' methods they made the arch the correct length and attached it under the bridge concluding that it did not now wobble. They had moved on in the modelling process to *validate* their completed model.

Both groups had completed their bridges by working through the whole modelling process and using mathematical techniques.

ASSESSMENT

For young children we are advocating the teaching of mathematics through modelling, not the teaching of modelling. Fundamentally, it is the ability to apply mathematical skills and knowledge that is important at this age, so that children can successfully use the process to provide a context for mathematical development. This is amplified by the Association of Teachers of Mathematics (1991) saying that the mathematics needs to "come from context" in order to avoid the "inevitable consequence of simply teaching mathematics in order to apply it".

The extent to which the activity of bridge building fulfilled the initial aim is evident when analysing the degree to which the children were stimulated to work towards the points cited earlier. They demonstrated their understanding of mathematical skills and concepts by using and applying them to the modelling situation. Throughout the study they used the language of measurement to consider and discuss concepts of weight, strength and length. When considering useful materials the children identified the need to apply their knowledge of sets to classify the materials. The children of both groups applied the concept of balance to determine the stability of their bridges. Group 1 used the concept of balance and equal weightings when balancing the towers and this also involved ideas of shape, size and symmetry.

The process involved estimation of relative strengths and dimensions of the bridges, along with the proportions of materials required. Skills of direct comparison were used to determine dimensions of the bridges, and again estimation of measurements was employed. Laura demonstrated her understanding of a half when her group were constructing their arch.

The nature of the starting point stimulated the children into discussing their thoughts and ideas and we were surprised by the length of time the children could sustain a productive discussion. The children demonstrated their ability to think and reason logically, accepting criticism and giving reasoned responses when generating variables and selecting and generating relationships. They were able to follow arguments through, responding to each other's thoughts with an attitude of open mindedness. Piagetian theory suggests that at this age children are egocentric, finding it difficult to follow such arguments, particularly when sustained over a long time period. The evidence from this study suggests, as does Donaldson (1978), that there are weaknesses in theory.

When selecting relationships the children were making choices and decisions, drawing on their reasoned arguments from the initial discussion. This involved the children working collaboratively and participating as equal members. Despite working in two groups on separate designs, the nature of the activity indicated by the sub-skills of Burkhardt *et al.* provoked discussion and co-operation at all stages. The children assumed roles of responsibility throughout the study. The roles of Nicola and Alex were particularly evident when they were presenting their results to their peers, as was Alex's position of responsibility when designing the mechanics of the bridge.

As their confidence of working within the modelling process grew the children developed autonomy. They were solving problems identified within the modelling process by applying their mathematical knowledge and ability. Children generally apply the skills and concepts they are familiar with when asked to solve problems, hence the modelling process, by its very nature, is a useful vehicle for assessing a child's ability to use and apply their mathematical knowledge. As the modelling activity progressed the children demonstrated their ability to use and apply a wide range of developing mathematical skills, concepts and language. Communicating their work orally in a presentation to their peers, they also displayed their understanding of the mathematics they had applied, using appropriate language.

CONCLUSION

The adoption of a modelling activity could stimulate work across most areas of the curriculum - indeed the children were considering ideas related to geography, science and design technology, while developing ideas of social and economic growth. The evidence accumulated here is not conclusive, since there are a number of limitations to the study. The fact that the study involved only six children for a full morning each week made the situation artificial. From the study it is evident that the children worked through the modelling process using the seven modelling sub-skills of Burkhardt. This supported the development of the children's mathematical ability. By presenting real problems, at the appropriate level, children of all ages can participate in the activity of mathematical modelling. All the children involved enjoyed the modelling activity, finding that it offered them a small taste of the wonder and satisfaction that mathematics can provide.

The children developed skills of applying mathematics, discussing thoughts and ideas, making choices and deciding, participating and collaborating as a team and developing autonomy. The process of mathematical modelling enabled children of differing abilities, working at different levels, to co-operate together to solve a real problem.

Since children generally apply the skills and concepts they are familiar with when asked to solve problems, mathematical modelling provided a vehicle through which it was possible to overcome some of the inherent problems of assessment at different levels. The children had different starting points and, through completing the modelling process, produced different, valid solutions for the same initial problem. By working co-operatively they were able to work at their own confidence levels, applying the skills and concepts with which they were familiar. It is this very nature of the mathematical modelling process that allows assessment at different levels to take place.

Precisely these same points allow for effective assessment of students within higher education through application of the mathematical modelling process. Students show their ability to apply their mathematical knowledge and understanding in contexts in which they have a genuine interest. Though not as straightforward to assess as more traditional methods, the benefits far outweigh this perceived disadvantage.

Burkhardt *et al.* indicate the importance of a modelling approach in their seven modelling sub-skills, with Naylor (1989) highlighting its importance in helping facilitate a child's mathematical ability. If mathematical modelling is valuable for the development of mathematical ability in the school curriculum, it follows that teacher training students need themselves to be aware of the nature of mathematical modelling and of its potential in providing a context for mathematical development.

The ideas behind modelling should also be inherent in curriculum study courses for all students, though not defined as such due to a lack of time for detailed analysis. Perhaps, with increased hours becoming available for a new College degree pattern commencing in 1996, serious consideration should be given to more focused consideration of mathematical modelling in these courses.

Incorporating mathematical modelling in a curriculum produces a natural influence on teaching styles. Its usefulness is evident in the way it accommodates the learning of mathematics at different levels and also in its value for assessment at these levels. This is as true in the reception class as it is for students at degree level, so providing an extremely valuable unifying element between students' own mathematical activity and understanding and that of the children they intend to teach. Practical school-based research has a very important role to play in the development of understanding of such links, and therefore on course provision for intending teachers.

REFERENCES

Association of Teachers of Mathematics, 1991, **Using and Applying Mathematics**, Leicester.

Burkhardt H, *et al.*, 1985, **Beginning to Tackle Real Problems**, Nottingham University.

Cockroft W H, 1982, **Mathematics Counts**, HMSO, London.

Donaldson M, 1978, **Children's Minds**, Fontana Press.

Kapadia R and Kyffin M, 1985, **Modelling for Schools/Colleges**, South Bank Polytechnic, London.

Naylor T. 1989, *The Unifying Effects of Mathematical Modelling in a College of Higher Education*, in Blum W *et al.*, (eds.), **Applications and Modelling in Learning and Teaching Mathematics**, Ellis Horwood , Chichester.

The Spode Group, 1986, **Mathematical Modelling for Teachers: Short Course Notes**, University of Exeter.

DES, 1989, **Non-Statutory Guidance for Mathematics**, HMSO, London.

29

An Introductory Course on Mathematical Models and Modelling: A Constructivist Approach for Middle School Teachers

Don Cathcart and Tom Horseman
Salisbury State University, 1101 Camden Avenue, Salisbury, MD 21801, USA
e-mail: dccathcart@sae.ssu.umd.edu **or** *tohorseman@sae.ssu.umd.edu*

ABSTRACT

This paper describes a course developed at Salisbury State University for freshmen and sophomore elementary education majors who are interested in teaching middle school (5th-8th grade or ages 10-13) mathematics or science. The course is designed to focus on problem-solving processes used in mathematics and science, and uses perspectives, knowledge, data-gathering skills and technological tools relevant to those disciplines. Specific problem assignments are suggested, and one is covered in detail.

The course was developed under the auspices of the Maryland Collaborative for Teacher Preparation (MCTP). The MCTP prescribes that instruction in mathematics and sciences be integrated using a constructivist, activity-based approach. The MCTP also prescribes that students use technological tools such as calculators and microcomputer-based laboratories (MBLs). Accordingly, students in this course will use these tools in collecting their own data, generating graphs, and analysing their results. Working together, students will be encouraged to construct physical concepts from their observations and make connections between science and mathematics.

Activities in the course focus on helping students:
(a)   see connections between mathematics and other disciplines,

(b)   *represent and analyse real-world phenomena using a variety of mathematical representations,*

(c)   *develop strategies and techniques for applying mathematics to solve real-world problems, and*

(d)   *explain and justify their reasoning, using appropriate mathematical and scientific terminology, in both oral and written expression.*

INTRODUCTION

The mission of the Maryland Collaborative for Teacher Preparation (MCTP) is to promote reform in the teaching and learning of mathematics and the sciences by improvement in the preparation of elementary school (ages 5-13 years) science and mathematics educators (MCTP, 1995). The mathematical modelling course described in this paper is the first mathematics course in the sequence of MCTP courses taken at Salisbury State University by prospective middle school (ages 10-13) teachers of mathematics and science; it is taken concurrently with an interdisciplinary science course. The following discussion is based on the Instructor's Guide (Cathcart and Horseman, 1995) that is used to introduce the instructor to the MCTP philosophy and to provide a framework for the course.

The need to improve the mathematical preparation of pre-service elementary school teachers is well documented in recent reports such as On the Mathematical Preparation of Elementary School Teachers (Cipra and Flanders, 1992), and Curriculum and Evaluation Standards for School Mathematics and Professional Standards for Teaching Mathematics by the National Council of Teachers of Mathematics (NCTM, 1989). The NCTM Standards call for increased emphasis on thinking, reasoning and problem solving. The NCTM Board of Directors, in a 1995 statement on interdisciplinary learning stated "...the curriculum must encourage the use of the perspectives, knowledge, and data-gathering skills of all disciplines...".

Currently the mathematics requirement for elementary education majors varies widely from college to college (Cipra and Flanders, 1992). In many cases, prospective teachers take a single, one-year course covering a variety of topics such as number systems, a little number theory, a superficial look at probability and descriptive statistics, and informal geometry. While some colleges require more than a single, two-course sequence, few prospective teachers are placed in situations where they are expected to discover and express relationships found in the world around them. The course described in this paper is intended to integrate mathematics and science, while placing special emphasis on the concept of a function as a model and on model building as a process. Few prospective elementary school teachers are currently exposed to such a course in their undergraduate training. The course is not intended to serve as a primary vehicle for teaching specific mathematics or scientific content. It is assumed that pre- or co-requisite mathematical or scientific content has been, or will be, addressed in other courses.

The Maryland Collaborative for Teacher Preparation (MCTP) has prescribed that instruction in mathematics and sciences be integrated using a constructivist, activity-based, approach. Accordingly, it is intended that students in this course will be engaged in active investigation and co-operative problem solving. The MCTP also prescribes that students have opportunities to use technological resources such as calculators, computers and microcomputer-based or calculator-based laboratories (MBLs or CBLs). Therefore it is expected that graphing calculators or computers with appropriate software (such as spreadsheets) and MBL or CBL equipment is available to students throughout the course.

Primary expected outcomes

In completing the exercises and activities it is expected that students will:
- see and value connections between mathematics and other disciplines,
- represent and analyse real-world phenomena using a variety of representations, use their own observations, experimentation, computers or calculators, and references to find answers to questions,
- develop strategies and techniques for applying mathematics to solve real-world problems,
- work with others while solving problems and learning,
- become skilful in explaining and justifying their reasoning, using appropriate mathematical and scientific terminology, in both oral and written expression, and critically evaluate their own work and the work of others.

COURSE CONTENT

Part I of this course focuses on functions as models for phenomena and on the development of a repertoire of techniques to be used in modelling. Functions are used to represent relationships of some particular interest. A change in the value of one variable causes a change in the value of another variable. We investigate the manner of that change in both qualitative and quantitative ways. Change and rate of change are illustrated and given meaning. Additional mathematical topics introduced in exercises include difference equations, difference tables, difference quotients, recursive functions, secants, least-distances criterion, and least-squares criterion.

The focus of Part II is on the concept of a mathematical model and on mathematical modelling as the essential element in applying mathematics to real problems (*eg* studies of motion, heat loss and light intensity). The phrase 'applications of mathematics' usually means useful connections between mathematics and other fields. Although the application of mathematics in different fields requires a variety of mathematical techniques, there is a common unifying element in applying mathematics to real-world problems. That unifying element is the concept of a mathematical model. The activity of constructing, analysing, interpreting, and evaluating mathematical models is the central process in applying mathematics to real problems. Our view of the modelling process is based largely on that of Maki and Thompson (1973), Roberts (1976), and the Curriculum and Evaluation Standards for School Mathematics (NCTM, 1989). This modelling process is illustrated in idealised form below.

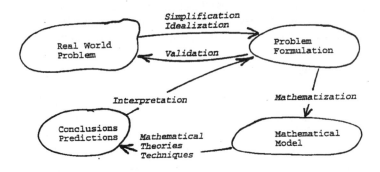

Figure 1: the Modelling Process

Step 1: Problem formulation

The modelling process usually begins when one is faced with a situation and is concerned about some aspect of that situation. Often the situation is very 'fuzzy' and the concern is vague. Thus the initial step in modelling is to define the situation or problem as carefully as possible. The formulation of a specific problem is important, but often difficult, and sometimes requires creative effort. Attempts at precision frequently involve making idealisations, simplifications, and approximations, because most real problems are too complex or nebulous to be treated mathematically in a way that includes all real-world aspects of the problem (Maki, 1975). In this problem-formulation step, essential variables are isolated and the critical questions identified.

Step 2: Mathematical translation

The real problem is translated into the language of mathematics. Symbols and mathematical operations are used to represent real quantities and processes. This process of mathematisation produces a mathematical realisation of the real problem called a mathematical model.

Step 3: Mathematical conclusions or predictions

Once the problem has been expressed in mathematical form, we study the resulting mathematical system using concepts and techniques of mathematics. If we are skilful, persistent, or fortunate, we will produce mathematical conclusions or predictions. Our mathematical conclusions or predictions must then be translated back from the language of our model to the language of the real world and interpreted as real-world conclusions or predictions (see Roberts, 1976).

Step 4: Evaluation

The conclusions and predictions are compared with the real-world phenomena being considered. Only rarely will our results from the mathematical theory agree completely with real-world outcomes. Usually a mathematical model will not reflect important aspects of the situation being studied. Only when the results of our mathematical study compare favourably with real-world data will we have some confidence in our model. Otherwise we must either make do with a model we suspect is inadequate, or we must refine the existing model by retracing the modelling process until an acceptable model is found (see Maki, 1975).

Disciplines integrated

The intent of the MCTP has been to integrate mathematics and the sciences. Applications have been selected from both physical and life sciences. A few applications from sports and management science have been included. Some MBL activities are co-ordinated with an integrated science course taken concurrently with this course.

Prerequisite knowledge

The emphasis in the exercises and activities is for the students to use mathematics and science they already know in approaching a task. Students with two years of algebra, one year of life science and one year of physical science should be able to perform and understand these activities and exercises. Of course some review and study of forgotten material may be required. The mathematical content requires knowledge of the following terms and concepts: function, table of values for a function, graph of a function, inverse function, linear function, power function, polynomial function, rational function, exponential function, logarithmic function, ratio, proportionality relations, variation, slope and average (mean).

Students' preconceptions

Students enter mathematics courses with many preconceptions concerning mathematics, its usefulness, and their own abilities. Some preconceptions for the instructor of this course to attend to are the following.

- Mathematics, beyond arithmetic, is of little value to me.
- There is only one correct answer to a math problem, and there is only one correct method for finding that answer.
- I can only solve easy math problems.
- The only important thing in solving a math problem is the answer.
- In functions, x and y are just numbers; not anything real.
- I cannot tell much about the graph of a function simply by looking at its rule.
- I cannot tell much about a function's rule simply by looking at its graph or table of values.
- We can easily tell who has the best answer to a math problem.
- In drawing a graph, I must use the same scale on each axis.
- Science is important, but lots of school math is not of much use in science.
- The math that is useful in science is beyond my capacity to learn.
- I don't know enough math to be able to solve real-world problems using math.

Activities and equipment

Many of the course's activities require students to gather or develop their own data, display that data in tables and graphs, and then try to find a functional model fitting that data. Some of the activities require specialised equipment. In Part I for example, Cuisenaire rods are used for one activity, the Tower of Hanoi puzzle in another; MBL or CBL equipment (motion detector, photogate, and microphone) is used in several activities. In Part II also, some activities require the use of equipment usually available in a chemistry or physics lab, and MBL or CBL equipment (light sensor, temperature probe). It is assumed that graphing calculators or computers with spreadsheet software are available to the students. Also, students are encouraged to use other resources that can be found in a library or on the Internet.

Pedagogical approach and assessment

As indicated above, MCTP prescribes a constructivist approach for teaching integrated mathematics and science. Therefore, as students are involved with activities in this course, they must be actively engaged in confronting their misconceptions and creating their own knowledge. Thus the instructor's role is that of facilitator or guide. The degree to which students meet the expected outcomes and alter their preconceptions is determined by examining their written work, listening to their oral explanations, and observing them during working sessions. For example, when students are trying to determine which of several functional models best fits the situation being studied, are appropriate criteria being utilised?

Case study: summary of some student work

The following is a summary from a written report submitted early in the course by a group of two students - Rita and Reggie. The students were asked to develop a model that could be used to predict the population of the United States in the year 2000. It is important to note that these students had not been exposed to the concept of best fit in any course. Instead, they developed models they intuitively understood by methods of their own invention.

Initially, in their problem formulation step, Rita and Reggie collected no data and consulted no references, but made a guess concerning the current US population and tried to estimate the number of couples of child-bearing age. They then assumed each couple would have one child during the next five years. On that basis, their first prediction was 250 million. They then obtained census data from The World Almanac and Book of Facts for 1995. The US population at the beginning of each decade from 1900 to 1990 was entered into a spreadsheet and plotted. The students invented curve-fitting techniques to fit models to these data. Since the year 2000 was only five years away, the students decided to extend their predictions to the year 2030.

Moving to the mathematical translation step, the students decided to develop a pair of models - a linear model and an exponential model. They developed models for the growth of the US population from 1900 to 1990. Letting $P(t)$ denote the US population, in millions, t decades after 1900, in one case they found values for m and b so that the graph of $P(t)=b+mt$ would fit their data; in the other case they found values for A and B so that the graph of $P(t) = A(B)^t$ would fit their data.

Next, Rita and Reggie proceeded to the step where they developed their conclusions and predictions. They suggested two methods for analysing the data. The mean of the nine decades average rates of change in population (in millions per decade) was calculated in what they called their 'rate of change method'. Taking that mean as the slope in their linear model and population of the US in 1900, in millions, as intercept , they proposed $P(t)=76+19t$ as their linear model. They also calculated the percent growth for each decade and determined that the average percentage growth per decade was 14.1% in what they called their 'percent growth rate method'. In this case, they proposed the model $P(t)=76(1.141)^t$. Letting $t = 10$ in each model, they estimated the US population in the year 2000 would be 266 million using their 'rate of change method' linear model, and would be 284 million using their 'percent growth method' exponential model.

In evaluating their models, the students were not sure which model they wanted to use. An excerpt from an email message one of the students sent to her instructor reflected her uncertainty. "We have a graph with the actual data and the data for our two models plotted. Our predictions for the year 2000 are close ... we don't know which one to use. Do you have a suggestion?" Rather than providing a direct answer, her instructor replied "Try your models for the years 2010, 2020, and 2030. I think you will find their graphs start to diverge considerably". Later the students stated that the exponential model was better because it more closely follows the actual growth graph. However, they offered no quantitative evidence that the exponential model offered a better fit to the data than the linear model.

This problem required students to use references to gather data. Although Rita and Reggie were able to develop a pair of models, they offered little support for their claim that the exponential model was better. If the instructor had emphasised the importance of establishing qualitative and quantitative criteria for goodness of fit before assigning this problem, the students might have thought of a compound-interest interpretation for the situation, and they may have compared the deviations of the models' population estimates from the actual population data. In the future this problem will be assigned after criteria for best fit have been developed by the students.

CONCLUSIONS

This course attempts not only to address the expected outcomes discussed earlier, but also to follow the guiding principles stated by Brooks (1993) by
(a)   posing problems in such a way that kindle the students' interests and therefore are made relevant,
(b)   presenting broad primary concepts that the students must 'break into parts that they can see and understand',
(c)   seeking, addressing, and valuing students' points of view, and
(d)   adapting the learning environment to address the students' preconceptions.

Our ultimate goal is to provide prospective teachers with an understanding of the problem-solving processes used in mathematics and science and a workable teaching/learning model for their own subsequent teaching careers. By de-emphasising the teaching of specific algebraic and statistical content, special emphasis can be placed on allowing students to construct personal knowledge and understanding of the concept of a function as a model, and on model building.

REFERENCES

Abruscato J, 1993, *Early Results and Tentative Implications from the Vermont Portfolio Project*, **Phi Delta Kappan**, 74, 6, pp. 474-477.

Brooks J and Brooks M, 1993, **In Search of Understanding: the Case for Constructivist Classrooms**, Association for Supervision and Curriculum Development, Alexandria, VA.

Cathcart D and Horseman T, 1995, **Mathematical Models and Modelling for Middle School Teachers: Instructor's Guide**, MCTP, College Park, MD.

Cipra B and Flanders J, 1992, *On the Mathematical Preparation of Elementary School Teachers*, **Report of a Conference Held at the University of Chicago**, The University of Chicago, Chicago.

Maki D and Thompson M, 1973, **Mathematical Models and Applications**, Prentice-Hall, Englewood Cliffs, NJ.

MCTP (Knight G and Layman, eds.), 1995, *Conceptual Framework for the Project: MCTP Frostburg Manifesto*, **Maryland Collaborative for Teacher Preparation**, The University of Maryland, College Park, MD.

Meir S, 1992. *Evaluating Problem-Solving Processes*, **The Mathematics Teacher**, **Vol. 85**, No. 8, pp. 664-666.

NCTM, 1989, **Curriculum and Evaluation Standards for School Mathematics**, National Council of Teachers of Mathematics (NCTM), Reston, VA.

O'Haver T, 1994, *CHEM 121/122 Chemistry in the Modern World (MCTP Physical Science)*, **Course Syllabus**, The University of Maryland, College Park, MD.

Roberts F, 1976, **Discrete Mathematical Models**, Prentice-Hall, Englewood Cliffs, NJ.

Schaufele C and Zumoff N, 1993, **Earth Algebra: College Algebra with Applications to Environmental Issues** (Preliminary Version), Harper Collins, New York .

Shannon K and Curtin E, 1992, *Special Problems: An Alternative to Student Journals in Mathematics Courses*, **PRIMUS**, **II**, 3, pp. 247-256.

30

Simulation Modelling for Undergraduate Mathematicians

Andrew Fitzharris
University of Hertfordshire, College Lane, Hatfield, Hertfordshire, England, AL10 9AB
e-mail: matqaf@herts.ac.uk

ABSTRACT

Mathematical modelling has been a feature of the mathematics degree at the University of Hertfordshire since the early 1960s. In September 1994 a new option was introduced into the final year of the degree in simulation modelling. This is unlike any other modelling course on the degree, and has become the most popular option in the final year. This paper will describe how mathematical modelling has featured in the mathematics degree in the past and how it is expected to develop in the future. It will also discuss the reasons for introducing the simulation modelling option, the structure of the course and the benefits gained by its introduction.

INTRODUCTION

Mathematical modelling first appeared in the mathematics degree at the University of Hertfordshire, formally Hatfield Polytechnic, in the early 1960s. Initially, it was taught as part of a traditional course in applied mathematics. By the late 1970s a significant number of students were performing poorly on the degree because of their inability to cope with the mechanics. It was therefore decided to introduce two separate modelling streams which would run throughout the taught part of the course. The first of these, called Applied A, considered modelling applications in the areas of economic and industrial systems. The second, called Applied B, considered modelling applications in the areas of physics and engineering.

By the late 1980s student numbers on the degree had fallen to a level where it was no longer viable to run two separate modelling streams. Furthermore, fewer and fewer students were choosing to follow Applied B, largely because it was perceived to be more difficult than Applied A. The modelling courses were therefore combined into a single modelling stream which considered modelling applications in all of these subject areas and many others - sport, politics, population dynamics and so on.

The combined modelling courses continue to run on the mathematics degree. These courses incorporate a significant coursework component in which the students are required to investigate open-ended modelling problems, often using IT aids such as computer algebra (Davies and Fitzharris, 1994), spreadsheets (Davies, 1994) and modelling packages such as Stella (Richmond *et al.*, 1990).

SIMULATION MODELLING

In September 1994 a new option was introduced into the final year of the mathematics degree in simulation modelling. This is essentially an advanced course in discrete event simulation. Courses of this kind are usually taught on masters degree courses and are rare on undergraduate programmes. The main reasons for introducing this course were to introduce stochastic modelling into the degree, an omission which had been noted by one of the external examiners and to provide more variety for the students and the personal interest of the member of staff involved.

Outline Course Structure

The course content is based upon the stages in the simulation modelling process (see Fig. 1).

Stage 1 considers the planning activities which take place at the beginning of a simulation study - identifying the goals of the study, forming the project team, organising the hardware and software required, planning the completion of the study.

Stage 2 looks at defining the boundary of the system (*ie* deciding which parts of the system to model) and the techniques used for simplifying large-scale systems such as omission (*ie* omitting unnecessary detail), aggregation (*ie* lumping multiple processes into a single process) and substitution (*ie* replacing complex processes with simplified ones).

Stage 3 discusses data collection (*ie* identifying the data required to complete the study, sources of input data, the techniques used for collecting data, *etc*), data representation (*ie* issues such as whether to sample directly from the data or to fit a statistical distribution to the data and to sample from this), statistical distributions and their applications in simulation modelling, the procedures used for fitting statistical distributions to data and the methods used for verifying the goodness-of-fit (the graphical method, the chi-square test, the Kolmogorov-Smirnov test).

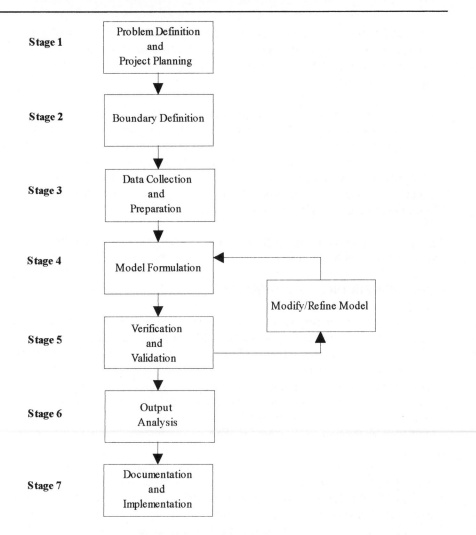

Figure 1: the Simulation Modelling Process

Stage 4 is concerned with building and running simulation models. The modelling component of the course will be discussed in greater detail later in this chapter.

Stage 5 considers the methods used for verifying that the simulation model behaves as intended and validating that it is an adequate representation of the real system. The verification techniques include walkthroughs (*ie* manually simulating the operation of the model), test runs (*ie* running the model with different input values), using an interactive debugger (*ie* stepping through the model line by line and observing queue lengths, variable values) and animation (*ie* animating the model and then observing its behaviour visually). The validation techniques include using a third party to confirm that the model has the correct structure and that it contains all of the important features of the real system, running the model to ensure that the data produced is consistent with the data from the real system and that changes in the inputs produce the expected changes in the outputs..

Stage 6 looks at experimentation (*ie* scenario analysis) and using the output from a simulation model to estimate the parameters of the system (resource utilisations, flow times, queue lengths*)*. The procedures used for estimating these parameters depend upon whether the system is terminating (such as a shop, an office) or non-terminating (such as a hospital, a telephone exchange). However, in outline, the procedures used involve storing the required output from the model in a file and then using these data to calculate confidence intervals for the parameters of interest.

Stage 7 is concerned with presenting the results of a simulation study and making recommendations to the customer for implementation.

THE MODELLING COMPONENT

The modelling examples used on the course are drawn from a variety of application areas such as manufacturing systems, shops and services and administration.. The students are introduced to the modelling techniques in the lectures, and then practice these by modelling similar systems in the practical exercises and the coursework assignments. At the beginning of the course the students are expected only to model small systems for which the input data are provided. However, as the course progresses they are required to model larger and more complex systems which eventually require them to work through all the stages in the simulation modelling process.

The larger models are built using the principle of parsimony, which involves starting with a simple model and then gradually adding detail until an adequate representation of the system is produced. For example, when modelling a large-scale manufacturing system the students would start by producing a simple model which consists of an arrival, a single machine and a departure. In the subsequent refinements they would then add in additional machines, inspection areas, material handling systems (*eg* fork-lift trucks, conveyor belts) and then build in machine breakdowns, operator breaks and suchlike.

SIMAN/Cinema

The modelling component is taught using the SIMAN/Cinema discrete event simulation language (Pegden *et al*, 1995). This package provides facilities for modelling and animating discrete, continuous and combined discrete/continuous systems. During the practical part of the course the students are introduced to all the main features of the package.

- Submodels - independent model segments. These have a variety of applications - for example, they can be used to establish priorities between customers competing for service at a ticket counter,

- Changes in resource capacity - modelling operator breaks, machine breakdowns, and suchlike.

- Station submodels. These remove duplication from a model by allowing the operations at a number of similar stations (*eg* machines, service areas) to be modelled using a single generic model segment,

- Station visitation sequences - modelling systems in which different types of component or customer visit the stations in a different order,

- Transporters and conveyors - modelling fork-lift trucks, conveyor belt systems,

- The output processor - analysing the output produced by a simulation model and estimating the parameters of the system,

- Animating simulation models using Cinema.

SIMAN Basics

A SIMAN model consists of two components - a MODEL frame and an EXPERIMENT frame. The MODEL frame describes the logical flow of the entities as they pass through the system. The statements within the MODEL frame are called blocks. The EXPERIMENT frame describes the environment in which the simulation takes place (*eg* the attributes, queues, resources, statistics, run length,). The statements within the EXPERIMENT frame are called elements. A complete SIMAN/Cinema model is constructed, compiled and run within a Windows - type environment. The software is user friendly and is therefore ideal for educational use.

The main elements of a SIMAN model are as follows

- **Entities**. These are the objects which flow through the system being modelled - customers, parts, documents, *etc*,

- **Attributes**. These are the values associated with each entity,

- **Queues**. These are waiting areas in which entities are held while their movement is suspended temporarily,

- **Resources**. These are objects which can be allocated to entities, such as machines and workers.

SIMAN provides a wide range of built-in functions for sampling values from distributions and for collecting run-time statistics.

- **Exponential(10)** returns a value from an exponential distribution with mean 10.

- **NR(*Res*)** returns the number of busy units of resource *Res*. This is sometimes called the **resource utilisation**.

- **NQ(*Que*)** returns the number of entities in the queue *Que*.

THE SINGLE MACHINE PROBLEM

To illustrate the content of the modelling component an example will be considered. This is similar to the small models considered at the beginning of the course, and is typical of the initial models constructed when modelling large-scale systems.

The Problem

A workshop containing a single machine processes two types of job. (See Fig. 2).

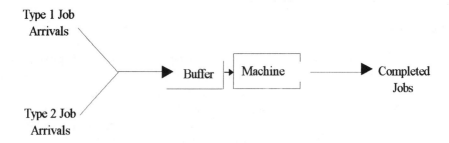

Figure 2: the single machine problem

- The inter-arrival times of type 1 jobs are exponentially distributed with a mean of 10 minutes. The processing time for each type 1 job is uniformly distributed between 2 minutes and 6 minutes,

- The inter-arrival times of type 2 jobs are also exponentially distributed with a mean of 6 minutes. The processing time for each type 2 job is uniformly distributed between 1.5 minutes and 4.5 minutes.

The problem here is to determine the number of jobs of each type which are processed during one 8-hour shift and to collect statistics on the machine utilisation, the number of jobs waiting for processing, the job flow times and the times between job exits.

Modelling strategy

The entities flowing through the system will be the jobs being machined. Each entity will have three attributes.

Job Type   This will be assigned either 1 or 2 depending on the type of the job and will be used to pick out the counter for each job type.

Process Time   This will be assigned a value from the appropriate uniform distribution and will be used to model the time taken to machine the job.

Arr Time   This will be assigned the arrival time of the job in the system so that the flow time statistics can be recorded later in the 'model'.

Buffer will be a queue. The queue length statistics will be calculated using NQ(Buffer). Machine will be a resource. The resource utilisation statistics will be calculated using NR(Machine). All times will be measured in minutes.

The blocks required in the MODEL frame will be :-

CREATE   to create an arriving job and record the arrival time in the attribute ArrTime.
ASSIGN   to assign values to the attributes JobType and ProcessTime.
QUEUE   to hold the job in the waiting area Buffer until the machine becomes available.
SEIZE   to hold the machine while it machines the job.
DELAY   to model the machining time of the job.
RELEASE   to free the machine.
TALLY   to record the flow time statistics.
COUNT   to count the job completed by type.
DISPOSE   to remove the completed job from the system.

The elements required in the EXPERIMENT frame will be as follows

PROJECT   to record the title of the model, the name of the analyst, *etc.*
ATTRIBUTES   to define the attributes of each entity - Job Type, Process Time and Arr Time.
RESOURCES   to define the resource Machine.
QUEUES   to define the queue Buffer.
COUNTERS   to create the counters used for counting the processed jobs by type.
TALLIES   to define the tallies used for collecting the flow time statistics.
DSTATS   to record the machine utilisation and queue length statistics.
REPLICATE   to define the run length of the model - 8 hours (or 480 minutes).

A Fitzharris

The SIMAN 'Model'

The MODEL frame for this problem is shown in Fig. 3 and the EXPERIMENT frame in Fig. 4.

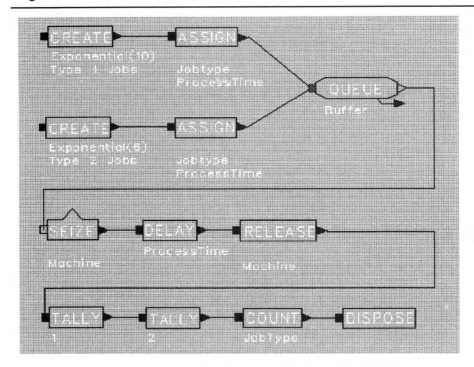

Figure 3: the MODEL frame for the single machine problem.

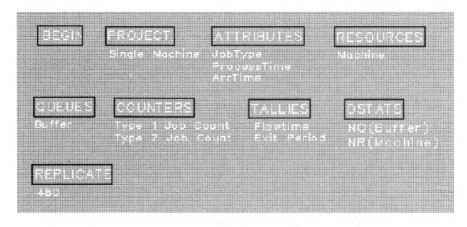

Figure 4: the EXPERIMENT frame for the single machine problem

The summary report produced by running this model is shown in Table 1.

TALLY VARIABLES

Identifier	Average	Variation	Minimum	Maximum	Observations
Flowtime	7.6548	.56697	1.5940	18.866	116
Exit Period	4.1138	.61447	1.5089	18.022	115

DISCRETE CHANGE VARIABLES

Identifier	Average	Variation	Minimum	Maximum	Final Value
Queue Length	1.0583	1.1847	.00000	5.0000	1.0000
Machine Util.	.80701	.48902	.00000	1.0000	1.0000

COUNTERS

Identifier	Count	Limit
Type 1 Job Count	43	Infinite
Type 2 Job Count	73	Infinite

Table 1. Summary report for the single machine problem

It can be seen from this report that during the 8-hour shift :

- 43 type 1 jobs and 73 type 2 jobs were machined,

- on average, the machine was busy for 80% of the time,

- on average, there was 1 job waiting in the queue for the machine,

- on average, each job took around 7.7 minutes to pass through the machine shop,

- on average, one job left the machine shop every 4 minutes.

ASSESSMENT

The course is assessed by coursework and examination and the marks are combined in the ratio 50:50. During the first presentation of the course the coursework component consisted of five modelling exercises in which the students modelled the flow of customers around a regional airport. In the first exercise they constructed a basic model of the system. In the subsequent exercises they embellished this model by allowing the customers to select the shortest queue, the ticket agents to take breaks for tea and lunch, adding-in security stations and departure gates, allowing some of the customers to travel to their departure gate on a transporter and allowing the customers to bring baggage which was then transported to the departure gates on a conveyor belt system.

The coursework exercises served their intended purpose in that they allowed the students to develop their modelling skills. However, most of the students produced near perfect solutions to these exercises, and obtained very high marks - the average coursework mark was 88%. This caused a problem at the Board of Examiners meeting, where it was difficult to discriminate between the students' ability. In future the students will be given a wider variety of coursework exercises in order to overcome this problem.

FUTURE PLANS

In the future it is hoped to expand the course content to include the following

• The use of the input processor for fitting statistical distributions to data,

• The advanced manufacturing features of SIMAN such as guided transporters, queue manipulation, parallel queues and resources, modelling entity sets (*ie* groups of entities), resource pre-emption and suchlike.

• Variance reduction techniques - statistical techniques for producing accurate estimates of system parameters using simulation output.

• An introduction to a visual interactive modelling system such as *Arena* (Pegden *et al*, 1995).

EVALUATION

The simulation modelling course has become the most popular option in the final year. Although the course was being presented for the first time, 31 of the 49 students in the final year chose to include this course in their study programme. A survey of student opinion conducted at the beginning of the course revealed that most of the students had chosen this option because it looked 'interesting' and offered something 'different'.

The main differences between this option and the other modelling courses on the degree are as follows.

• The option considers 'real' modelling applications. In the other modelling courses the problems being considered are simplified, often to an unrealistic level, so that the students can cope with the underlying mathematics. On this course it is not necessary to make concessions of this kind - using SIMAN it is possible to model discrete systems of almost any size and complexity. The course also exposes the students to the problems which arise when modelling 'real' systems - working from imprecise specifications, dealing with data values which do not follow a known or expected distribution and so on.

- Simulation modelling combines mathematical modelling with statistics. In the other modelling courses the statistical content is limited largely to fitting laws to data using the method of least squares.

- The course considers a completely different range of applications.

The introduction of this option has produced a number of significant benefits.

- The course has increased the breadth of the modelling courses on the degree and has therefore provided more variety for the students.

- The course has allowed the students to develop their professional skills as mathematicians. For example, when undertaking a simulation study the students must liaise with the customer, exercise diplomacy (particularly when collecting timed data from operators), manage employee relationships, use a variety of presentation skills and provide professional advice. The modelling and professional skills provided on this course are highly applicable, and will be extremely useful to the students in their future careers.

- The course has opened up a new range of undergraduate projects. During the 1994/95 academic year two final-year students completed a project in which they were able to model and animate the operation of the restaurant facilities at the university. During their investigation they identified a number of efficiency improvements which they subsequently forwarded to the catering manager for implementation.

- The course has inspired a number of new research ideas. Areas currently under consideration include the application to modelling traffic flow and the application to modelling the flow of information around computer networks.

- The course has lead to income - generating short courses in simulation modelling and has opened up the possibility of consultancy in this area at some time in the future.

CONCLUSIONS

The mathematical modelling courses at the University of Hertfordshire continue to develop. New topics are introduced each year so that a different mix of applications can be included in each presentation. This provides variety for the students and prevents the coursework exercises and the examination papers from becoming repetitive.

A major area of development is in the use of IT aids for modelling. For example, during the 1994/95 academic year the students were introduced to environmental modelling in *Stella*, solving mechanics problems using a spreadsheet *(Excel)* and solving the Leslie matrix model using a computer algebra package (Maple). The use of IT aids is likely to increase in the future.

Mathematical modelling is the ideal vehicle for teaching applications of mathematics and for drawing together the various components of a degree. It will therefore continue to play a major role in the mathematics degree at the University of Hertfordshire for the foreseeable future.

REFERENCES

Davies AJ, 1994, *The Spreadsheet as a Medium for Teaching and Learning Undergraduate Mathematics*, presented at ICTCM7, Orlando, Florida, USA.

Davies AJ and Fitzharris AM, 1994, *DERIVE in Undergraduate Courses for Mathematicians and Engineers*, presented at the First International DERIVE Conference, University of Plymouth, UK.

Pegden CD, Shannon RE and Sadowski RP, 1995, **Introduction to Simulation Using SIMAN** (2nd ed.), McGraw-Hill, USA.

Richmond B, Peterson S and Boyle D, 1990, *Stella II* **User's Guide**, High Performance Systems Inc, Hanover, New Hampshire, USA.

31

Experiences With System Modelling in a Social and Business Context

MJ Hamson and MAM Lynch
Glasgow Caledonian University, Glasgow, G4 0BA.
e-mail : mjha@ gcal.ac.uk

ABSTRACT

In this paper the mathematical modelling of social and business problems is considered. The setting is a module in the third year (level) of the BSc Mathematics for Business Analysis degree programme at Glasgow Caledonian University. One main objective in the module is to show how a System Dynamics approach provides an effective tool in helping to formulate models of social and business problems. The resulting models are normally systems of differential equations. In the paper a brief explanation of the systems dynamics approach is given, together with an illustrative example. The work done by the students in the sessions 1994-95 and 1995-96 is then described, with comments on their achievements.

INTRODUCTION

Students undertaking business mathematics programmes are unlikely to devote much time to physical models, where differential equations traditionally play a large part. The students on the business mathematics programme at Glasgow Caledonian are no exception in this respect and, although they have studied the mathematical techniques for solving differential equations, their use of differential equations in, say, mechanics is not on the syllabus. However there are many contexts (for instance waste disposal, manpower planning, population control), where processes can be adequately modelled using differential equations. At Glasgow, students attend modules entitled Communication and Modelling in levels 1 and 2 and so have already experienced the modelling process, producing written reports and giving oral presentations. In this level 3 module, there is the opportunity to extend the modelling experience to more open-ended problems found in business applications, where there are often inter-related dynamic effects between various problem variables and their rates of change. These inter-relations will normally be empirical, but nevertheless can lead to a model in the form of a system of differential equations. Source material which helped provide the module tutors with examples and ideas for suitable student assignments was found in System Dynamics Review and in the New Scientist.

The study of systems of differential equations may appear to be a barrier for business mathematics students, however appropriate software packages can be used to solve systems numerically (without the user needing to follow a detailed course in numerical methods for differential equations). The Glasgow students were taught to use the package Interactive Simulation Language (1986) and, in addition, material on equilibrium points and stability analysis of a system of differential equations was covered in the module (see Huntley and Johnson, 1983, for technical details). Students had to judge the behaviour of their model system and its effectiveness in representing the real problem. Providing both theoretical and practical solution tools helped to motivate student understanding of the mathematics as well as allowing the model to be analysed.

The modelling activity described above forms a major project within the level 3 module. The particular projects were selected for open-endedness, and hopefully address general issues of interest to the students - for example a large fruit market could be the basis for such a project . In such a market there will be a daily supply and demand feature, seasonal availability problems, daily sales profiles and waste disposal issues. This example may seem somewhat distant from that which conventionally leads to differential equation models but, by constructing input-output relations and modelling the associated rates of change, a plausible model can be set up. The precise form of the relations between problem variables may require library data or other information concerning the project topic.

The key part in mathematical modelling is, as always, the success of the model formulation. To aid this process students used the methodology of System Dynamics as a formal modelling tool , as described in texts such as that by Roberts (1983). This proved to be a powerful ally in developing models for the type of projects the students were asked to tackle. Problem variables (or state variables) are identified as 'levels' and these are linked through associated 'rate functions'. This is done as a 'directed graph' (digraph) or 'Influence Diagram'. Such a diagram is an attractive way of pursuing the eventual model since changes , additions and corrections are easily made to the diagram. Students like the uniformity of the Influence Diagram language which is unambiguous in nature, and helps beginners to contribute quickly in group discussions. The process is described briefly in the next section .

SYSTEM DYNAMICS METHODOLOGY

To introduce the notation of Influence Diagrams, first consider two coupled differential equations:

$$x'(t) = f\{ x(t), y(t) \}$$
$$y'(t) = g\{ x(t), y(t) \}.$$

The state variables (or levels) are x and y and the rate functions are f and g. Thus the time derivatives of levels are defined as f and g respectively and f and g in turn depend on the levels. This can be conveniently represented as an Influence Diagram as in Fig. 1.

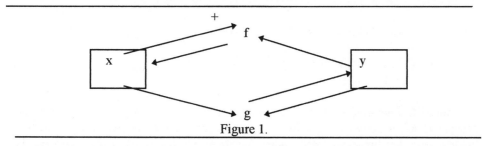

Figure 1.

An arrow into a level indicates that the time derivative of that variable depends on (or is influenced by) the rate function at the tail of the arrow. An arrow into a rate function indicates that that rate function has that level at its tail as an argument.

A 'plus' attached to the arrow, say, from x to the rate f (see Fig. 1) indicates that $\partial f/\partial x \geq 0$, whilst a 'negative' sign would indicate $\partial f/\partial x \leq 0$. An arrow from a rate to a level can also be signed, and is interpreted as causing the level to increase or decrease, - a birth rate causes a population level to increase, whilst a death rate causes a decrease. This procedure is demonstrated in Fig. 2 below

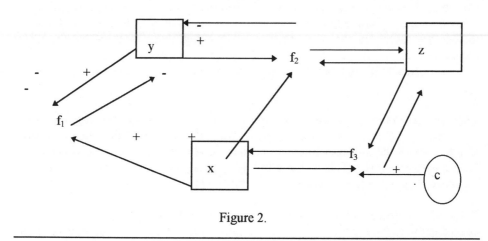

Figure 2.

In this case the Influence Diagram, showing three levels and three rates, gives the structure for a system of three differential equations:

$$x' = f_3, \quad y' = f_1 - f_2, \quad z' = f_2 - f_3$$

The functional dependence of the rates can be read from the diagram:

$$f_1 = f_1\ (x,\ y)\ ,\ \ f_2 = f_2\ (x,\ y,\ z)\ ,\ \ f_3 = f_3\ (x,\ z;\ c)$$

where c denotes some capacity parameter which may be required in the model.

Finally constraints on the signs of the derivatives are

$$\partial f_3 / \partial x \geq 0\ ,\ \ \partial f_3 / \partial z \geq 0\ ,$$
$$\partial f_1 / \partial y \leq 0\ ,\ \ \partial f_2 / \partial x \leq 0\ .$$

The model formulation is not yet complete since the precise nature of the functional dependencies indicated for f_1 , f_2 and f_3 has to be constructed This would need recourse to data collection, or some other empirical feature, so that simple expressions can be tried , at least initially.

SYSTEM DYNAMICS EXAMPLE

Typical of some of the tutor-led problems used in the module to illustrate the System Dynamics method was the following.

Set up a model to represent Scottish Tourism and resulting Bed and Breakfast Bookings in Glasgow.

Consider a typical summer season in which visitors are attracted to Scotland, and spend a few days in Glasgow before moving off around the country. If it is assumed that the tourists are casual, in the sense that no pre-bookings have taken place, then their requirement is probably for bed and breakfast accommodation. Bookings in Glasgow will be affected by the number of tourists coming into the city and by the availability of rooms.

There is cause and effect between the number of tourists and room bookings. Bookings will rise at the start, but hotel capacity will limit the rate of admission later. Construction of the Influence Diagram requires the selection of suitable levels and rates for a simple first model. This is shown in Fig. 3, where there are two linked parts in the Influence Diagram, one for tourism and one for hotels.

From the Influence Diagram, it is seen that three levels have been chosen. The connecting lines with attached arrows are in place and the functional dependencies can be listed:

$$r = r(T,\ N;\ c)\ ;\ b = b(N)\ ;\ a = a(T,\ S)\ ;\ e = e(S).$$

Units are sensibly fixed in this example by taking the time variable (t) in days, and the variables N,T and S all as 'number of people'.

There are three differential equations, corresponding to the three levels

$$N'(t) = r - b$$
$$T'(t) = a - r$$
$$S'(t) = e.$$

Tourist Industry

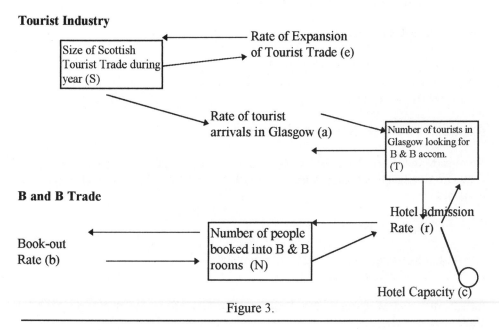

Figure 3.

The model is not fully defined until particular functional forms are substituted into the rates r, b, a and e. This will require some modelling skill to ensure that suitable expressions are chosen. We examine each expression in turn.

(i) $r(T, N; c)$ is the hotel /B & B admission rate.

      Select $r = k_1 N (1 - N/c)T$ since

      (a) the rate is taken as directly proportional to T,
      (b) the logistic model is convenient for the dependency on N with at
             first a growth due to popularity and later a brake as full capacity is
                 approached.

(ii) $b = k_2 N$ is the booking rate - it seems reasonable to take this as directly proportional to N

(iii) $a = k_3 (1 - T/T_{sat}) T S$ (logistic model used)

(iv) $e = k_4 (1 - S/S_{sat}) S$ (again logistic model)

In the latter two formulae the logistic terms T_{sat} and S_{sat} are included to represent the maximum values of tourist numbers and tourist trade respectively. Direct proportionality is used in (ii) for the inclusion of S. Proportionality constants k_i , $i = 1,2,3,4$ have to be introduced and require fixing before a numerical solution can be obtained. Also initial values for N,T and S have to be suitably set.

The final model is three coupled nonlinear differential equations:

$$N'(t) = k_1 TN(1 - N/c) - k_2 N$$
$$T'(t) = k_3 S T (1 - T/T_{sat}) - k_1 TN(1 - N/c)$$
$$S'(t) = k_4 S (1 - S/S_{sat}).$$

(1)

RESULTS FOR THE TOURIST PROBLEM

As already indicated, the constants $\{k_i\}$ present in (1) have to be selected before a computer simulation is tried. Students find such a selection difficult even when data are available.

Here the following values were taken:

$k_1 = 0.01$, $k_2 = 0.5$, $k_3 = 0.001$, $k_4 = 0.0005$.
$c = 500$, $T_{sat} = 1000$, $S_{sat} = 8000$.

Initial values (say at the start of the season) were set at $N(0) = 50$, $T(0) = 100$, $S(0) = 1000$.

As an illustration of assessing the value of the parameters, consider $k_1 = 0.01$. While it is not possible to state that k_1 will be 0.01 exactly, some attempt at setting it can be seen from the first term in the N' equation. Writing this in the form

$k_1 \cong \delta N / \{ \delta t N T (1 - N/500) \}$,

look at the representative values of N and T and also the expected increment in N , δN per δt. Take N and T = 50, take $1 - N/500 \cong 1$, and then with 25 people (say) taking up B & B places per day, the value $k_1 = 0.01$ is obtained.

Thus the system of three differential equations to be solved is:

$N' = N T (1 - N/500)/100 - 0.5 N$
$T' = T S (1 - T/1000)/1000 - N T (1 - N/500)/100$
$S' = 0.0005 S (1 - S/8000)$.

An equilibrium point can be identified at $N \cong 474$, $T \cong 969$, and $S \cong 8000$.

Stability analysis can be used to confirm that the point is a stable equilibrium point. The time taken, measured in days, to reach this equilibrium point is however much too large to be valid for a summer season. A graph of the behaviour of N and T against time over a 100 day period is given below (Fig. 4), showing that there is some preliminary oscillation before equilibrium is approached.

It was noted that:

(i)   only a simple model has been constructed for the tourist problem; a number of features have been ignored, notably that tourists would not wait in Glasgow for accommodation to become available but would 'leak' away elsewhere,

(ii)   the particular solution given here is dependent on the parameter values chosen. Whilst these are hopefully of the correct order of magnitude in each case (as illustrated above), there is nevertheless some flexibility.

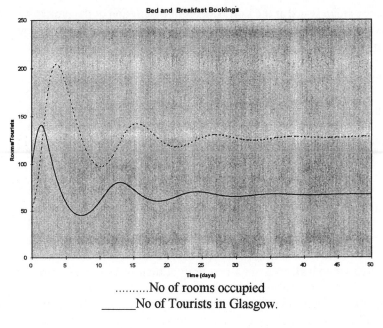

..........No of rooms occupied
_____No of Tourists in Glasgow.

Figure 4.

Students worked on other similar class examples to gain experience of the System Dynamics approach before their major modelling projects were issued.

EXPERIENCES FROM WORK DONE BY THE STUDENTS

The module was taught in the style described above in 1994-95 and in 1995-96. Students worked in groups of five on the modelling projects, and had about six weeks to carry out the work. On completion a single report was produced and oral presentations were also held. Usual modelling practices were adhered to in which groups kept a log book of their progress and reported any problems with group harmony as well as modelling difficulties. The topics chosen were selected for their interest to business mathematics students and also for their suitability to the System Dynamics approach.

Sample modelling projects for 1994-95 and 1995-96

- CO_2 pollution from fossil fuel consumption,
- Vehicle tyre trade in the UK, with attention to imports and exports of second hand tyres due to differing tread regulations,
- Supply chain management,
- Modelling population by age category,
- Spread of HIV in the intravenous drug user population,
- Spatial variations in fish populations.

Each student group issued at the start with a full description of the project was including source material from journals or magazines. This gave them important background focus to the subject material of the project and helped direct their subsequent library searches.

Comments on outcome

- These projects were all quite open in context and each group was expected to read around its topic and gather background information. The time span allowed for the completed work was about six weeks. It may be thought most were too wide to be usefully modelled, even by third-level students. It should be remembered however that a systems approach - with emphasis on cause and effect - can give rise to effective models; complicated mathematical formulae within the subsequent equations were not expected.

- In the event students did find the projects quite difficult and much discussion was necessary before a preliminary list of levels and rates could be drawn up. At an early stage each group had to submit a first draft of their Influence Diagram. This meant a check could be kept and serious errors or misconceptions corrected. It was found that a sense of 'model ownership' developed and generally all group members were able to contribute, particularly since there were no physical technicalities that required special knowledge.

- Students did not always relate all their 'levels' appropriately, but learned considerably by checking if quantities increased or decreased as they should when the numerical solution was run. The main numerical problem was in choosing effective units and then scaling the differential equations appropriately before solving. This was a particularly useful exercise since , for example, an initial rate of change could not be set at 10^9.

- The fact that students had a personal interest in the problem encouraged a further interest in the associated mathematics which perhaps would not have occurred if , say, an isolated treatment of systems of differential equations had been given . Stability analysis methods were examined with renewed enthusiasm.
- Students struggled initially to find appropriate functional relations to insert into their 'rate' equations. This key modelling skill needed some tutor-led work first, on such matters as logistic growth.

- Emphasis was put on interpretation of graphical output so that the behaviour observed could be related back to the Influence Diagram, and perhaps corrections made.

- The main problem with the chosen projects was the absence of reliable real data. Students found it difficult to use representative or surrogate data and help from tutors was sometimes needed with this aspect.

CONCLUSIONS

Modelling non-physical situations using the method of System Dynamics was an attractive task for business mathematics students. Not only were the contexts of the projects attractive for their course, but sufficiently interesting mathematical analysis was present to provide quite a demanding task. Whilst not every difficulty was adequately solved, students found the tasks stimulating. Tutors look forward to another challenging run of this module in 1996-97.

REFERENCES

Huntley ID and Johnson R, 1983, **Linear and Nonlinear Differential Equations,** Ellis Horwood, Chichester.

Interactive Simulation Language (ISIM), 1986, Salford University Business Services, Manchester.

New Scientist, IPC Magazines Ltd., England.

Roberts N, *et al.*, 1983, **Computer Simulation - a System Dynamics Approach,** Addison Wesley, Reading, Massachusetts, USA.

System Dynamics Review, Wiley, Chichester.

32

Using Critical Reviews in Mathematical Modelling Courses

Bryan A Orman
University of Southampton, Southampton, SO17 1BJ, UK.
e-mail: bao@maths.soton.ac.uk

ABSTRACT

Critical reviews in mathematical modelling courses are a means of generating, developing and moderating a critical potential in students towards the use of mathematics in areas of application.

In mathematical modelling courses students acquire knowledge of existing models together with an understanding of the characteristics of the modelling process through direct instruction. In addition, they can perform the modelling themselves either by applying known models to related novel situations or by building or modifying existing models. These aspects of mathematical modelling have to be coupled with some knowledge of report writing in order to communicate the essential results of the activities.

This paper describes the author's experience in engendering an appreciation of good report writing in students through the critical analysis of selected mathematical modelling reports, both existing ones taken from relevant publications and ones that they or other students have written. This activity allows students to exercise personal judgement on issues raised within the reports in relation to the methodologies developed in the course. It appears to instil in students an awareness of the need to employ correct modelling methodologies and, from comments made in feedback questionnaires, the critical analysis component of the modelling course is favourably received.

INTRODUCTION

The only mathematical modelling courses available to undergraduate mathematics students at the University of Southampton are in the final year of study. The Introduction to Mathematical Modelling course in the first semester is a prerequisite for the Applications of Modelling second semester course and, taken together, they represent a quarter of the programme of study for a final year student. The existing modular structure of the many undergraduate degree programmes means that students with diverse backgrounds and varied expectations register for these courses, since no previous knowledge of mathematical modelling is assumed. This package of courses has become very popular with students and, in this last academic year, the two courses attracted 37 and 11 students respectively from a cohort of 105 students.

Since both courses are assessed entirely by coursework, one of the main concerns of the students is their ability to write their mathematical modelling activities in an acceptable format. Whilst many of the students would have experienced highly structured and directed coursework assignments in second year courses they would not have encountered the requirements associated with the methodologies occurring in the final year Mathematics Education type courses. The major feature that distinguishes the two types of coursework is that, in the former courses, submissions are generated through formal and sometimes quite didactic instructions whilst, in the latter courses, submissions arise through an informal and less authoritarian regime of student-centred inquiry.

RATIONALE

As with most undergraduate mathematical modelling courses, a well-balanced programme of conventional model-building exercises is developed after the introduction of some apparently trivial problem-solving exercises at the beginning of the course. These early exercises are necessarily elementary, since their main purpose is to expose the students to the underlying modelling methodologies upon which the whole philosophy of the course is built. In addition, many students are unable to cope with mathematics that is no more sophisticated than that experienced in their early calculus, algebra and geometry courses. Their mathematical naiveté has to be acknowledged, and this results in the internal models associated with these exercises having to be wholly transparent and quite undemanding, thus ensuring that the methodology acquires prime importance at this stage of instruction. Indeed an awareness of acceptable modelling methodologies is certainly the earliest requirement in any mathematical modelling course, since students who understand what will be expected of them in this intellectual area tend to be more at ease with both the individual and the group modelling activities as and when they occur.

The gradual progression, from the simple problem-solving exercises at the beginning of the course to the quite demanding open-ended case studies of the students' own choosing towards the end of the year-long course, allows the early emphasis on the methodologies employed to be coupled with an exposure to the requirements of sound report writing. Students are aware that the assessment is based entirely on written reports, and they are quite naturally concerned to know what might be expected of them both in the content and the style of the written reports. More generally, in any course that involves communication of work undertaken by students, whether it be through oral presentations, video presentations, written reports or the like, it is essential that the students are fully aware of the expected format of these submissions. There is therefore an obligation to expose the students to an accepted style, manner and content of the submissions which describe both the activity and the results obtained from work undertaken. Since the format is more often than not highly course dependent, the exposure of students to what is considered good practice in these modes of communication is therefore of paramount importance.

Students thus need to develop sound report-writing techniques, in conjunction with general mathematical modelling skills, and they also need to proceed from the early model-presentation stage of the instruction to the model-construction stage of the assessment. Models are presented to students to try to involve them in the process of modelling so that they become competent and independent modellers. The students' transition from passive modelling to active modelling is probably the most difficult aspect of the instructor's task. Before the introduction of critical reviews, a significant number of students admitted at the end of their course that they did not fully understand what was ultimately expected of them.

The old rigidity in the presentation of the modelling process and report writing required overhauling since many students requested practice in producing modelling reports before using the acquired skill in the submission of both their individual and group reports. They also requested that these practice reports be formatively assessed since they were naturally apprehensive about any form of new coursework assessments - even if such assessments were afforded the peer protection of group submissions!

COMPREHENSION TESTS AND CRITICAL REVIEWS

The benefits of using comprehension tests in the teaching of mathematical awareness have been thoroughly reported by Houston (1993a, 1993b, 1995) in a series of papers in the Teaching Mathematics and its Applications journal and in the proceedings of ICTMA-6 held at Delaware in 1993 (Sloyer *et al.*, 1995). In these communications, a rationale was offered for the introduction of comprehension tests, together with its aims and objectives, and the difficulties encountered in the implementation of such tests were well documented. The rationale associated with critical reviews has much in common with that of comprehension tests; however, there are some distinct additional aims and objectives to the critical review activity. Furthermore, being entirely coursework based, its implementation is quite different, whereas comprehension tests are by their nature conventionally assessed.

To engender an appreciation of good report writing the students should critically analyse and assess mathematical models, both existing ones and models that they or other students have constructed, with respect to given report writing guidelines. If the existing mathematical models are taken from those published in the mathematical modelling and related literature, then a comprehension test scenario could be employed. If, on the other hand, the models are ones that have been produced by students who have attended a modelling course at the same institution in previous years, then a quite different form of critical analysis will be required. These reports will have too many points of departure from the accepted guidelines and will require either a partial or, in some extreme cases, an almost total rewrite. It is this process of critically analysing given reports with a view to rewriting them that forms the cornerstone of critical reviews.

The major aims of this activity are:
- to enable students to learn from the written mathematical modelling work of others,
- to encourage students to discuss openly aspects of mathematical processes relevant to the modelling of real problems,
- to provide students with an opportunity to appreciate and tolerate differences of opinion between their own and others' interpretations of modelling methodologies,
- to enable students to acquire experience in analysing and assessing critically mathematical models and processes which are appropriate to their level of understanding,
- to instil in students an awareness of the need to employ proper modelling methodologies,
- to motivate students to acquire the necessary skills of sound report writing.

The assessment submissions associated with this activity are designed to promote in students the following specific objectives:
- to distinguish between the relevant and the irrelevant aspects of written reports,
- to detect implicit assumptions not mentioned in reports,
- to locate digressions and inconsistencies in reports,
- to identify incorrect mathematical calculations and to incorporate the corrections and the consequences in their revision of written reports,
- to establish the appropriateness of the data and to analyse it in the most convenient manner,
- to write mathematical reports that are technically unambiguous,
- to write clear and comprehensible reports that have been systematically and logically developed.

IMPLEMENTATION AND ASSESSMENT

The critical review activity supplements the conventional model-building exercises and group project work associated with the first semester course, and it is introduced alongside the elementary model-building examples at the beginning of the course. Students are given a copy of a modelling report written by a contemporary student which is to be discussed quite openly in a tutorial class. The report tends to be written quite subjectively (containing many misconceptions and overstatements, not to mention irrelevant information and sophistication) and in a style that does not conform with the taught modelling methodology of the course. The report is analysed line-by-line, page-by-page, equation-by-equation, in relation to a general modelling methodology developed for the elementary model-building examples at the start of the course. At this stage a general requirement for good report writing is established by common consent. The overall structure of the report is then discussed, paying attention to issues like the elucidation of the technical particulars, the appropriateness of the data, and the benefits of diagrams and appendices. Students are encouraged to annotate the report in the light of the discussion. This first activity normally takes at least two 45 minute tutorial classes to complete. Two further reports are treated similarly. Finally a trial report is handed out for overnight annotation, and this allows the subsequent tutorial class discussion to be lively and highly focused. Examples of good annotated reports are also handed out to the students, together with critiques and subsequently revised reports. These reveal to students the norms for the assessment submissions.

The assessment of this activity must be viewed in relation to the overall assessment strategy of the course. The assessment of the first-semester course is based on the main components of the course, and the assessment falls into three main parts.

(i)   A critical review of two written modelling reports produced elsewhere is undertaken. Students are expected to discuss these reports using the established criteria and in three quite distinct ways. The importance students give to each mode is entirely up to them. Firstly a report is annotated with margin comments, queries and corrections. The report should not be obliterated by these annotations and they should form the basis for the further collective observations. Secondly a written critique is attached to the report, which forms the constructive criticism aspect of the submission. No more than 1000 words are expected. Finally a revised version of the given modelling report is submitted, based on the student's annotation and critique of it. The report is expected not to exceed 2500 words in length. For an example of a typical submission under this assessment heading see Houston and Orman (1996). This part of the assessment contributes 30% to the total course assessment.

(ii)   Two specific model-building exercises are undertaken from a selected list of five elementary real problems that had already been introduced and discussed briefly in class. The students are expected to identify an aspect of the problem worthy of further discussion, and the written report is expected to contain an abstract which includes the actual conclusion reached and an evaluation of the significance of the problem. Each report has a 3000 word maximum length, and this part of the assessment contributes 30% to the total course assessment.

(iii)   A group project is justified on the grounds of practicality in that there are problems which cannot be adequately addressed by an individual student in the time available. In addition there are problems for which data collection can only be achieved by the co-operation of several individuals. The group project involves many aspects of communication skills, and individual members of the group can contribute some special expertise to the group project activity. Peer assessment is used to moderate the work and no limit is placed on the length of the final report. This part of the assessment contributes the final 40% to the total course assessment.

CONCLUSIONS

It should be noted that peer assessment of the group project submission has only rarely been used to moderate the mark assigned to individual students under the final assessment heading. In the main the groups are self selecting and the topics for investigation are of the students choosing (subject to the course co-ordinator's approval), so little conflict arises when it comes to the assessment. For those few groups that have a disparate membership the activities are strongly overseen, with the aim of encouraging co-operation and of eliminating unequal workloads. The peer assessment strategy advertised to the students has much in common with the models described by Conway *et al.* (1993) and by Goldfinch and Summers (1996).

The use of critical reviews gives the student a better understanding of mathematical modelling methodology and the essential requirements of sound report writing. The activity helps to remove the subjective nonconformity often found in the unregulated submissions of many students.

A number of important issues should be acknowledged before any implementation of this activity is contemplated namely:-

(1)   the activity requires open discussion of the reports by the whole class and this often results in the instructor having to exercise resolute control over the proceedings, even though many of the digressions reveal interesting features worthy of further investigation. The problem of the occasional domineering and opinionated student also requires careful handling. The instructor has to guide the discussions, without inhibiting or suppressing the students' freedom of expression. It is also true that the instructor's criticism might not be the students' criticism and, more significantly, students themselves differ enormously in their perception of what is good or bad in a report.

(2)   the feedback to students is also extremely important, since it is the formative nature of this part of the assessment that is of significance - especially with respect to the further work undertaken by the students, both individually and in a group. The critical review assessment has an early and indispensable deadline. This allows the work to be graded prior to detailed marking conflation with the other assessments at the end of the course. The assessment of the submission requires careful consideration, since the activity concerns an individual student's interpretation of a written report in the light of an agreed modelling methodology. The assessor has to look for self-consistency in the submission, in addition to the recognition of a sufficient number of known and significant defects in the given report, to generate a satisfactory rewrite.

Dissimilar rewritten reports often attract the same overall grade. The submission is discussed privately with each student. Although this would appear to be extremely time consuming it is executed with no additional workload during the week when the students commence their group modelling work. The course does not loose continuity through this feedback mechanism, and the reassurance it gives the students does much to encourage them in their further work.

REFERENCES

Conway R, Kember D, Sivan A and Wu M, 1993, *Peer Assessment of an Individual's Contribution to a Group Project*, **Assessment and Evaluation in Higher Education, 18,** 45-56.

Goldfinch JM and Summers R, 1996, *Peer and Self Assessment after Groupwork*, in Haines CR and Dunthorne S, (eds.), **Mathematics Learning and Assessment: Sharing Innovative Practices**, Edward Arnold, London, pp 5.10-5.20.

Houston SK, 1993a, *Comprehension Tests in Mathematics*, **Teaching Mathematics and its Applications**, **12**, 22-34.

Houston SK, 1993b, *Comprehension Tests in Mathematics II*, **Teaching Mathematics and its Applications**, **12**, 60-73.

Houston K, 1995, *Assessing Mathematical Comprehension*, in Sloyer C, Blum W and Huntley I, (eds.), **Advances and Perspectives in the Teaching of Mathematical Modelling and Applications**, Water Street Mathematics, Yorklyn, Delaware, pp 151-162.

Houston K and Orman BA, 1996, *Comprehension*, in Haines C R and Dunthorne S, (eds.), **Mathematics Learning and Assessment: Sharing Innovative Practices**, Edward Arnold, London, pp 4.37-4.59.

Sloyer C, Blum W and Huntley I, (eds.), 1995, **Advances and Perspectives in the Teaching of Mathematical Modelling and Applications**, Water Street Mathematics, Yorklyn, Delaware.

Albion Books on Mathematics and its Applications

FUNDAMENTALS OF UNIVERSITY MATHEMATICS
COLIN McGREGOR, JOHN NIMMO and WILSON W. STOTHERS, Department of Mathematics, University of Glasgow

A unified course for first year mathematics, bridging the school/university gap, suitable for pure and applied mathematics courses, and those leading to degrees in physics, chemical physics, computing science, or statistics.

Contents: Preliminaries, Functions & Inverse Functions; Polynomials & Rational Functions; Induction & the Binomial Theorem; Trigonometry; Complex Numbers; Limits & Continuity; Differentiation - Fundamentals; Differentiation - Applications; Curve Sketching; Matrices & Linear Equations; Vectors & Three Dimensional Geometry; Products of Vectors; Integration - Fundamentals; Logarithms & Exponentials; Integration - Methods & Applications; Ordinary Differential Equations; Sequences & Series; Numerical Methods.

ISBN: 1-898563-10.1   540 pages   1994

DECISION AND DISCRETE MATHEMATICS: Maths for Decision-making in Business and Industry
THE SPODE GROUP, Truro, School, Cornwall

A complete coverage in the Decision Mathematics Module (Discrete Mathematics) of the syllabuses of English A level examination boards. Also suitable for foundation first year and undergraduate courses in qualitative studies and operational research, or for access courses for students needing strengthening in mathematics or those moving into mathematics from another subject.

Contents: An Introduction to Networks; Recursion; Shortest Route; Dynamic Programming; Flows in Networks; Critical Path Analysis; Linear Programming (graphical) Linear Programming: Simplex Method; The Transportation Problem; Matching and Assignment Problems; Game Theory; Recurrence Relations; Simulation; Iterative Processes; Sorting and Packing; Algorithms.

ISBN: 1-898563-27-6   256 pages   1996

MATHEMATICAL ANALYSIS
DAVID S.G. STIRLING, Senior Lecturer in Mathematics, University of Reading

This fundamental and straightforward text for first and second year degree courses in the UK addresses a weakness observed among students, namely a lack of familiarity with formal proof. Dr Stirling begins with the need for mathematical proof, developing associated technical and logical skills. This lucid analysis and development reads naturally and in straight-forward progression by indicating how proofs are constructed. The text emphasises
   (1) the need for familiarity with long mathematical arguments and manipulation,
   (2) the importance of the ability to construct proofs in analysis

Contents: Setting the Scene; Logic and Deduction; Mathematical Induction; Sets and Numbers; Order and Inequalities; Decimals; Limits; Infinite Series; Structure of Real Number System; Continuity; Differentiation; Functions Defined by Power; Appendix: Decimal Expansion of Integers.

ISBN 1-898563-36-5   250 pages   1997

Albion Books on Mathematics and its Applications

SIGNAL PROCESSING IN ELECTRONIC COMMUNICATIONS
MICHAEL J. CHAPMAN, DAVID P. GOODALL, and NIGEL C. STEELE, School of Mathematics and Information Sciences, Coventry University

This text for advanced undergraduate and postgraduate courses in electronic engineering, applied mathematics, and computer science, deals with signal processing as an important aspect of electronic communications in its role of transmitting information, and the language of its expression. It develops the required mathematics in an interesting and informative way, leading to confidence on the part of the reader.

Contents: 1 Signal and Linear System Fundamentals 2 System Responses 3 Fourier Methods 4 Analogue Filters 5 Discrete-time Signals and Systems 6 Discrete-time System Responses 7 Discrete-time Fourier Analysis 8 The Design of Digital Filters 9 Aspects of Speech Processing Appendixes: (a) The Complex Exponential (b) Linear Predictive Coding Algorithms (c) Answers to the exercises.

ISBN 1-898563-23-3   *ca.* 300 pages   1997

CALCULUS
Introduction to Theory and Applications in Physical and Life Science
R.M. JOHNSON, Department of Mathematics and Statistics, University of Paisley

A lucid text conveying clear understanding of the fundamentals and applications for first year undergraduates in applied mathematics, computing, physics, electrical, mechanical and civil engineering, chemical science, biology and life science.

Contents: Prerequisites from Algebra, Geometry and Trigonometry; Limits and Differentiation; Differentiation of Products and Quotients; Higher-order Derivatives; Integration; Definite Integrals; Stationary Points and Points of Inflexion; Applications of the Function of a Function Rule; The Exponential, Logarithmic and Hyperbolic Functions; Methods of Integration; Further Applications of Integration; Approximate Integration; Infinite Series; Differential Equations.

ISBN: 1-898563-06-3   336 pages   1995

LINEAR DIFFERENTIAL AND DIFFERENCE EQUATIONS:
A Systems Approach for Mathematicians and Engineers
R.M. JOHNSON, Department of Mathematics and Statistics, University of Paisley

An advanced text for senior undergraduates and graduates, and professional workers in applied mathematics, and electrical and mechanical engineering.

"Should find wide application by undergraduate students in engineering and computer science ... the author is to be congratulated on the importance that he attaches to conveying the parallelism of continuous and discrete systems" - *Institute of Electrical Engineers (IEE) Proceedings*

Contents: PART I: CONTINUOUS SYSTEMS: An Approach to the Laplace Transform; Solution of Linear Differential Equations; Steady State Oscillations; Piece-Wise Continuous Functions; PART II: DISCRETE SYSTEMS: From the Ideal Sampler to the Z-transform; Solution of Linear Difference Equations; Digital Filters; Tables: z-transforms of (X_n); z-transforms of $f(t)$; Comparison Between Continuous and Discrete Systems.

ISBN: 1-898563-12-8   200 pages   1997